biosaline
RESEARCH A LOOK TO THE FUTURE

ENVIRONMENTAL SCIENCE RESEARCH

Recent Volumes in this Series

---

biosaline
RESEARCH
A LOOK TO THE FUTURE

Edited by **Anthony San Pietro**

Indiana University
Bloomington, Indiana

Library of Congress Cataloging in Publication Data

International Workshop on Biosaline Research (2nd : 1980 : La Paz, Mexico)
  Biosaline research.

  (Environmental science research ; v. 23)
  Proceedings of the Second International Workshop on Biosaline Research, held in
La Paz, Mexico, Nov. 16-20, 1980.
  Includes bibliographical references and index.
  1. Biosaline resources—Congresses. 2. Salt-tolerant crops—Congresses. 3. Mari-
culture—Congresses. 4. Halophytes—Congresses. 5. Micro-organisms, Halophilic—
Congresses. I. San Pietro, Anthony Gordan, 1922-   . II. Title. III. Series.
S620.I57   1980   631   81-19885
ISBN 0-306-40892-9   AACR2

Designed by Alice San Pietro from a suggestion by R. N. Ondarza.

Proceedings of the Second International Workshop on Biosaline
Research, held in La Paz, Mexico, November 16-20, 1980

© 1982 Plenum Press, New York
A Division of Plenum Publishing Corporation
233 Spring Street, New York, N.Y. 10013

Printed in the United States of America

FOREWORD

The concept that one could successfully grow plants and trees on soil which is high in saline concentration could be of very great importance to the future agricultural and economic development of many countries where food products are scarce and such areas are vast. All of whom are gathered here are interested in basic research and its application to practical purposes – especially to remove some of the handicaps that exist in taking full advantage of soil, brackish water and other heretofore unproductive areas.

The first workshop was organized at Kiawah Island just three years ago. A relatively small group of us discussed the different possibilities of what could be done to take greater advantage of our knowledge and basic research to develop this field. A short report was published which was later extended by the addition of a number of chapters formulating the volume entitled The Biosaline Concept: An Approach to the Utilization of Underexploited Research.

The Kiawah Island workshop was followed by one in Kuwait and another in Egypt. All were important in their contribution toward the accumulation of knowledge in this area. Then an extremely interesting workshop followed which was held at La Paz, Baja California, Mexico, during November of 1980. The proceedings of this workshop demonstrate very well the progress which has been made in these different areas in recent years. It has also brought out the issues about which investigators from developing countries have been actively studying in their own countries and trying to overcome the limitations which increased saline concentration present to their agricultural development. Their results are most interesting.

It is so obvious to us that several excellent possibilities exist if we pursue our knowledge of plant breeding and our understanding of the adaptation of plants to tolerate saline environments. Some of the basic techniques were brought forth in a

v

symposium which was held at the Brookhaven National Laboratory and later published under the title <u>Genetic Engineering of Osmoregulation</u>. On the basis of such efforts, we are trying not only to develop the basic work which is fundamental to fully utilizing various environments, but also to apply this information to productive benefit.

Especially interesting at the La Paz meeting was the broad range of countries represented. This illustrates a deep interest shared by many countries in the southern and eastern hemispheres where problems of salt effect in soil and along coastlines are common. The outline of subjects discussed at this workshop demonstrates the wide variety of solutions to this situation. It is clear that we need additional basic research to overcome some of these genetic barriers.

Unfortunately, I was not able to attend this workshop; however, I have carefully studied the abstracts with interest. I would like to express thanks to the host country which did such a marvelous job in arranging for this meeting. We also express our gratitude to the participants who promptly submitted their manuscripts so that the publication of this workshop could be possible in efficient time. Of course, the organizers of this workshop should be especially commended; Dr. San Pietro's capability in pulling this together has been most impressive and I think the volume will be an especially interesting one.

Alexander Hollaender

PREFACE

Biosaline research will provide, hopefully, the experimental basis for the Biosaline Concept which envisions the harmonious interplay of high solar insolation, high temperature and saline (or brackish) water availability as the foundation of a unique renewable resources program for desert lands focussed on biogrowth in a saline environment. Within this concept, sunny deserts, inland brackish water, lengthy coastlines, and the oceans themselves, are viewed as climatic and geographic resources rather than as detriments. The potential for increased future exploitation of these resources for food, fiber, energy and fine chemicals production, utilizing marine organisms, marine-adapted plants, and arid and semi-arid plants, is almost limitless provided the necessary technology becomes available through basic research.

The IInd International Workshop on Biosaline Research was convened in La Paz, Mexico, from 16-20 November 1980. The major topics selected for discussion were: Regional Reviews; Food and Economic Plants; Potential Uses of Microalgae; Stress Biology; and Present and Future Applications. The Regional Reviews focussed on current biosaline research efforts in various parts of the world and underscored the international scope of the overall endeavor. For each of the other topics, one or more acknowledged experts presented an up-to-date assessment of the "State of the Science"; their own research, as well as that of other investigators, was described. Additionally, some twenty-six contributed papers were presented.

In retrospect, the workshop provided an excellent forum for communication and discussion among the international community of investigators. The proceedings are provided in toto in this volume.

On behalf of all attendees, I thank most sincerely the Governor, Lic. Angel Cesar Mendoza Aramburo, for his gracious hospitality in arranging a special luncheon for us at the State Residence. Further, I extend sincere appreciation to Ing. Alfonso Gonzalez Ojeda, State Development Secretary, representing the

Governor, and to Lic. Norberto Corella, on behalf of Raul N. Ondarza as commissioned by Dr. Edmundo Flores, Director General of Consejo Nacional de Ciencia y Tecnologia, for their participation in the opening ceremony of the Workshop.

It is a sincere pleasure to acknowledge with gratitude the excellent collaboration of Dr. Felix Cordoba-Alva, the Mexican Co-Organizer whose tireless efforts assured the excellence of the Workshop. Thanks are due Mrs. Blanca Herrera de Gregoire and the staff of the Centro de Investigaciones Biologicas for the excellent local arrangements in La Paz.

Special thanks are due Dr. Alexander Hollaender for the Foreward to this volume and Mrs. Lillian Cooney and Ms Christina Phillips for excellent organizational and secretarial assistance.

The financial assistance of the Office of Problem Analysis, National Science Foundation (USA), and Consejo Nacional de Ciencia y Tecnologia (Mexico) is acknowledged most appreciatively.

A. San Pietro
June 1981

CONTENTS

WELCOMING ADDRESSES

REGIONAL REVIEWS

FOOD AND ECONOMIC PLANTS

WELCOMING ADDRESSES

WELCOMING ADDRESS

George E. Brosseau, Jr.

National Science Foundation
Washington, D. C.   20550   USA

The National Science Foundation is proud of its continuing role in the development of Biosaline research through joint sponsorship with CONACYT of the IInd International Workshop on Biosaline Research. It is appropriate that the workshop be held in La Paz, Mexico. Important biosaline research is centered in Mexico, for whom the salt affected environment is an important problem. Further, it emphasized the international nature of saline environments and of the scientific community who work on biosaline problems.

Much has happened since the first Biosaline Workshop was held on Kiawah Island in 1977. It is time to examine the progress made since then. The NSF recognizes the value of periodic assessment of a developing field and of the importance of that assessment to the further development of relevant research. An overview can identify gaps to be filled and opportunities for future research. By summarizing the status of today's research, it gives impetus and focus to what needs to be done tomorrow.

The workshop also functions to facilitate communication between scientists of many nations. Publication is slow, but, face to face exchange allows scientists to benefit from the latest information. Hopefully, more international collaborative work will spring from these exchanges. Another benefit is the transfer of useful technology from one nation to another.

What does NSF anticipate as the outcome of this workshop? A funding agency wishes to make the best use of its resources. The workshop can be useful to program planners by identifying gaps and work that needs to be done next. This report, by bringing the results of diverse work together in one place and by introducing investigators to each other, can lead to better coordination of

3

research. Most important of all, it can stimulate more research by
attracting new investigators, especially young people, to biosaline
research.

The NSF is pleased to be a participant in this continuing
activity.   We are confident that this workshop will   point the way
to exciting new knowledge on how living systems deal with   salt and
simultaneously open up new sources of resources vital to   the needs
of future generations.

WELCOMING ADDRESS

R. N. Ondarza

CONACYT
Mexico City 20, D. F. MEXICO

Ladies and Gentlemen: On behalf of the National Council for
Science and Technology, and   in the   name of the   General Director,
Dr. Edmundo Flores and myself, I wish to extend the warmest welcome
to all participants in the IInd International Workshop on Biosaline
Research.   At the   same time, I wish to   thank the Governor of the
State of Baja California Sur, Angel Cesar Mendoza Aramburo, for his
determined support of all scientific activities in this   region and
especially of this International Workshop.

To sponsor an event like this means to CONACYT the accomplish-
ment of one of its functions and its interest goes beyond a simple
responsibility, since Biosaline Research is a   topic   of   great
importance for Mexico.

As is well known, half of our territory is considered within
the arid and semi-arid zones. For this reason the topics discussed
here will have advantageous effects on Mexico's development.

The general topic of this Workshop is extremely relevant from
a worldwide point of view, since the Biosaline concept relates
biological systems to the saline environments. Taking this into
account, scientists have been researching to understand mechanisms
used by the different organisms in adapting to adverse environments
and a great deal of research has been done on exploitation of
natural resources such as energy, food, fertilizers, fibers,
chemical compounds and drugs, in order to satisfy the primary needs
of the people living in these regions.

In Mexico, as in other parts of the world, arid zones comprise
a good portion of the total land area. They occupy 36% of the
total area of the Earth and are located between 15^{0} and 40^{0}
latitude North and South of the Equator, including the areas that

5

receive the highest ratio in solar power, up to 200 KCal/cm^2/year.
They are located in the southern part of the Mediterranean Sea,
Arab and Red Seas, the Oman and Persian Gulf, the Atlantic Coast of
North Africa and the Pacific Coasts, and in Mexico, the Peninsula
and Gulf of Baja California.

The Mexican Government has been supporting a great deal of
research in the arid zones.   An example of this is the National
Council for Science and Technology, which has made a considerable
effort to increase research in this field.   Also, it is worthwhile mentioning the creation of the Biology Research Center of
Baja California and the Applied Chemistry Research Center in
Saltillo, Coahuilla.   These two institutes carry out good standard
scientific research within aspects directly linked with the topics
that will be discussed here.

The reason for CONACYT proposing that the IInd International
Workshop on Biosaline Research should take place in La Paz, Baja
California, was the intention of linking our already existing
research substructure with that of other countries with similar
interests.   We do hope that all scientists, national as well as
international, benefit from this meeting.

In Mexico we have great possibilities for studying biosaline
topics, especially those concerning the basic fields.   As an
example, we have the work undertaken in the Figueroa Lagoon or
Mormon Lagoon, located near Guerrero Negro here in the Peninsula,
where they have found micro-organisms that produce Stromatolites
capable of profiting from a hypersaline environment.

We are certain that the biosaline concept is a challenge for
all scientists interested in making good use of the already scarce
natural resources for a growing population demanding a higher
quality and quantity of products.   For this reason there is an
urgent need to use all our means and intelligence.

I thank you all for your attendance and I hope that the IInd
International Workshop on Biosaline Research will be a successful
one.

REGIONAL REVIEWS

BIOSALINE RESEARCH IN THE UNITED STATES AND CANADA

J. C. Aller*

National Science Foundation
1800 G Street, N. W.
Washington, DC   20550   USA

SUMMARY

The biosaline concept depends upon the observation that net primary production of biological systems operating at salinities characteristic of the oceans is three orders of magnitude higher than naturally on deserts. This leads to summarizing research as a function of salinity levels showing that maximum research interest is in the salinity level found in irrigation waste waters and the other maximum is related to sea water salinity. Despite the fact that two microorganisms found in terminal lakes living at near precipitation levels show commercial promise, there are very few research efforts in this salinity region. Overall, the research efforts identified span all possible parts of the biosaline research spectrum.

INTRODUCTION

Sometimes the obvious escapes us. The obvious in this case is that salt water is potentially valuable and provides us with an opportunity to alleviate some of our materials shortages. Recently, the word "biosaline" was coined to describe the activity needed to alleviate this ever-increasing problem (1, 2).

The "biosaline concept" envisions the harmonious interplay of biological systems with saline environments for the ultimate benefit to man. In particular, the desire is to provide alternative options to critical material needs from renewable resources

---

*Opinions are those of the author and are not official.

9

in an environmentally acceptable manner. The essential elements of
the concept are biological systems and a saline or marine environment. Biological systems represent animals, plants or microorganisms or their individual cellular constitutents, especially enzymes
for these biological catalysts can be used for a host of transformations. Thus, the boundary conditions for biosaline research are
biological organisms, or their essential constituents, living in or
being able to adapt to and tolerate saline or marine environments.
Arid lands become an intrinsic part of the biosaline concept
because they comprise a resource that is currently underutilized
and which have, at times, available saline water.

The biosaline concept is both simple and complex. It is
simple in that most of man's activities to convert the enormous
resources offered by desert areas to productive uses are based on
modifying the environment to be suitable for plants and animals
that are already known and used. One example of such an environmental change is the removal of salt and minerals from brackish
water and the use of desalinated water on plants which accept only
fresh water. Nature does not levy this fresh water requirement.
In fact, throughout the world--in the seas, brackish waters, and
the salt ponds and lakes of the deserts--many plants and organisms
exist which not only tolerate the presence of minerals, including
salts, but actually relish them.

The fact that the oceans support a varied life is self
evident. What is not so obvious is that the capture of primary
energy from the sun can be nearly the same for both fresh water and
marine organisms. Black (3) points out the difficulties in
measuring primarily productivity, i.e., mass per unit over a period
of time. Estimates of primary productivity vary but the analysis
of Lieth and Whittaker (4) can be accepted. Primary productivity
data are plotted in Figure 1 to illustrate two points: first, both
deserts and the open ocean are low in productivity compared to
agricultural land and productive marine systems; and second,
although the open ocean productivity is limited by a shortage of
nutrients, it is still over 10 times the productivity of deserts.
One clear conclusion is that both nutrients and water are limits in
higher productivity.

Net productivity is an important consideration and in this
case nature demonstrates to us that salt is not an insurmountable
barrier to the storage of solar energy in the form of fixed carbon.
Experience in management of terrestrial systems shows that mean
values can be greatly exceeded under appropriate conditions. On
the other hand, harvesting of a desirable product such as a cereal
grain means that only a fraction of the available net productivity
is used. Similarly, if one continues the effort to identify and

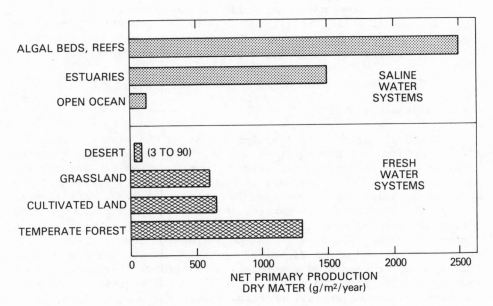

*NOTE: MEAN VALUES-MAXIUM LEVELS ARE MUCH HIGHER. PRIMARY PRODUCTION
IS NORMALLY MANY TIMES HARVESTED VALUE.

Figure 1.   Comparison of Saline and Fresh Water Productivity (3).

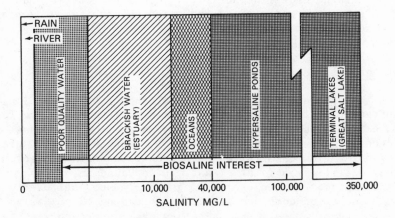

Figure 2.   Examples of Saline Environments.

adapt existing salt water organisms, the final harvest output in
the form of various materials will be the criteria for utility.
What is known is that the fundamental step of capturing the solar
energy is not only possible, but also potentially at least as
rewarding as on presently managed grazing, agricultural, and forest
systems.

Salinity as a Variable

Water is itself not a limiting factor on our planet. Epstein
(5) has pointed out however that on both land and in sea that the
apparent ease in water acquisition - when available - is acquired
only at high metabolic cost and through complicated structural and
functional adaptions. One should expect several variables to
control the adaptions by organisms. Among these variables can be
cycles of wet and dry weather, high and low temperatures, and a
number of other factors. In this paper, however, a single
variable - the amount of dissolved salts - is used for analysis of
the type of research in progress. The reader is referred to a
recent article by Epstein et al (6) for a discussion of the
complexities of the overall problem.

Water that is recycled through precipitation contains a small
amount of dissolved materials. For conventional agriculture,
however, it is considered fresh water or irrigation water of good
quality up to 1,000 parts ppm corresponding to an electrical
conductivity of less than 2 mmho/cm. While the boundary is not
precise, the shift from fresh water to marginal quality into what
can now be termed salty agriculture is about 2,000 to 3,000 ppm
dissolved salts. This is shown on Figures 2 and 3 as the boundary
for describing biosaline activity. It also corresponds to water
that is presently rejected to the ocean as unsuitable for agricultural use.

Salty Agriculture

By reference to the recently published International Directory
of Current Biosaline Research Projects (7), we find the following
crops under investigation: alfalfa, cowpeas, mung beans, melons,
and other cucurbits, tomatoes, wheat, lettuce, dates, grapes and
nitrogen fixation on alfalfa. Research is underway in seven United
States localities and at least three Canadian sites. Various
research methods are in use varying from tissue culture techniques
through selection combined with plant breeding to emphasize desired
salt tolerant properties. The general approach is to extend
present agricultural crops. The differences between Canadian and
U.S. scientists appear to be primarily associated with crops of

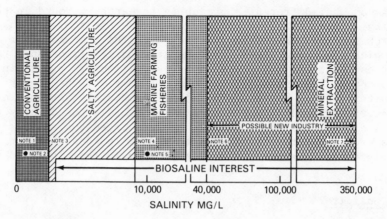

Figure 3.   Examples of Activity.

    (1)   Irrigated water of more than 1,050 mg/l considered
           marginal;
    (2)   Irrigation up to 3,000 mg/l by crop selection;
    (3)   Research area on salt tolerant plants--Barley, Rice,
           Pomegranate;
    (4)   Research on Native Estuarine Sea Coast Plants;
    (5)   Alga, Marine Bacteria, Sea Grasses, Mangroves;
    (6)   Dunaliella, other possible organisms for chemical
           production;
    (7)   Limits of Salinity-abundant life, but minor diversity.

regional interest rather than differences in research approach. As
pointed out by Niemarr (8), previous research has shown the
importance of climatic and edaphic factors as well as method
irrigation. The diversity of research in these various locations
may therefore reveal new methods for predicting plant response to
salt stress thereby improving methods of selecting and breeding
to improve tolerance.

Estuary Salinity

A somewhat different approach in research is found in the next
salinity levels which is called estuarian research. Much of this
research is directed toward understanding the processes of estuaries which are now known to be critically important to many
fisheries. It is a fact that investigation of plants in these
environments has identified several that may be important as
monoculture forage or feed crops that can resist salt to levels as
high as 25,000 mg/l. Somers et al (9) have discussed early results
of this approach. Somers (10) also has pointed out that ornamental
shrubs could be derived this way to reduce residental water use and
possibly become a source of exotic flowers in the floral industry.

Ocean Salinity

We can observe the scope of this research at salinities of the
oceans by reflecting that all of oceanography is involved. There
are some very important types of research that can be highlighted.
For example, consider reforestation with trees such as mangroves as
described by Teas (11). A portion of the NSF program on International Decade of Ocean Exploration which focused on sea grass
ecosystem study (12) is also of interest. Overall the research is
too varied to summarize except to note that research results-
-particularly with microorganisms--elucidate salt tolerance mechanisms and may apply to other environmental nitches.

There is also another type research that is of potential
interest to the biosaline research community. If one visualizes
that time when large areas are suited for various applications of
biosaline technology, it is possible that one final step would be
return of concentrated brine to the ocean. If this is accomplished
through membranes, the energy stored by water evaporation can be
recovered as electric power. Nevertheless, the result locally
would be hypersalinity. A series of research efforts have
addressed the question of the impact on the environment from brine
release (7). The motivation was derived from the U.S. strategic
oil reserve program which required excavation of salt domes. The
research results give insight into problems that might derive from

use of research results.

Hypersaline

As shown in Figures 2 and 3, biosaline research activity extends to saturation levels. There are two points to be made. Research is very sparse in this region. As pointed out by Keck and Hassibe (13) however, life exists at these salinities and has commercial use.

"Due to the high salt content of the water, most forms of aquatic life cannot live in the lake. Certain species have adapted, however, and, in fact, have flourished. Forms of algae (primitive plants without true root systems), bacteria (single-cell microorganisms that are either free living or parasitic), and protozoa (single cell organisms that are the most primitive form of animal life) all exist in the lake. Due to their primitive nature, the algae, bacteria and protozoa have been able to adapt to the lake's high salinity.

The more advanced lifeforms in Great Salt Lake are the brine fly and brine shrimp. Both exist in large numbers. The larvae (newly hatched wingless form of an insect) of the brine flies live beneath the waters and feed upon the algae and bacteria. Eventually they change into pupae (the dormant form of an insect) and float to the surface where the wind blows them to shore. There, after a third transformation to the adult stage, the brine flies gather in large numbers.

The brine shrimp are the largest form of life living in the lake, some being almost half an inch long. Great numbers of adults are in the water during the summer, and the shrimp eggs are sometimes so numerous that they cover large areas of the lake and drift with the currents.

Both the brine fly and shrimp have been used for unique purposes. The Paiute Indians formerly used the brine fly pupae as dough, which they baked into bread called Koo-chah-bie. The adult brine shrimp are netted, frozen, and sold as food for tropical aquarium fish. The shrimp eggs are also collected in large numbers, dried, sold, and eventually hatched by fish fanciers throughout the world as living food for their tropical fish."

Still in the recently published directory (7) there are only two references to studies of hypersaline lakes (Mono and Great Salt Lake). We should note that while both lakes are saline lakes their chemistry differs. The sparse research in this region is even more noticeable when this fact is considered. Neither of these research efforts has federal suport. There are, of course, related research efforts not in the directory. For example, Ho et al (14) recently reported on work involving Spirulina maxima. Other hints lie in Szalay and R. E. Macdonald (15) reporting on a 1979 workshop which pointed out research results by Dr. Hellebust (Canada), Dr. Kushner (Canada), Dr. Fitt (Canada) as well as Dr. Yaguchi, also from Canada, addressed to a range of salinity stress. It is probable that considerable microorganism research is underway both in the United States and Canada which spans the entire range of salinity from zero to salt saturation.

Other Research

An experimental program for research is conducted by NSF under its small business research program (16). Table 1 shows four awards that address the salinity ranges of salty agriculture, ocean salinity, through hypersaline. Because this research is in commercial concerns, it is possible that relatively rapid use of research can take place. The abstract of one of these awards is quoted as follows:

"The Chinese tallow tree, Sapium Sebiferum Roxb., is an introduced species of semi-tropical Euphorb which has become naturalized in the coastal wetlands from South Carolina to Texas. It is a potential high-yielding source of fat and oil which could substitute for petroleum derived lubricants and chemical intermediates.

This project is making a preliminary survey of the economic potential of the Chinese tallow tree to formulate detailed research plans for its development and exploitation."

One should also note the patent (17) issued to apply Duniella to production of glycerol as an example of applying salt tolerance directly.

Conclusions

From the viewpoint of salinity, it is clear that biosaline research in the United States and Canada spans the entire possible spectrum. Combined with research from other countries and regions,

TABLE 1. Examples of Biosaline Research Awards in Small Business

Ocean Pond	Fishers Island, NY	Enrichment of Brackish Water Pond for Shellfish
Ecoenergetics, Inc.	Vacaville, CA	Microalgae Production of Glycerol and Related Chemicals on Saline Waters.
Native Plants, Inc.	Salt Lake City, UT	Investigation of New Source of Natural Rubber in the United States.
Simco, Inc.	Weston, CT	The Chinese Tallow Tree as a Source of Petroleum Substitutes.
Neushul Mariculture, Inc.	Goleta, CA	Experimental Macroalgae Mariculture.

we can hope that research results will lead in time into useful endeavors. Several possibilities in applications from salty agriculture through hypersalinity microorganisms can be visualized. Because of the paucity of research in the hypersaline region combined with evidence of early commercial exploitation, this region may be fruitful for more research. The variety of research at lesser salinity levels shows that extremely varied and difficult research questions may exist in this little known region.

References

1. A. San Pietro, (ed.) "Summary Report of International Workshop on Biosaline Research," Indiana Univ., Bloomington (1977).
2. An expanded account is found in: "The Biosaline Concept: An Approach to the Utilization of Underexploited Resources,"

(A. Hollaender, J. C. Aller, E. Epstein, A. San Pietro and
O. R. Zaborsky, eds.) Plenum Press, N. Y. (1970).

3. C. C. Black, Jr., in: "Handbook of Biosaline Resources," Vol.
1B, Fundamental Principles, (A. Mitsui and C. C. Black,
Jr., Vol. 1B eds.) CRC Press, Boca Raton, FL (In Press).

4. H. L. Lieth and R. H. Whittaker, "Primary Productivity of the
Biosphere," p. 306, Springer-Verlag, N. Y. (1975).

5. E. Epstein, in: "Genetic Engineering of Osmoregulation--
Impact on Plant Productivity for Food, Chemicals and Energy"
(D. W. Rains, R. C. Valentine and A. Hollaender, eds) pp.
7-21, Plenum Press, N. Y. (1980).

6. E. Epstein, J. D. Norlyn, D. W. Ruch, R. W. Kingsbury,
D. B. Kelley, G. A. Cunningham and A. F. Wrona, Science
210:399-404 (1980).

7. International Directory of Current Biosaline Research Projects
S.S.I.E. (available through NTIS) (July 1980).

8. R. H. Nieman, in: "Genetics Engineering of Osmoregulation: Impact on Plant Productivity for Food, Chemicals, and Energy,"
(D. W. Rains, R. C. Valentine and A. Hollaender, eds.)
pp. 355-357, Plenum Press, N. Y. (1980).

9. F. Somers, M. Fontes and D. Grant, "Proc. of Inter. Arid Lands
Conf. on Plant Resources," Lubbock, TX (1978).

10. F. Somers (pers. comm.)

11. H. J. Teas, in: "The Biosaline Concept," pp. 117-163, op. cit.

12. International Decade of Ocean Exploration Progress Report,
Vol. 7, April 1977 - April 1978, Govt. Print. Ofc.
Stock No. 003-017-00431-7.

13. W. G. Keck and W. Hassibe, "The Great Salt Lake," U.S. Govt.
Print. Ofc., 0-277-036, Stock No. 024-001-03132-9 (1978).

14. Kwok Ki Ho, W. L. Ulrich, D. W. Krogman and C. Gomez-Lojero,
Biochim. Biophys. Acta 545:236-248 (1979).

15. A. A. Szalay and R. E. McDonald, in: "Genetic Engineering of
Osmoregulation Impact on Plant Productivity for Food,
Chemicals and Energy," pp. 321-329, op. cit. (1980).

16. Summary of Awards, NSF (1980).

17. U. S. Patent 4,115,949, Aaron et al, "Production of Glycerol
from Algae," U.S. Patent 4, 199, 895 (1980); "Production
of Glycerol, Carotenes and Algae Meal," U.S. Patent,
Plant 4, 511 (1980) "Algae Strain."

BIOSALINE RESEARCH IN LATIN AMERICA

Felix Cordoba Alva and Albertina Cota

Centro de Investigaciones Biologicas
de Baja California, A. C.
La Paz, B. C. S., Mexico

Biosalinity is a new concept comprising the study and development of the natural resources of Arid Zones surrounded by oceans. It is necessary, therefore, to understand the adaptative mechanisms whereby living organisms resist soil and water salinity. Thus, biosalinity research includes studies of halobacteria, marine algae and seaweeds as well as marine fauna up to marine mammals. Further, biosalinity research is concerned with the mechanisms which have evolved that allow them to survive in hot dry and salty environments. This research is directed also to a better utilization of desert and marine natural resources for the benefit of mankind in less developed countries, where these ecogeographical conditions prevail.

The biosalinity concept contains fundamental ecological implications as concerns the distribution and frequency of biological population of saline environments. One expectation is a decreased emphasis on fossil fuel consumption by employing solar energy and traditional renewable fuel sources. Photosynthesis is very much involved, as is the interaction of high temperatures -due to solar energy absorption --in marine and desert areas.

Latin America contains varied coastal regions, from equatorial and tropical zones, down to cold austral waters, sometimes with great deserts close to the marine environments. We can mention the deserts of northern Chile and southern Peru, of the Brazilian northeast, of the Guajira peninsula in Colombia, of the Samara beaches of Costa Rica, of the dry valleys from the Chitre in Panama, as well as the whole Baja California peninsula in Mexico.

We shall describe briefly some of the biosaline investigations underway currently in Latin America; this review is by no means exhaustive but selective in a few examples.

The scarcity of fossil fuels in Brazil, and the various approaches taken to increase their fuel supply by utilization of vast forests and jungles, is well known. One typical case is a "fuel plant," the cassava (Mandioca or tapioca) fruit, that is exploited for the production of a brazilian flour, called "farinha d'agua" (water flour), consumed by people as bread. Cassava contains 20 to 35% starch and 1 to 2% protein and has a most efficient photosynthetic physiology. It is resistant to various plant diseases and - which is particularly interesting now - shows good tolerance to saline soils and drought. In Brazil alone, cassava production amounts to about 28 million tons with a yield of 40 to 60 tons per hectare; most is used for flour.

Professor Barreto de Menezes, of the Institute of Food Technology in Campinas (Sao Paulo), is developing microbiological techniques for the optimal utilization of cassava. The flour can be saccharified to yield glucose which, by fermentation, produces the ethanol used in engines as a 10% mixture with regular gasoline (1). Glucose can be transformed into unicellular protein for cattle fodder or syrups for human diets. Finally, by means of selective fermentation, it is possible to obtain amino acids, antibiotics and solvents, all of which increases markedly cassava root utilization.

Following this example, it is interesting to consider cassava as a crop suitable for development in the saline areas of Latin America since these territories contain poor saline soils and receive only sparse precipitation.

Another interesting idea for developing the biosaline areas of the world is that put forward by Dr. Mario Gutierrez of the International Tropical Agriculture Centre in Cali, Colombia (2). He has demonstrated that the great losses in grains are due to fungal and bacterial infection. Focussing on beans - which are consumed abundantly in Latin America - Gutierrez has found that seeds from temperate areas suffer from about 95 diseases in comparison with 280 different diseases for seed beans from tropical areas. Also, crop yields increased 300% when clean seeds were employed. In other words, the overall productivity of bean fields would increase considerably if healthy seeds, rather than the genetically selected but infected beans that are normally planted, were used. This is a general problem and is reflected in several countries, including Mexico, that need to import grain to feed the increasing population.

Nevertheless, as pointed out by Gutierrez, it is increasingly

difficult to find areas for seed production since traditionally bean
fields, and those fields employed for other commercial crops, do not
meet the sanitary conditions required, and are already contamined.
As a consequence the author points out that potentially useful areas
for seed production could be located in the biosaline areas of the
Continent free of bean pathogens and endowed with high solar
irradiation (for rapid plant growth), temperatures of 28 to 35°C,
and healthy soils. One drawback to this interesting proposition
concerns the water resources needed for irrigation. However,
Gutierrez points out that this idea is suggested for clean seed
production only, an operation that requires less than 200,000 liters
of water daily. This water could be obtained from passive solar
distillation taking advantage of the high solar irradiation available in the deserts of the world. The economic benefit and investment recovery would be obtained in the increased (3-fold) yields
using disease free seeds instead of the short yields with bean seeds
from over exploited (and thus contaminated) soils. As before, this
technical proposal becomes particularly interesting for several
important biosaline areas of Latin America, and especially to Baja
California, comprising great virgin deserts that fulfill the
characteristics mentioned by Gutierrez.

Perhaps the most typical biosaline plants are mangroves. These
marine forests, widespread in the tropical coastlines, are composed
of several species with particular physiologies adapted to tolerâte
high salinities.

The group coordinated by Federico Pannier (3), of the Central
University in Venezuela, has done complete studies of Latin American
mangroves. The trees grow in thick forests in intimate contact with
marine water partially diluted by fresh water from rivers or rain.
In some instances, like Baja California where there are no rivers
and rain precipitation is low, the mangroves depend almost
exclusively on sea water with saline concentrations above 3%.

The adaptive mechanisms developed by mangroves to cope with
high salinities are of great interest in the scope of plant
physiology and molecular biology. Pannier and his associates are
beginning to perform research in this area and their most recent
results indicate that protein synthesis in mangrove (R. mangle and
Avicennia nitida) is related with the substrate saline gradient up
to 1% salt concentration; A. nitida is a species better adapted to
saline environments than R. mangle since it appears to tolerate salt
by means of a secretory mechanism. This finding helps one to understand the mangrove natural zonation; that is, the characteristic
distribution of mangrove species as observed in mangrove forests.
For example, in Baja California, R. mangle is implanted in the ocean
front followed by A. germinans and Laguncularia racemosa to the
back, close to the land.

Mangroves form an important marine resource and they have been utilized as a source of firewood (in places where no fossil fuel is available), as a source of tannins and other pigments, as cattle food and for other less important uses.

Perhaps the best way to understand mangroves is to consider the great ecological role that they play in desert coastlines; mangroves are rich ecological niches with great productivity comparable to coral reefs. Marine detritus is produced by abundant leaf decay and there is capture and formation of new land by the intricate mangrove root system in the estuaries and coastal lagoons. Plentiful marine and terrestrial organisms use mangrove forests as a suitable vital niche; from fish, crustaceans and molluscs to various insects and birds —mostly migratory species like ducks, geese and heron that occupy periodically the mangroves. This is the case of the mangrove forest in front of this Hotel.

In connection with the biosalinity concept, mangroves fulfill the requirement that salt tolerance is directly related to a very effective photosynthetic process in the high solar irradiated areas where these plants thrive. Due to the fact that mangroves constitute the only true marine forest, they represent interesting biological models for studying the biochemical mechanisms involved in salt resistance in higher plants.

Another plant typical of biosaline areas and developing in lands close to oceans is jojoba (Simmondsia chinensis). Originally from Baja California, and transplanted and developed successfully in other places, it is a well known desert specimen. Jojoba is becoming an economically important crop due to the fine oil (wax) present in the seed and the increased world demand for fatty products. Jojoba grows in poor saline soil, needs very little water and withstands prolonged drought and high temperatures (4).

Interest in, and development of, jojoba is concentrated at present in areas around the Gulf of California (both USA and Mexico). However, the particular characteristics of this plant, and the high profits expected from increased oil demand, has attracted the attention of Government agencies and private enterprises. They are studying the feasibility of transplanting and developing jojoba fields in other biosaline areas of the World. At present, jojoba grows well in Israel, Sudan, Australia, Costa Rica, and there are good indications that it will develop in some desert areas in Argentina. Thus, jojoba is becoming a typical biosaline - high profit crop donated by Latin America, where it is native, to other biosaline areas of the world. It is expected in the near future that jojoba will represent a main crop growing in most biosaline areas including South America. Nevertheless, much basic research is required before this goal is achieved. It is necessary to understand natural jojoba varieties in relation to high productivity stocks. Also, the factors which determine the appearance of certain suitable oil characteristics must be elucidated. The question about jojoba diseases is pending, as well as other questions related to massive jojoba cultivation. All these questions must be answered adequately before jojoba crops become a reality.

In Latin America many other investigations related –directly or indirectly – with biosalinity interests are underway. In Mendoza, Argentina, Brucher (5) is interested in developing new crops by adaptation of wild potatoes of the arid western part of the country. In Balcarce, Torres and Boelcke (6) are supplementing low quality pasture with corn to feed cattle, and in Sta. Rosa, Lavado and Reinaudi (7) are studying the effects of salinization in the soil of the Argentina plain. In Mar del Plata, De Ciechomski's group (8) has begun research on fish culture and Charpy-Roubad's group (9) is investigating the primary productivity of the northern Gulf of Patagonia.

In Brazil, in the Escuela de Agricultura de Piracicaba near Sao Paulo, several important projects on arid soils and agricultural problems are under investigation. In Brasilia, plant communities in calcereous tropical soils are analyzed by Furley's group (10).

Yoneshique et al (11) studies the production of seaweed for human consumption in the upsurgings that are present in Cabo Frio, near Rio. In the same place, Paschoa and Baptista (12) investigate, by way of isotopes, problems of plankton physiology.

In Santiago de Chile, several laboratories deal with the topic of biosalinity; Montenegro et al (13) analyze the dynamics of growth of the Chilean brush and Avila et al (14) direct physiological research on those plants which are resistant to drought and to salinity. In the Universidad Austral, located in Valdivia, Campos et al (15) are finishing a physical-chemical study related to the biotic resources of Lago Renihue, and, in Valparaiso, Finney (16) studies the saline "Stress" in copepods.

In Colombia, research done by the Centro Internacional de Agricultura Tropical, located in Cali, has received much attention. Here, they study nutritional aspects of wheat, cassava, etc., as well as the characteristics of soils dedicated to agriculture. Also, the fattening of cattle with "Elephant grass". In Bucamaranga, Gentry (17) studies the diseases of plants of economic interest, such as local flora, Steyermak (18) compiles the floristic inventory of the Guayana and Pannier (19) accomplishes a complete research of the Venezuelan mangroves of Orinoco. In Puerto Cabello, Luckhurst's group (20), in the Universidad Simon Bolivar, analyzes the substrates of fish communities in the coraline reefs.

In Havana, Cuba, there is marked interest to study the nutritional potentiality of raw honey and molasses in chickens and ducks. Martinez-Viera and Page (21) study the possibility of enriching the ferralitic soils of Cuba by the incorporation of nitrogen fixing organisms, such as Azotobacter.

In Mexico, aside from the projects of this Center (elsewhere in this volume), outstanding research is being done in the Centro de Investigaciones en Quimica Aplicada, in Saltillo, where the group directed by Campos (22), studies guayule, the rubber plant, as an advantageous substitute for Hevea rubber. They are interested also in agricultural and industrial aspects of jojoba, and other economically profitable desert plants like "governadora".

The University of Mexico (Centro de Ciencias del Mar) is performing various marine science projects. Of special interest is the work of Mandelli's group (23) concerning pollution in the Gulf of Mexico, and that of the Centro Internacional for the Improvement of Corn and Wheat (CIMMYT) which include in their studies, cereal varieties collected in the Americas (24).

REFERENCES

1. T. J. B. Menezes, Process Biochem 13:24-26 (1978).
2. M. Gutierrez, Agricultura de las Americas 28(1):36-41 (1979).
3. D. Mizrachi, R. Pannier and F. Pannier, Botanica Marina XXIII: 289-296 (1980).
4. La Jojoba. Mem. II Conf. Internacional Jojoba, Ensenada, B.C. Mexico, CONACT, CONAZA (1978).
5. H. Brucher, Angew Botanik 53:1-14 (1979).
6. F. Torres and C. Boelcke, Anim. Prod. 27:315-321 (1978).
7. S. R. Lavado and N. Reinaudi, Fluoride 12:28-32 (1979).
8. J. D. De Ciechomski and M. C. Cassia, J. Fish. Biol. 13:521-526 (1978).
9. C. J. Charpy-Roubaud, L. J. Charpy, S. Y. Maestrini and M. J. Pizarro, Comptes Rendu Acad. Sc. Paris 287:1031-1034(D) (1978).
10. P. A. Furley and W. W. Newey, Journal of Biogeography 6(1):1-16 (1979).
11. Y. Yoneshigue-Braga, S. Y. Maestrini and E. Gonzalez, Comptes Rendu Acad. Sci. Paris 288(D):135-138 (1979).
12. A. S. Paschoa and G. R. Baptista, Health Physics 35(2):404-408 (1978).
13. G. Montenegro, O. Rivera and F. Bas, Oecologia (Berl.) 36:237-244 (1978).
14. G. Avila, S. Araya, F. Riveros, and J. Kummerow, Ecol. Plant 13:367-373 (1978).

15. H. Campos, J. Arenas, W. Steffen and G. Aguero, "Physical and Chemical Limnology of Lake Rinihue (Valdivia, Chile)."
16. C. M. Finney, 2(2):132-135 (1979).
17. P. Genty, Oleagineus 33(8-9):421-427 (1979).
18. J. A. Steyrmark, Taxon 28(1,2/3):45-54 (1979).
19. F. Pannier, Environmental Management 3(3):205-216 (1979).
20. B. E. Luckhurst and K. Luckhurst, Marine Biology 49:317-323 (1978).
21. R. Martinez-Viera and H. Page, Landwirtschaft 16:137-144 (1978).
22. E. Campos, Chem. Tech. 9(1):50-57 (1979).
23. M. T. C. Rosales, A. V. Botello, H. Bravo and E. F. Mandelli, Mexico Bull. Environm. Contam. Toxicol. 21:652-656 (1979).
24. P. G. Coertz, W. G. Rollmer, E. Villegas, and B. S. Dhillon, Maydica XXIII:221-232 (1978).

SOME EUROPEAN CONTRIBUTIONS TO BIOSALINE RESEARCH

R. G. Wyn-Jones

Dept. of Biochemistry and Soil Science
University College of North Wales
Bangor, Gwynedd, Wales
UNITED KINGDOM

SUMMARY

Although biosaline problems are not a major preoccupation of European agriculture, there is an enormous interest in many aspects of this topic. While no uniquely European approach to biosaline research can be delineated, many significant contributions to fundamental aspects of salt tolerance and toxicity in plants, fungi and algae have been made by European laboratories. These are, however, part of the international scientific effort and close collaboration with other laboratories, particularly in Australia, has been and remains an important feature of this work.

Since little purpose is served by cataloging the European groups active in the field, the author has selected for brief discussion the major themes to which European laboratories have made outstanding contributions. These are: Ion compartmentation in relation to tolerance; Organic solute accumulation and the compatible cytosolute hypothesis; and the Relationship of respiration to osmoregulation. Emphasis will be placed on work with higher plants, again reflecting the author's prejudices.

The importance of integrating fundamental laboratory studies with field work and associating work in Europe with that in less developed countries where "Biosalinity" is a practical problem is also discussed. The bilateral link established with ourselves in Bangor and colleagues in the University of Agriculture, Faisalabad, Pakistan is discussed as an example of how such cooperation could be achieved.

INTRODUCTION

Soil salinity and alkalinity are not significant problems in the context of European agriculture and overall productivity is such as to generate little pressure for research into the exploitation of saline and brackish waters. Indeed, within the framework of the European Economic Community there is more concern about excess production - milk and wine lakes and butter and beef mountains - than about producing food or fuel from saline habitats. There are, of course, specific local problems such as those experienced in the Netherlands during the reclamation of the polders which had led to great technical expertise in the reclamation of such salt affected soils but this is not strictly relevant to the "Biosaline Concept." Therefore in Europe biosaline research has grown out of fundamental physiological and ecological studies rather than being a response to agronomic or economic pressures.

Although lacking a strong external stimulus, many European laboratories have been prominent in studying the ecology, physiology and biochemistry of salt stress in plants and microorganisms. Indeed it must be pointed out that the late Professor Marcel Florkin of the University of Liege was the father of modern comparative physiology and biochemistry and his group contributed substantially to our knowledge of the adaptation strategies of marine invertebrates. Unlike highly applied or agronomic studies which are concerned with exploiting the productive potential of a specific region, most European work in this field is not 'regional' but is best regarded as part of the international scientific effort. Often the country of origin is irrelevant to such studies and indeed the cooperation between laboratories in different countries, particularly between numerous European laboratories and colleagues in Australia, has proved particularly productive. With this caveat a substantial European contribution to fundamental biosaline research can be recognized. I shall not attempt the near impossible and, I believe, counter productive task of listing all the relevant work in progress in Europe. Rather, I shall subjectively select four strongly but not exclusively European themes of work which have made important contributions to our understanding of salt toxicity and tolerance, particularly in higher plants. Without doubt, another author would have chosen with equal or perhaps greater justification to discuss other work. Finally, I feel that the extensive Australian contribution to these themes of work must be emphasized.

Ion Compartmentation

An increasing body of evidence has shown that plant cells selectively accumulate K^+ in their cytoplasms (cytosols) while occluding most of the Na^+ and Cl^- in their vacuoles. Evidence for this proposition, which is clearly of fundamental importance to our understanding of the mechanisms(s) of salt tolerance comes from several laboratories and is based on a variety of techniques.

Based on Professor Pitman's original method, a sophisticated system of flux analysis has been developed by Professor Jeschke in Wurzburg using initially roots of barley (1) and more recently of Atriplex hortensis (2). This has allowed cation selectivity at the plasma membrane, the tonoplast and the symplasm/xylem boundary to be assessed and semiquantitative measurements of ion concentrations in the major compartments to be made. A summary of the cation selectivities observed in barley is shown in Figure 1.

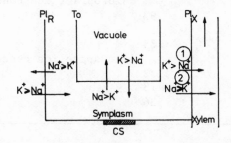

Figure 1. Specificities of cation fluxes in barley and
           A. hortensis roots. (1) Barley, Jeschke (1);
           (2) A. hortensis, Stelter (2).

X-ray microprobe analysis of plant samples has been developed by Professor Andre Lauchli and his colleagues in Hanover and by Drs. Diana Harvey and John Hall in Brighton. Some data from the latter on subcellular ion distribution in Suada maritima is presented in Table 1 to illustrate the potency of this technique. However, difficulties still remain, particularly with regard to the cytoplasmic ion levels (concentrations) and the K^+ levels reported in this table are open to some doubt as they do not account for the total tissue K^+ content.

Complementary to microprobe analysis has been the microchemical analysis of root tissues of differing degrees of vacuolation by

TABLE 1.   Intracellular ion cencentrations in salt grown
           Suaeda maritima (3)

Inorganic ion	% of cell occupied by cell type	
	60%	30%
	vacuolar ion concn. (mM)	
Na	565	422
K	24	11
Cl	388	301
	cytoplasm/cell wall ion concn. (mM)	
Na	109	146
K	16	28
Cl	21	112
	chloroplast ion concn. (mM)	
Na	93	75
K	16	29
Cl	85	86
	free space ion concn. (mM)	
Na	132	22
K	13	0
Cl	36	38

Jeschke's group.   Again, the evidence indicated a high degree of
K^+-specificity in tissues of low vacuolation (4).

The comparative biochemistry of the effects of salts on
enzymes and organelles from glycophytes and halophytes has been
developed by Dr. Flowers in Sussex as well as   several Australian
groups and more recently by my own colleagues in Bangor (5-7). Such
studies have also provided indirect evidence and a   theoretic basis
for ion compartmentation in plant cells (see 5, 7, 8).

While there are problems with each of these techniques, taken
together they indicate strongly that cytoplasms have a high K^+
selectivity, even in halophytes and that Na^+ and Cl^- are largely
vacuolar. However, many doubts remain; for example, it may be that
ion discrimination is much higher in meristematic cells and that
cytosolic ion discrimination is less marked in more mature vacuolated cells of lower and more restricted   metabolic activity.
Unfortunately, detailed discussion of the evidence is impossible in
this presentation, nevertheless it is interesting to consider two
other recent pieces of work as they serve to underline the significance of ion compartmentation to whole plant physiology and the

exploitation of saline habitats. Professor Raven (9) speculated on the close relationship of the phloem to cytosol and this hypothesis would lead to the expectation that the phloem would exercise a similar ionic discrimination to cytoplasm. While studying aphid nutrition on Aster tripolium, Dr. Dowling in Cambridge (10) obtained dramatic evidence to support this theory. Some of his data are shown in Table 2, which also includes some of our data on Aster petals. Since these organs are supplied by the phloem, we reasoned that their chemical composition would reflect that of the phloem and be clearly different from that of highly vacuolated leaves. Such a proposal seems borne out by the data in Table 2. Thus, the phloem-fed tissues such as fruits and seeds even of halophytes would be expected to have relatively low Na^+ and Cl^- contents. Such a conclusion has important practical implications in the development of cereal grain crops able to withstand salinity and of course is consistent with the observation that the fruits from the tolerant tomato crosses obtained in Davis still require table salt to eat! (Rush and Epstein, pers. comm.)

TABLE 2: Chemical composition of Aster tripolium phloem sap, florets and leaves.

Osmal Kg^{-1}	Leaves[1]	Florets[1]	Phloem Sap[2] Fresh Water 599 $\pm$ 20	Sea Water 1544 $\pm$ 50
K	72 $\pm$ 6	133 $\pm$ 21	86	106
Na	360 $\pm$ 18	56 $\pm$ 3	0.35	31
Cl	320 $\pm$ 15	51 $\pm$ 9	5.4	28.3
Amino acids	12 $\pm$ 5	58 $\pm$ 10		
Proline	N.D	21 $\pm$ 1		
Glycinebetaine	18 $\pm$ 2	82 $\pm$ 6		
Total soluble sugar	53 $\pm$ 6	493 $\pm$ 69		
Sucrose equivalents			302	672

[1]Data from Gorham et al. (11) analyzing field collected sample.

[1]Data from Downing (10) analyzing phloem from Aster plants grown in fresh or sea water.

An interesting problem arising from the compartmentation model is the need to rationalize the evidence for relatively low Na^+ levels in the root symplasm with the ability of halophytes to accumulate this ion in their leaves even at relatively low external Na^+ level (12,13). Recently Dr. Stelter (2), in collaboration with Professor Jeschke, observed the reversal of K^+ and Na^+ selectivity at the xylem:symplasm boundary of A. hortense compared with barley. In this halophyte discrimination at the boundary favored Na^+ over K^+, thus it is possible to rationalize the apparent inconsistency noted above (Fig. 1).

Membrane Chemistry and Biochemistry

The study of the changes in membrane lipid composition in response to salt stress and the comparative lipid chemistry of halophytic and glycophytic plants has been pioneered by Professor Kuiper and his colleagues in Groningen (14, 15). The Dutch team has over the years worked closely with Professor Kylin who, with collaborators in both Denmark and Sweden, investigated extensively the comparative biochemistry of the ATPases of species from different habitats (16, 17). It is only possible to illustrate this work very briefly. In a recent study the Dutch group compared the lipid compositon of membrane extracts from three Plantago species differing in salt tolerance and in the fertility of the ecological niche. It is apparent from the data reproduced in Figure 2 that there are both salt-induced changes in many lipid components and difference in the basic composition of the related species from differing habitats (14). In other work this group has compared the lipid composition of bean, barley and sugar beet, the former being much more sensitive than the two other species, and observed marked differences (15). For example, the more tolerant species were found to have higher root sterol and sulpholipid levels.

In much the same way Kylin and his colleagues reported differences in the ATPase activities from sugar beet cultivars of different tolerances (17) and in this work an association of (Na^+ + K^+) - ATPase activity with sulpholipid was observed (16).

This is a significant line of work which will, I'm sure, produce important information in future.

Compatible Organic Solutes

A major preoccupation of several European laboratories has been the accumulation of organic solutes in halophytic plants. The original work on proline accumulation in halophytes was published

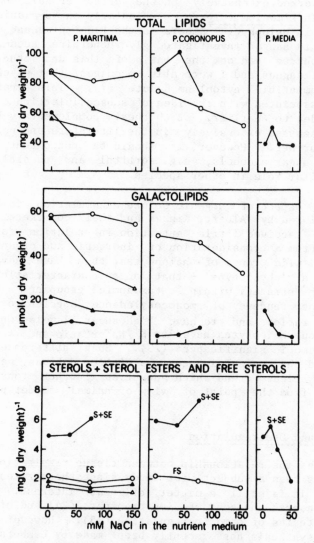

Figure 2. Lipid composition of the roots and shoots of Plantago maritima, P. coronopus and P. media as affected by salinity. (●) roots and (o) shoots from nutrient-rich conditions; (▲) roots and (△) shoots from nutrient-poor conditions. Data from (14).

in French by Professor Goas (18) and these observations were extended by Drs. Stewart and Lee in Manchester (19) and Treichel in Darmstadt (20). Dr. Larher, also a member of the French group at Rennes, has worked extensively on the effect of salt on nitrogen metabolism in Limonium vulgare, particularly on β-alanine betaine accumulation in halophytic members of the Plumbaginaceae (21). Dr. Storey (22) in Bangor investigated glycinebetaine accumulation in various halophytes and on the basis of this and other data, a number of colleagues and I were able to suggest that glycinebetaine acts as a compatible cytoplasm solute (7). This hypothesis is intimately associated with the idea of subcellular ion compartmentation alluded to earlier. Both these complementary hypotheses have been discussed extensively in the literature and need not be discussed further. However, it should be noted that there is evidence of sugar alcohols, e.g. sorbitol and possibly pinitol, having a similar role in other species.

Another aspect of organic solute accumulation in halophytes has been explored by Albert, Kinzel and their coworkers in Vienna (23) and has received little attention in relation to Biosaline Research. From a consideration of inorganic and organic solute levels in a wide range of halophytes, they have advanced the concept of a 'physiotype' - that is a characteristic solute -accumulation strategy within a taxonomic grouping. They have emphasized the tendency of monocotyledenous species to have high leaf K^+/Na^+ ratios and to have high sugar levels whereas the dico-tyledenous halophytes have low K^+/Na^+ rtatio and sugar levels of little osmotic significance (23). Such differences are not clear cut and their significance is a little dubious, nevertheless there is a clear trend which deserves further consideration, particularly from the point of view of animal fodder production.

Respiration and Osmoregulation

Although the relationship between tissue respiration and ion transport has been a subject for study since the earliest work of Lundegardh, it is still a matter of great interest. There are attempts to assess the cost of ion transport (24) and of the different strategies of osmotic regulation (25). However, an interesting new synthesis has recently been made by Lambers (26) who surveyed the role of the alternative pathway of respiration (SHAM-sensitive-cyanide-resistant) in several aspects of plant physiological chemistry. In P. coronopus the alternative pathway is inhibited by external NaCl and this is correlated with the accumulation of sorbitol, the probable compatible solute in this species. Alternatively, excess sugars are thought to be oxidized by this pathway if not required for osmoregulation or other metabolic link. Clearly, Lambers' concept requires further exploration but it is another interesting example of the integration of stress physiology into the basic metabolism of plants.

Discussion

The examination of these four themes is far from comprehensive and does less than justice to the contributions from many other laboratories such as those of Professors Mengel and Marschner in Germany studying a range of agronomic and physiological problems; Dr. Rozema in the Netherlands emphasizing ecophysiological topics; Professor Jennings in England working on marine fungi as well as higher plants; and Professor Heller in France who has pioneered work on salt tolerance using single cell suspension cultures. Such a list is itself quite inadequate and omits many important highly relevant contributions and entirely neglects work in progress on macro and micro-algae from saline habitats.

Despite obvious limitations, I hope that the review has demonstrated the enormous contribution made by European laboratories, with the colleagues and collaborators all over the world, to our fundamental understanding of salt tolerance in plants. Through further basic research I believe that European laboratories can continue to make major contributions to biosaline research. I would illustrate these possibilities by reference to some other recent developments. If intracellular ion compartmentation proves as crucial to the salt tolerance of halophytic plants as currently anticipated then the study of the biochemistry and biophysics of the tonoplast from different species, and possibly different cell types, is an urgent priority. The comparative work on the lipid chemistry and ion-specificity of ATPases from different plant species developed by Kuiper and Kylin must be refined to the level of specific membranes. An indication of what can be achieved comes from recent work by Walker and Leigh (27) in Cambridge who have demonstrated the presence of an anion-stimulated ATPase on the tonoplast of red beet cells. Dr. Walker's presence in this conference presenting a regional report on behalf of Australia also emphasizes the role of international cooperation in this work.

Other exciting developments should follow from the application of the micro-pressure probe, developed by Professor Zimmerman and his colleagues (28), to halophytes. Currently, our understanding of the interrelationships of the osmotic and ionic stresses on higher plants is very inadequate. However, this new device should allow water relations on a cellular basis to be explored in some detail.

In addition to pursuing these fundamental problems, there is a strong case for more direct cooperation between European laborator-

ies and colleagues in less developed countries where salinity is an
immediate problem.   Such cooperation can be envisaged on several
planes.   While   in this review I have tried to show how our
fundamental knowledge is developing, it is also apparent that more
and more sophisticated biochemical, biophysical and micro-analytic
techniques are being applied to these problems. Clearly there is,
therefore, an important educational role for European laboratories.
In my opinion conventional graduate studies are inadequate as the
return of the graduate to his home country leads to a high
'scientific casualties' rate.   A continuing link between institutions and individuals in the two countries is required to effect
technological and scientific interchange, to allow a scientific
outlet for the scientist from the developing countries, who might
otherwise be overwhelmed by his problems and to bring European
scientists into direct contact with field problems.   Such a link
has recently been established by the Overseas Development Adminis-
tration of the United Kingdom Government between myself and
colleagues at the University of Agriculture.   This contact has also
brought me into contact with very interesting Pakistani work on the
use of the grass, Diplanche fusca, to exploit saline and alkaline
soils (29).   This very tolerant grass is particularly interesting
and has a high leaf K^+/Na^+ ratio unlike the Chenopodiaceae and thus
may be more favorable in maintianing the salt balance of the
animals.   In general much more consideration should be given to the
chemical compositional differences underlying Albert and Kinzel's
concept of physiotypes in selecting possible species for agronomic
exploitation.

    The importance of such international contacts is underlined by
the development of a genetic approach to the exploitation of saline
soils and waters pioneered by Professor Epstein and his colleagues
in Davis.   So far European laboratories have not figured prominent-
ly in this work but a notable exception is the cooperative work on
screening rice for salt tolerance being carried out by Dr. Tim
Flowers in Sussex in collaboration with IRRI.

References

1. W. D. Jeschke, in: "Recent Advances in the Biochemistry of
        Cereals," (D. L. Laidman and R. G. Wyn Jones, eds.),
        pp. 37-61, Academic Press, London-New York (1979).
2. W. Stelter, Doctoral Thesis, Universitat Wurzburg, West
        Germany (1979).
3. D. M. R. Harvey, T. J. Flowers and S. L. Hall, unpub. data.
4. W. D. Jeschke and W. Stelter, Planta 128:107-112 (1976).
5. T. J. Flowers, P. F. Troke and A. R. Yeo, Ann. Rev. Plant
        Physiol. 28:89-121 (1977).
6. H. Greenway and R. Munns, Ibid, 31:149-190 (1980).

7. R. G. Wyn Jones, R. Storey, R. A. Leigh, N. Ahmad and
    A. Pollard, in: "Regulation of Cell Membrane Activities in
    Plants," (E. Marre and O. Ciferri, eds.) pp. 121-136,
    Elsevier/North Holland (1977).
8. R. G. Wyn Jones, C. J. Brady and J. Speirs, See Ref. 1,
    pp. 63-104 (1979).
9. J. Raven, New Phytol. 79:465-480 (1980).
10. N. Downing, Doctoral Thesis, Cambridge University, England
    (1979).
11. J. Gorham, H. Hughes and R. G. Wyn Jones, Plant Cell Environ.
    3:309-318 (1980).
12. R. Collander, Plant Physiol. 16:691-720 (1941).
13. R. Storey and R. G. Wyn Jones, Ibid 63:156-162
    (1979).
14. L. Erdei, C. E. E. Stuiver and P. J. C. Kuiper, Physiol.
    Planta. 49:315-319 (1980).
15. C. E. E. Stuiver, P. J. C. Kuiper and H. Marschner, Ibid,
    42:124-128 (1978).
16. G. Hansson, P. J. C. Kuiper and A. Kylin, Ibid, 28:430-435
    (1973).
17. J. Karlsson and A. Kylin, Ibid, 32:136-142 (1974).
18. M. Goas, Bull. Soc. Fr. Physiol. Veg. 11:309-316 (1965).
19. G. R. Steward and J. A. Lee, Planta 120:279-289 (1974).
20. S. Treichel, Z. Pflanzen Physiol. 76:56-68 (1975).
21. F. Larhes, Doctoral Thesis, Universite de Rennes, France (1976)
22. R. Storey, Doctoral Thesis, University of Wales, Cardiff, Wales
    (1976).
23. R. Albert, M. Popp, Oecologia 27:157-170 (1977).
24. R. G. Wyn Jones, in: "Physiological Processes Limiting Plant
    Productivity," (C. B. Johnson, ed.), pp. 271-292,
    Butterworth Publ. (1981).
25. B. W. Veen, in: "Genetic Engineering of Osmoregulation,"
    (D. W. Rains, R. C. Valentine and A. Hollaender, eds.)
    187-195, Plenum Publ. Corp., NY (1980).
26. M. Lambess, Plant Cell Environ. 3:293-303 (1980).
27. R. A. Leigh and R. R. Walker, Planta 150:222-229 (1980).
28. D. Husken, E. Steudle and U. Zimmermann, Plant Physiol.
    61:158-163 (1978).
29. G. R. Sandhu, Z. Aslam, M. Salim, A. Sattar, R. H. Qureshi,
    N. Ahmad and R. G. Wyn Jones, Plant Cell Environ. 4
    In Press) (1981).

BIOSALINE RESEARCH IN ISRAEL: ALTERNATIVE

SOLUTIONS TO A LIMITED FRESH WATER SUPPLY

Dov Pasternak

Applied Research Institute
Ben-Gurion University of the Negev
P.O.Box 1025, Beer-Sheva 84110, Israel

SUMMARY

Biosaline research can significantly contribute to an increase in the agricultural production of Israel, a country in whch all available fresh water resources are already being utilized. The following biosaline research projects are reviewed: (1) Introduction, selection and development of dryland fodder shrubs for regions of marginal ($\approx$200 mm) rainfall; (2) Production of animal feed and other products from microalgae; (3) Introduction and development of new water-sparing industrial crops such as jojoba; (4) Utilization of geothermal and solar energy for the production of out-of-season export crops; (5) Irrigation of agricultural crops with brackish water; (6) Utilization of sea water for the production of fodder crops, fish and algae.

A detailed bibliography on biosaline research in Israel is included.

INTRODUCTION

Israel is a relatively small country with a total area of 20,325 km^2 and a population of 3.8 million. About 60% of the country is classified as desert. The climate is typically eastern Mediterranean, and rainfall occurs over the five winter months. The summer is hot and dry, and daily evaporation rates range between 8-20 mm according to the region and the elevation. These facts, together with the country's high rate of economic development, have led to intensive utilization of water for irrigated agriculture.

Today, practically all the known fresh water supplies of
Israel are being fully utilized. Some facts and figures on water
consumption are given in Table 1. Most of the fresh water
resources are concentrated in the northern part of the country.
Much of the water is transported to the dry south, over long
distances, via the National Water Carrier. Accordingly, the cost
of fresh water in the Negev desert in the South is about four times
higher than that in the area of lake Tiberias which lies in the
north (Table 1).

TABLE 1. Water in Israel - Some Facts and Figures[*]

a. Water production and consumption in 1977

Parameter	Amount $(m^3 \times 10^6)$
Total water production	1,740
Total water consumption	1,670
Total agricultural consumption	1,230
Agricultural consumption of brackish water	104
Estimated potential for brackish water development	120 (+ 10%)[**]

b. Estimated cost/m^3 of water in various regions, March 1980

Region	Cost (US$/$m^3$)
Lake Tiberias	0.06
Coast	0.11
Negev	0.22

[*]The data are from ref. 1 and 2.

[**]Salinity range: 1,000-4,000 ppm TDS.

TABLE 2. Agricultural Production and Trade in Israel (1979/80)

Some Facts and Figures[*]

Total agricultural area	–	429,700 hectares
Total area under irrigation	–	200,000 hectares
Total value of agricultural exports	–	$783.6 million; – 38% of total production
Total value of agricultural imports	–	$755.2 million
Total value of imported grains	–	$133.0 million; 18% of agricultural imports
Quantity of grain imported	–	1.1 million tons
Area required to produce this quantity	–	275,250 hectares[**]
Water required to irrigate this area	–	1,100 million m^3 [***]

[*]The data are adapted from ref. 1.

[**]Estimated on the basis of 4 tons grain/ha.

[***]Estimated on the basis of 4,000 m^3/ha.

Some facts on agricultural production and the agricultural trade balance in 1979/80 are given in Table 2. Israeli agriculture is concentrating more and more on the production of lucrative cash crops. Last year about 40% of the country's agricultural input was directed to the export market. This has more than offset the costs of the imported agricultural products. The major agricultural import item is grain for animal feed. Practically all the livestock produced is for local consumption. If Israel were to produce its total needs for feed grain locally, it would require an area of 275,000 hectares and 1,100 million m^3 of water. In other words, it would have to double not only the total area currently under irrigation but also the quantity of the available irrigation water.

Biosaline research can help Israel to overcome some of the handicaps which stand in the way of further agricultural development. The ways in which the country is trying to increase its agricultural input (Fig. 1) may be divided into the following categories:

I.   Increased water use efficiency of crops;
II. Increased returns per cubic meter of water applied; and
III. Development of new alternative water sources.

The first two divisions are related but not necessarily identical. Within each category, agriculture is based on two approaches: (a) Changing the environment to suit the plant, i.e. technology and management; and (b) Changing the plant to suit the environment, i.e., introduction and genetic manipulation.

Figure 1 illustrates the subdivision of each of the three alternatives into the technological and the plant-manipulation approaches.

I. INCREASED WATER USE EFFICIENCY OF CROPS

Water use efficiency is defined as the yield divided by the amount of water applied. Water use efficiency may be increased either by decreasing the amount of water applied (without decreasing the yield proportionately) or by increasing the yield.

A. Technology

Two relatively new systems of irrigation are under development - drip (trickle) irrigation and soiless culture.

Drip irrigation was first introduced in the mid-sixties. It has since received international recognition as the system through which water and fertilizers can be applied most accurately (both in space and in time) and as an irrigation method in which water losses due to evaporation or inefficiencies are minimized. Irrigation with a drip system results in about 30% water saving as compared with sprinkle (3) or furrow (4) systems. One of the latest developments in drip technology is a mobile drip irrigation machine which is being developed by Mr. M. Farbman of the Faculty of Agricultural Engineering of The Technion - Israel Institute of Technology in Haifa and Mr. U. Shani of the Arava Regional Research Station at Yotvata.

Another recent achievement in irrigation technology is computerized irrigation. With computerized systems it is possible to control the timing and the quantity of irrigation water, taking

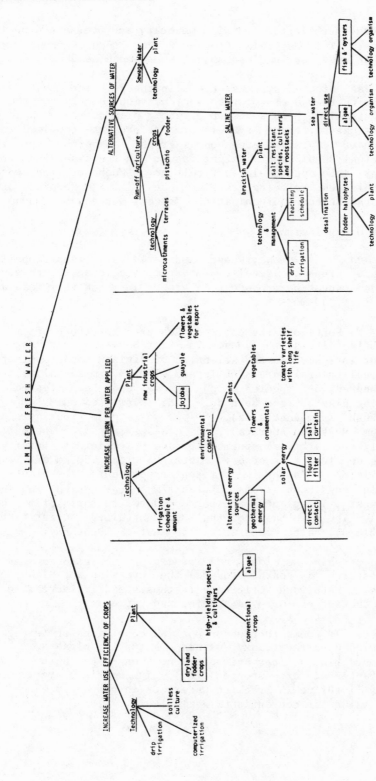

Figure 1: Alternative solutions to a situation of limited fresh water supply.

into account such factors as wind velocity and pressure drop in the line, etc. The Motorola Company which produces these systems claims that water savings can amount to as much as 30%.

Soiless culture systems are, in theory, the best systems through which the water, oxygen and nutrient supply to the plant can be controlled. As a result of their high costs and high level of sophistication, these systems are at present confined to the production of lucrative greenhouse crops. Nevertheless, future technological developments might result in the expansion of this technology into open-field agriculture. Much of the work on soiless cultures is being carried out by Dr. H. Sofer of the Agricultural Research Organization, Volcani Center, Bet Dagan.

B. Plant Manipulation

Dryland fodder crops. Wheat and barley are traditionally the major fodder crops of Israel's semi-arid zone, even though economical yields cannot be produced in areas in which the mean annual rainfall is $\leq$ 250 mm.

Mr. M. Forti of the Applied Research Institute of the Ben-Gurion University of the Negev in Beer-Sheva is carrying out a project of introduction and selection of dryland fodder plants for the marginal rainfall areas of Israel (180-250 mm annual rainfall). Out of hundreds of species which have already been tested, three outstanding plants have been selected: *Atriplex nummularia* (old man saltbush from Australia), *Atriplex canescens* (four wings salt bush from northern America) and *Cassia sturtii*. These plants, especially the *Atriplex* species are very efficient water users, producing up to 5 tons of dry matter/ha with 150 mm of rain per year (Table 3). In the most recent experiments, the plants were planted in dense populations (2 plants/m^2) over buried raw municipal garbage. The plants are harvested two to three times during the year. The feed value of the fodder obtained so far is fairly high (Table 4), being comparable with that of wheat hay.

High-yielding species and cultivars. Contemporary agricultural research is geared towards the creation of high-yielding cultivars. Water use efficiency is thus steadily increasing with the introduction of improved species and varieties.

Algae. It seems that under optimal conditions of temperature, radiation, and nutrient, oxygen and CO_2 supply, algae are able to produce more biomass (per unit of land area or per unit of water) than conventional crop species (6). In addition, various algal species will thrive in brackish water or in sea water which has a marginal value for conventional agriculture.

TABLE 3.   YIELD OF DRYLAND FODDER SHRUBS (TON/HA) FOR THE PERIOD
           DECEMBER 1978 - NOVEMBER 1979 AS AFFECTED BY RAW
           MUNICIPAL GARBAGE.* Values are means of four
           replications. The following conditions prevailed:
           Amount of garbage applied, 150 tons/ha; Planting date,
           5.5.1977; Planting density, 20,000 plants/ha; Number of
           harvests per year, 2; Mean annual rainfall, 200 mm; and
           Winter rainfall (1978-1979), 154 mm.

Atriplex nummularia		*Atriplex canescens*		*Cassia sturtii*	
Garbage	Control	Garbage	Control	Garbage	Control
3.87	2.78	5.37	3.83	0.85	0.62

*Adapted from ref. 5

       The development of production technologies and the selection
of algal species are under way at The Desert Research Institute at
the Sde Boker Campus of the Ben-Gurion University of the Negev.
The project is led by Prof. A. Richmond.   Two species of *Spirulina*
are being investigated -*S. platensis*, a brackish water thermophilic
alga, and *S. Subsalsa*, a marine algae with low temperature
requirements.

II.   INCREASED RETURN PER UNIT OF WATER APPLIED

A. Technology

       Irrigation regime. Some years ago, the definitive project on
water economy was carried out by a group of scientists, led by Dr.
J. Stanhill, of the Institute of Soils and Water, Agricultural
Research Organization. They succeeded in quantifying the amount of
irrigation water and in finding the optimal scheduling of irrigation which would give the highest return per cubic meter of water
applied.   The USA class A evaporation pan has been introduced as a
tool for estimating irrigation water requirements.   It has been
demonstrated that for many crops the highest biological water use
efficiency is not necessarily synonomous with the highest capital
return per cubic meter of irrigation water.

       Environmental control. By 1970, there were only a handful of
greenhouses in Israel.   Today, a decade later, the greenhouse area

TABLE 4. FEED QUALITY OF THREE DRYLAND FODDER SHRUBS MANURED WITH RAW MUNICIPAL GARBAGE*

Quality Component (dry weight basis)	*Atriplex nummularia*		*Atriplex canescens*		*Casia sturtii*	
	Garbage	Control	Garbage	Control	Garbage	Control
Protein (%)	13.7	10.6	12.5	9.6	10.7	12.1
Cellulose (%)	24.8	32.0	35.5	32.7	27.7	25.3
In vitro digestibility (%)	55.7	–	48.6	–	43.8	–
Metabolic energy (megacalories/kg)	2.0	–	1.7	–	1.5	–
Ash (%)	21.6	18.4	13.4	15.6	6.6	7.0

*Adapted from ref. 5.

is estimated to be about 500 ha, and it is still growing. Most of
the greenhouse   products   are   out-of-season   ornamentals   and
vegetables for export.   Within the next 10 years, Israel is
planning to build some   25 new   settlements in the   western Negev,
which will be based almost exclusively on out-of-season greenhouse
tomatoes.   At present, greenhouses are not heated, but, additional
heating could result in substantial increases in yields and in the
addition of new heat-loving crops.   Greenhouse heating could be
accomplished by means of two energy sources which are abundantly
available in the Negev desert -geothermal energy and solar energy.

     Geothermal energy: The vast brackish water aquifers which
underlie most   of the southern   part of Israel are deep, and the
water is   therefore warm (30-60°C).   For the past few years, the
Applied Research Institute of the Ben-Gurion University of the
Negev has been developing systems for the utilization of this heat
trapped in the water (7).   The water is pumped out of deep wells
and is used to heat the soil and the air of protected crops.   It
then flows to a storage pond from where it is pumped to irrigate
large fields of conventional crops (Fig. 2).   Energy, if used in
this way, is very cheap because it can be regarded as a by-product
of the irrigation water. This project is led by the author.

     Solar energy: The collection and storage of 20% of the solar
energy which falls onto a greenhouse during a typical winter's day
is sufficient for night heating.   Three different systems are at
present under development: (a) The direct-contact system in which
solar heat is absorbed or dispersed by direct contact of warmed air
with a wall of water drops ("cooling tower"); the heat is stored in
a water pond.   The system is being developed under the leadership
of Mr. N. Zamir, at the Institute of Agricultural Engineering of
the Agricultural Research Organization; (b) The liquid-filter roof
is being developed by Dr. J. Gale at The Desert Research Institute,
Ben-Gurion University of the Negev.   The system is based on the
filtration of the sun's longwave radiation through a liquid-filled
double-walled greenhouse roof which has been treated with selective
paints.   The heat is stored in water.   The heat is dissipated by
the liquid-filter roof (see Fig. 2); and (c) The salt envelope[*].
This system is based on the collection and storage of solar energy
in clear hydrate salts which are placed inside double-layered
plastic sheets from which the greenhouse roof and walls are
constructed.   The salts change phase from liquid to solid and vice
versa at about 20°C; in so doing they release or store large
amounts of latent heat, respectively.   A roof of 1.5 cm of salts
over the plants stores enough energy for its temperature to be
maintained at 20°C during a whole winter's night.   The system is

---

[*]Patent pending

TABLE 5.   SELECTION FOR HIGH YIELD IN A JOJOBA POPULATION OF 270
           PLANTS DURING THE PERIOD 1965-1975[*]

a. Average yield of six plants

High Yielders		Low Yielders	
Plant No.	Yield (g/plant)	Plant No.	Yield (g/plant)
1	1,500	1	300
2	1,300	2	240
3	1,200	3	90

b. Distribution of the yield in 1973

Yield Range (g/plant)	% of Population
1,000 - 1,500	7
1,500 - 2,000	2.2
above 2,000	1.9

[*]Data in the Table are obtained from ref. 8.

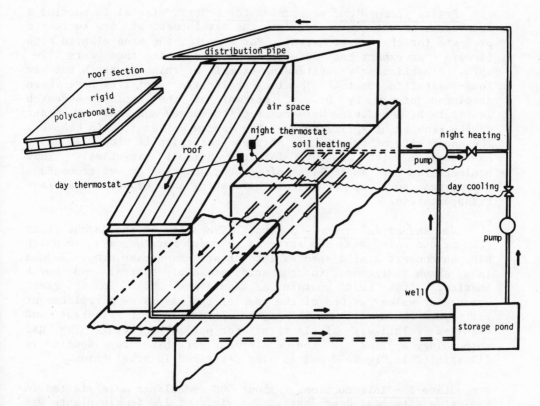

Figure 2.   The "water curtain system" for heating greenhouse space
with low-temperature water.   The water is pumped from a geothermal
(40°C) well and flows through a network of pipes buried in the
ground to heat the soil.   The water returns to a storage pond from
which it is pumped through a double-layered roof made out of rigid
polycarbonate.   The warm roof heats the plants by longwave radiation and prevents heat loss to the atmosphere.   The water then
returns to the storage pond from where it is pumped to irrigate
conventional field crops (Adapted from ref. 7).

being developed at the Applied Research Institute of the Ben-Gurion
University of the Negev.   The leader of this project is Prof. D.
Wolf of the University's Department of Chemical Engineering.

B. Plant Manipulation

Fruit, flowers and vegetables for export. Israel is putting a
tremendous amount of effort into the development of new lucrative
products for the export market. For example, the area planted with
flowers for export has more than doubled during the years 1976-
1979. A particularly outstanding example of this effort is the new
long-shelf-life tomato. The long-shelf-life varieties have been
developed jointly by Dr. Y. Mizrahi of the Applied Research
Institute of Ben-Gurion University of the Negev and Prof. N. Kedar
of the Faculty of Agriculture of The Hebrew University in Rehovot.
The shelf life of ripe fruits of these varieties is about 40 days,
as compared with 8 days for conventional varieties.   This
prolongation of shelf life offers the tomato grower tremendous
advantages, such as mechanical harvesting and export by surface
transportation.

New industrial crops - jojoba. The story of the jojoba plant
(*Simmondsia chinensis*) can serve as a model for the ways in which
R&D can convert a wild species into an arid-zone cash crop. Jojoba
is a shrub indigenous to the arid areas of central and north
America. Its fruit contains a liquid wax which is of great
commercial value.   (One of the uses of the wax is as a replacement
for sperm whale oil).   The jojoba plant is salt resistant and
requires relatively little water for optimal growth.   The R&D
methodology applied in the development of this crop species is
illustrated in Figure 3 and is also described in brief below.

Stage I - Introduction: About 500 seedlings were planted in
1961 at a site near Beer-Sheva. The yield of 270 female plants was
assessed over a period of ten years.   From these plants, five
outstanding specimens (including three high yielders) were selected
(Table 5), and they served as the basis of a hybridization program.
Preliminary trials showed that with proper techniques, the
outstanding plants could be propagated vegetatively. A feasability
study was then carried out and in the light of the promising
results a commercial company - the Negev Jojoba Company - was
established.

State II - Research & Development: The Company has three
definite, interrelated and equally important duties: funding, R&D
and commercialization (Fig. 3).   Research & Development is carried
out at the Applied Research Institute of the Ben-Gurion University
of the Negev (under the direction of Mr. M. Forti) and is divided
into industrial and agricultural aspects.   The agricultural side is

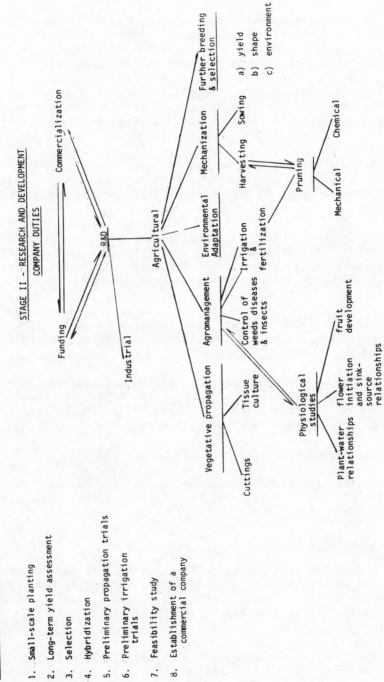

STAGE I - INTRODUCTION

1. Small-scale planting

2. Long-term yield assessment

3. Selection

4. Hybridization

5. Preliminary propagation trials

6. Preliminary irrigation trials

7. Feasibility study

8. Establishment of a commercial company

Figure 3. Procedures and methodology in the development of a new crop - jojoba (*Simmondsia chinensis*).

subdivided into the five categories described below.

1) Vegetative propagation. Jojoba may be propagated by means of cuttings or by tissue culture techniques. The latter method has the potential to yield an indefinite number of genetically identical seedlings within a short period of time (Fig. 4).

2) Agromanagement is divided into physiological studies; control of weeds, diseases and pests; pruning for purposes of yield control and plant shaping; and irrigation and fertilization trials.

3) Introduction. Jojoba has so far been introduced into 14 different localities in Israel, from the hot Arava valley (with water containing up to 5,000 ppm salts) to the mild coastal area (Fig. 5).

4) Mechanization. A special harvester for harvesting and collecting the jojoba nuts has been developed by the Institute of Agricultural Engineering of the Agricultural Research Organization. Direct sowing techniques have also been developed.

5) Breeding and selection trials have been expanded. At present, some 15,000 plants are being assessed for the following traits: yield, shape (for mechanical harvesting) and environmental adaptation.

The Negev Jojoba Company uses the information obtained from the R&D program as a basis for decision making in the commercialization of the jojoba plant. Since 1977, 100 ha of jojoba have already been planted in commercial groves.

III. ALTERNATIVE SOURCES OF WATER

In the absence of fresh water, Israel can look to at least three alternative water sources; run-off water, sewage water and saline water (Fig. 1).

Run-off water

The collection of run-off water into large reservoirs is feasible in the northern part of Israel but not in the Negev. This is because desert storms are scarce and erratic, and it is impossible to plan a viable agricultural system under conditions of uncertainty.

The small-scale run-off agricultural systems which are being developed by Prof. M. Even Ari of The Desert Research Institute can supply some solutions for arid zones where both labor and land are

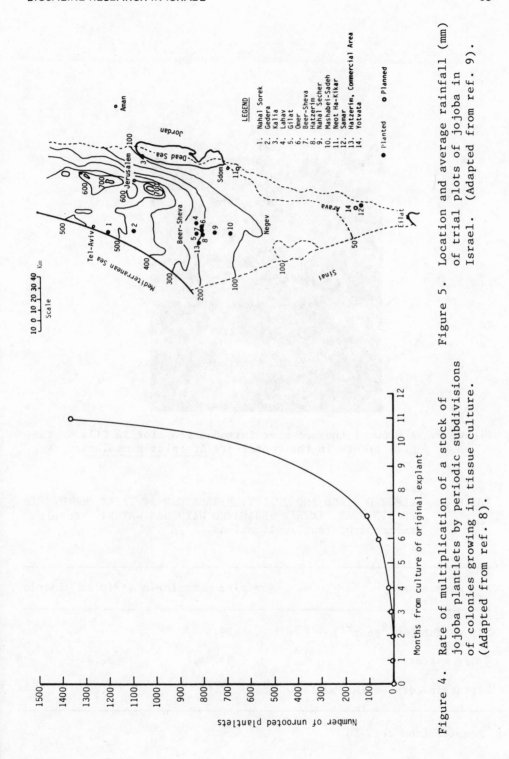

Figure 5. Location and average rainfall (mm) of trial plots of jojoba in Israel. (Adapted from ref. 9).

Figure 4. Rate of multiplication of a stock of jojoba plantlets by periodic subdivisions of colonies growing in tissue culture. (Adapted from ref. 8).

Figure 6. A view of the seawater introduction plot in Eilat. The
grey plants in the center are <u>Atriplex</u> <u>hummularia</u>.

TABLE 6. SELECTED YIELD AND QUALITY PARAMETERS *ATRIPLEX NUMMULARIA*
AND *ATRIPLEX CINEREA* IRRIGATED WITH SEA WATER.[*] Results
are means of four replications.

Parameter	*Atriplex nummularia*	*Atriplex cinerea*
Dry weight ($gm^{-2}year^{-1}$)	429.5	–
% dry matter	35.5	37.8
% protein (dry weight basis)	12.6	9.20

[*]Adapted from ref. 10.

abundant. His work is centered around optimal water harvesting techniques (large terraces and microcatchments) and around the trials with trees (notably the pistachio nut), hardy grasses and shrubs.

Sewage water

Treated sewage is becoming an important alternative water source. Israel can produce about 200 million cubic meters of treated sewage per year for agricultural and industrial uses. This will replace 200 million m^3 of fresh water which will be diverted to urban uses. The R&D on uses of sewage water is being carried out mainly at the Institute of Soils and Water.

Saline water

1) Brackish water: It is difficult to estimate the exact quantitites of brackish water available for further exploitation in Israel. The estimate of 120 million m^3/year given in Table 1 is possibly very conservative.

The salt content in the water of the Negev's brackish aquifers ranges between 1,000-10,000 ppmTDS. Brackish water research is being carried out at two research stations belonging to the Ramat Negev Regional Council and the Arava Regional Council in cooperation with the Agricultural Research Organization and the Applied Research Institute of the Ben-Gurion University of the Negev. The work at the Arava station is centered around the selection of salt-resistant rootstocks for avocado, mango, citrus, and deciduous fruit trees, on the mechanization of tomato harvesting and on the mechanization of drip irrigation. The Ramat negev Station concentrates on the selection of salt-resistant species and varieties and the development of agromanagement practices and technologies for brackish water irrigation.

2) Sea water: The utilization of sea water for the production of food and energy in the desert could become the ultimate solution to Israel's water shortage. The major part of the arable land of the Negev desert is situated within 40 km of the seashore. Because of high fuel prices, the desalination of sea water for large-scale agriculture is not feasible. Sea water will thus have to be used directly. Three main sea-water projects are, in fact, under way: the production of halophytic fodder crops, the production of single cell algae, and the production of fish and oysters in sea-water ponds.

Fodder halophytes. In 1977, a 400 m^2 plot was planted in Eilat with a collection of 38 halophytic plants which were then trickle irrigated with water from the Red Sea (4.2% salts). Some of the

plant species grew surprisingly well (Fig. 6): the yields recorded
for two of the species are given in Table 6. Based on this
successful preliminary work, a large-scale introduction scheme has
just been started whereby halophytic fodder species will be introduced, tried and selected for yield and feed quality. The work is
being carried out by the author at the Applied Research Institute
of the Ben-Gurion Unviversity of the Negev.

Fish and oysters. Two scientists of the Israel Oceanographic
and Limnological Ltd, Dr. H. Gordin and Dr. W. L. Huges-Games, are
responsible for overseeing a project on a polyculture system
centered around two marine organisms -the sea bream, *Sparus aurata*
and the Japanese oyster, *Crassostrea gigas*. The sea bream is fed
with a protein-rich diet which results in a high level of nutrients
in the water from the excretion of the fish. Phytoplankton thrive
in this environment and they serve as the feed for the oysters.
Fish and oyster yields are among the highest recorded in the world.
The detailed R&D which is being carried out is similar in basic
plan to the jojoba project (Fig. 3).

Algae. Research & Development is under way on a unicellular
alga *Dunaliella* which grows well in high salt concentrations even
higher than those of sea water. The products that can be extracted
from species of this alga include glycerol, β-carotene and protein.
Dr. A. Ben Amotz of the Israel Oceanographic and Limnological Ltd.,
and Dr. M. Avron of the Weizmann Institute of Science, Rehovot are
responsible for heading this R&D project. A pilot plant has
already been established near the Red Sea by a large commercial
enterprise.

IV. THE DESERT GOAT - AN EXAMPLE OF AN ALTERNATIVE SOLUTION

The desert goat fits into all three categories (Fig. 1) which
serve as the framework of this presentation. This animal is the
best known converter of dry desert plants into meat and milk. It
can recycle up to 90% of the residual nitrogen in its fodder and
produce up to 10% of its own weight in milk every day. The desert
goat can live without water for four successive days and can drink
an amount of water equal to 40% of its body weight. It can drink
water containing 15,000 ppm salts with no ill effects. All these
outstanding features make the desert goat a very promising object
for study and for development. The study of the desert goat is
headed by Dr. A. Schkolnick of the Tel Aviv University.

REFERENCES

1. Z. Greenwald, "Water in Israel," The Water Workers Association
      (In Hebrew) (1980).
2. "Economical Report on Agriculture in Israel for the Year

1979/80," The Authority for Planning and Development of Agriculture, The Ministry of Agriculture, Jerusalem (1980).

3. D. Goldberg, B. Gormat and D. Rimon, "Drip Irrigation. Principles, Design and Agricultural Practices," Drip Irrigation Scientific Publications, Kfar Shmaryahu, Israel (1976).
4. L. Bernstein and L.E. Francois, Soil Science, 115:73-86 (   ).
5. M. Forti, J. Levi, R. Padova, A. Ross and I. Leibovitz, "Utilization of raw city garbage for production of fodder and industrial crops in the Negev," Research & Development Authority, Ben-Gurion University of the Negev, Beer-Sheva, Report No. BGUN-RDA-279-80 (In Hebrew) (1980).
6. A. Richmond, and K. Preiss, Interdisciplinary Science Reviews, 5:60-70 (1980).
7. D. Pasternak, E. Rappeport, Y. De malach, M. Twersky and I. Borovic, in: "Conference on Alternative Strategies for Desert Development and Management," Sacramento, Calif., May 31-June 10, New York, United nations (1977).
8. M. Forti, pers. comm.
9. M. Forti, Research and Development Authority, Ben-Gurion University of the Negev, Report No. BGUN-RDA-158-77 (1977).
10. D. Pasternak, J. Ben Dov, and M. Forti, Ibid., Report No. BGUN-RDA-220-79 (1979).

Appendix - Addresses of Israel institutes mentioned in this review

1. Agricultural Research Organization, Volcani Center, P.O. Box 6, Bet Degan.
2. Applied Research Institute, R&D Authority, Ben-Gurion University of the Negev, P.O. Box 1025, Beer-Sheva, 84110.
3. Arava Regional Research Station, Yotvata.
4. Authority for Planning and Development of Agriculture, The Ministry of Agriculture, Hakiryah, Tel-Aviv.
5. Faculty of Agriculture, The Hebrew University of Jerusalem, Rehovot.
8. Negev Jojoba, c/o Delek, P.O. Box 1831, Tel-Aviv.
9. Ramat Negev Regional Research Station, Revivim, D.N. Negev.
10. Technion-Israel Institute of Technology, Haifa.
11. Tele-Aviv University, Ramat Aviv.
12. The Institute for Desert Research, Ben-Gurion University of the Negev, Sde Boker Campus, P.O. Box 2053, Beer-Sheva 84120.
13. The Weizmann Institute of Science, Rehovot.

BIOSALINE RESEARCH IN E. ASIA AND NEW ZEALAND

V.J. Chapman

Botany Department
Auckland University
Auckland, New Zealand

SUMMARY

An account is given of recent and current biosaline research in the following countries: India, Pakistan, Bangladesh, Malaysia, Thailand, Indonesia, Papua New Guinea, New Zealand and Oceania. In recent years there has been very considerable emphasis upon all aspects of mangrove vegetation. However, there has been research on other saline plants and also upon the marine algae, especially those capable of commercial usage.

INTRODUCTION

This survey is concerned with recent and current research on bio-saline problems in the following countries: New Zealand, Pakistan, India, Bangladesh, Burma, Malaysia, Indonesia, Thailand, Papua New Guinea and Oceania. In this part of the world, it is not surprising that the current major thrust of research is directed towards the species comprising mangrove swamps, largely because of their economic importance. Less attention has been given to the sea-grasses and marine algae, even those associated with the mangroves, though in Oceania considerable attention is being given to algae of commercial importance (cf, Neushul). In India some work is directed to plants that grow on interior saline soils, but such areas are not available in most of the other countries. Whilst there are some studies related to the physiology of the plants, much of the current work is essentially ecological. The extent of this will become evident in the accounts for the different countries.

59

New Zealand

At the present time in New Zealand biosaline research is concentrated primarily upon the mangrove vegetation and the marine
algae. Practically no attention currently appears to be being
given to the the salt marsh plants that occupy considerable areas
in the south of the North Island and the coast of the South Island.
One recent contribution (1) has dealt with auto- and heterotrophic
processes of marine algae and Zostera on sand flats in Delaware
Harbour near Nelson (South Island). Here the highest rates of
micro- and macroalgal production occurred on sandy sediments. In
the case of Euglena mats $^{14}CO_2$ fixation ranged from 20-200 mg C/m^2
x hr. Maximum rates of 190, 27 and 8 mg C/m^2 x hr were recorded
respectively for Ulva, Enteromorpha and Zostera. Microalgae on mud
flats exhibited the very much lower rate of 2-4 mg C/m^2 x hr. On
the other hand heterotrophic ^{14}C-glucose rates for microalgae on
mud flats were up to forty times greater than on sand flats. So
far as salt marsh plants are concerned, one study is currently
being undertaken at Auckland on the nutrient relationships of
Salicornia australis and Samolus repens.

So far as the New Zealand mangrove, Avicennia marina var.
resinifera, is concerned the present author has reported on a
survey of nearly all the known swamps (2-5) in the course of which
two widespread areas of a black sooty mould, Capnodium fuliginodes,
were reported on mangroves infested with the scale insect
Ceroplastis sinensis. This infestation gave the two areas a completely black appearance. Arising from this survey a number of
distinct communities have been recognized (6):

1. Avicennia marina var. resinifera community of tall or low
   bushes. This may have associated with it on the mud
   Zostera muelleri or the brown alga Hormosira banksii.

2. Juncus mauritimus-Leptocarpus similis savannah with
   scattered trees of Avicennia.

3. Juncus maritimus var. australiensis community. This may
   have Leptocarpus associated with it and at higher levels
   one may find Samolus repens (on better drained areas),
   Selliera radicans, Triglochin striatum, Scirpus cernuus,
   Muehlenbeckia complexa.

4. Leptocarpus similis community - monospecific or with the
   same associated species as in 3.

5. Salicornia australis community.

6. General salt marsh community of Selliera, Samolus,

Triglochin, Scirpus, Plantago coronopus and Cotula coronopifolia.

7. Juncus - Leptocarpus - Typha brachish marsh.

At the Leigh Marine Laboratory, Taylor is continuing his studies on the productivity and litter fall of the New Zealand mangrove and has recently discussed its biology (7). He has also queried the systems analysis of mangals and argues for more research on population dynamics and ecological tolerances of the various species (8). In the Gulf of Thames a fungus of the genus Phytopthora has caused some damage to the mangroves and the biology of this fungus has been studied over some years by Maxwell (9).

Turning to the marine algae, Dromgoole at Auckland will shortly have published an extensive study of research on the functional morphology of vesicles of brown algae, especially within the genus Carpophyllum. Another study nearing completion is that of Miss Begum at the Leigh Marine Station on the germination and establishment of Hormosira banksii. At the same station Choat has about completed his studies on the growth and decline of Ecklonia radiata forests and the impact of grazers upon these forests. The growth and reproduction of this potentially important commercial alga is also being studied by Mrs. Novaczek. Also at Leigh, Hawkes has under way life history studies of some of the local Rhodophyceae, including Pseudogloiophloia bergrenii, Scinaia firma and Apophloea Sinclairii, whilst Ms Bonin is working similarly on Asparagopsis armata and Delisea compressa. One of the more important pieces of work in progress at Auckland is concerned with the mariculture of Graciliaria secundata var. pseudoflagellifera at the Manukau sewage purification works by Johnson. Pilot plant culture work has proved very successful giving a yield of not less than 10 gr/m^2 x day and this alga has an extremely high potential for the production of agar-agar.

In Wellington, at Victoria University, Ms Ellis has about completed her study of the use of seaweeds as a means of indicating metal pollution in the water. In the South Island, at Canterbury University, MacRaild has been studying the structure of the gametangia of the brown alga Marginariella boryana, whilst a student (B. Cummack) has been examining aspects of the biology and ecology of the large brown alga, Macrocystis pyrifera. There is one significant study continuing by Parsons of Botany Division, Lincoln, on the various species of Gigartina and the type of agar that can be obtained from them.

Pakistan

Much of the country is low-lying and some 23 million acres out

of 33 million irrigated acres are affected by salinity problems.
The increase in saline areas has been brought about in the present
century by the increasing use of irrigation resulting in the redistribution of salts in the soil.   The entire salt affected area
has been divided into a number of salinity control and reclamation
areas of about 1 million acres each and regional plans and a master
plan prepared.   Whilst the plans generally involve a physical
approach to the problem, Quereshi (10) considers that a biological
approach to soil reclamation will be necessary to deal with these
saline sodic soils.   This will mean studies to determine species
that can be used agriculturally and which will grow under saline
conditions. This programme may now be under way.

India

      This country with its long coastline and arid interior regions
can experience many salinity problems and major studies on biosaline problems are carried out at coastal universities and at
interior institutions such as the Central Salt and Marine Chemicals
Research Institute.   Work in this country falls, then, mainly into
two categories: (a) plants of interior saline deserts; and (b)
mangroves.

      In the case of saline deserts much of the recent work was
reported at the INTECOL Conference at Jerusalem in 1978.   Thus
Basuchaudhary et al (11) gave an account of species that can be
used for the biological reclamation of salt affected soils in India
and this resulted in a recommendation for the use of Prosopis
spicifera.   Choudhury and Varshney (12) gave an account of the
productivity of the different plant communities in salt affected
(USAR) lands of the Indo-gangetic plain, whilst Sen and Rajpurohit
(13) described the distribution of various species in relation to
salinity in Indian desert areas.   Use of these lands is increasingly important because of the increasing pressure for production.
The primary production potential of these saline lands, especially
in the Indo-Gangetic plain, was also reported on at Jerusalem by
Varshney and Choudhuri (14).

      In India salt-affected soils limit or prevent crop production
over an estimated area of about 7 million hectares.   Of this some 4
million hectares are coastal soils of arid areas and there are 2
million hectares of alkaline soils in the Indo-gangetic plains.
The most promising way of using these soils is to develop and grow
crop tolerant varieties and the Central Government Salinity
Research Institute is currently evaluating locally adapted plant
species.   Efforts are also under-way to develop improved varieties
by breeding and induced mutations (15).

      At the present time the Central Salt and Marine Chemicals

Research Institute is conducting experiments to use some of the halophytes as an additive to fodder and also as a source for useful chemicals. This study involves mangroves as well as plants of interior saline regions. Among the mangroves of Saurashtra Avicennia marina is being exploited as a fodder for animals during periods of scarcity. Studies at the Institute are showing that this fodder is quite comparable with stocks of Sorghum and Pennisetum (Iyengar, pers. comm.). The biological evaluation of the feed is being made on a large scale using sheep and calves at both the veterinary college of Madras and the Rajasthan College of Agriculture. In order to increase Avicennia crops, experiments on vegetative propagation have been carried out by using air layering. Four to six month old air layers have shown very good establishment with a 60-80% success in the monsoon.

The halophytic shrub Aeluropus lagopoides is also being studied for its potential in preventing erosion on salt farm bunds (banks) and also as a source of fodder. Growth is greatly improved by the foliar application of nitrogenous fertilizers after pruning. Studies are also underway using exotic Atriplex species of known food value from the U.S.A., Australia and Israel. Species involved are A. cinerea, A. canescens, A. nummularia, A. paludosa, A. vesicaria, A. patula, A. rhagioides.

On a less applied basis, Joshi (16) has been carrying out physiological and ecological studies on the halophytes Salicornia brachiata, Suaeda nudiflora and Sesuvium portulacastrum. In the first species germination takes place in pure sea water and as well as sodium chloride accumulation in the plant tissues there is also excessive accumulation of free amino acids. With the second species increased germination can be obtained by scarifying the seed but 10,000 ppm salt is the upper limit for germination. A two-layered tunica is found in Suaeda in the transition from vegetative to flowering stages and marginal growth of the leaf lamina is of the middle submarginal type.

Turning now to the Indian mangroves, a recent summary of the area occupied and the utilization is given by Christensen (17). Two earlier studies were reported at the Jerusalem INTECOL Conference. Kerrest (18) described the mangrove communities in Pichavaram, including the differences between natural succession and that following felling by man. In this area of India grazing of mangrove, mainly Avicennia, by cattle leads to stunted growth and limited productivity. The bunds constructed for fishing also cause siltation by affecting the current pattern and they interfere with the movements of floating seedlings and so prevent new mangrove colonization (19). Datta and Chanda (20) reported on problems associated with mangroves in the Western Sunderbans. In this area the nature of the vegetation is influenced by adaptive and

hydrological peculiarities. As may be expected the succession
varies in different sites and at present the whole ecosystem is
under stress due to slow tilting of the Ganges delta, lack of fresh
water storage, lack of enriching alluvium and man's demands for
land use resulting in the phasing out of Nypa and Heritiera (21).
The authors report the rate of litter fall as averaging 9 tons/ha x
yr. A more recent account of the Sunderban mangroves in relation
to fresh water supply has been given by Mukherjee (21a). A rather
comparable study has been carried out by Bannerjee (22) for the
Mahanadi delta.

More ongoing research has been reported at the recent Wetlands
Conference in New Delhi, where Untawale et al (23) described the
use of remote sensing techniques to study the distribution of mangroves along the estuaries of Goa, where they occupy about 2000
hectares. Untawale (24) provides a more general account of the
West coast mangroves. This technique could well be extended further. A study of the mangroves near Bombay, especially of the
water and phytoplankton, indicates that mangrove swamps do have a
high capacity to degrade organic wastes present in sewage effluent
provided proper dilution is available (25). This, however, is not
the entire story and before mangroves are finally used to purify
sewage effluent a study needs to be made of the heavy metal component, especially where effluent comes from industrial sources.

Bhosale and Shinde (26) have been investigating the significance of cryptovivipary in Aegiceras corniculatnum. During the
active growth stage of the seedling on the parent tree maximum
translocation of the ^{36}Cl and products of photosynthesis takes
place. During maturation the seedling slowly acquires independence
from the mother plant, especially in respect of adjustment to
salinization, to the point of fall. There appear to be two sites
acting as a barrier to Na^+ and Cl^- translocation. One is between
the fruit wall and seed coat and the other between seed coat and
embryo. On the morphological aspects of mangroves Rao et al (27)
have reported on the significance of schlereids in leaves as a
means of distinguishing between different species of Rhizophora.
This can be very valuable in the absence of other features such as
flowers and seedlings.

Continuing work on photosynthesis by Joshi et al (28) and
Joshi and Waghmode (29) has confirmed the role of aspartate as the
initial product of photosynthesis in species of Rhizophora,
Avicennia, Bruguiera, Ceriops and Acanthus. Alanine also occurs as
one of the products of short term photosynthesis. This work leads
the authors to believe that in mangroves the carbon follows a
modified C4 pathway. This conclusion has been challenged for
species of Rhizophora by Andrews and Clough (30) who argue that
they photosynthesise through the C3 pathway.

Finally, in India work is also being carried out on aspects of algal biology in saline areas. Balakrishnam and Gunde (31) have demonstrated that Schizomeris leibleinii is an excellent indicator of water pollution. Somewhat surprisingly the role of Cyanophyceae does not appear to be significant in determining the nitrogen status of soils in the salt affected (USAR) habitats (32). Commercial uses of seaweeds are starting to assume significance in India as in other countries. The genus Sargassum is widespread and abundant and regarded as a very suitable source for fermentation and production of methane. Using alginate digesting bacterial strains Rao et al (33) got a substantial increase in gas output by the addition to the Sargassum of cow dung and Ulva as additional carbon source. As well as the production of biogas, alginates are also manufactured from Sargassum and Turninaria. The other major seaweed industry is agar production, mainly from Gelidiella acerosa and Gracilaria edulis. Because the agarophyte resources are limited the Central Salt and Marine Chemicals Research Institute has initiated pioneering work on aquaculture at Mandapam. Here Gracilaria has been cultivated using the long-line rope method at a constant level and has given a yield of 100 tons wet wt/ha. per year using three harvesting periods (cf also New Zealand). Experimental pilot farms of Gelidiella acerosa are also being developed as a result of co-operative work with M/S Cellulose Products of India Limited. Recently a technique for the cultivation of the red seaweed, Hypnea musciformis, a source for carrageen, has been developed. Vegetative fragmentation is used and a four-fold increase in biomass over 25 days has already been achieved (Iyengar, pers. comm.).

Bangladesh

In his survey of the Bangladesh forests Christensen (17) notes that extensive coastal afforestation is being undertaken to hasten land reclamation by stabilizing new accretions. For this purpose Sonnerati apetala is one of the most favoured species being raised in nurseries and transplanted when 30-60 cm high. At Chittagong Sonneratia is also associated with plantings of Avicennia spp. and annual growth data have been obtained for all species in each area. The growth of the different species appears to be related to rate of silt deposition, maximum growth being associated with maximum silt deposition (34). The exploitation and management of the valuable Sunderbans mangrove resources is so important for the well-being of the country that Ahmad (35) argues for multi-disciplinary research in the future on a sound scientific basis. One of the problems in the Sunderbans area is the current gradual climatic change in the region which is leading to more drought and desertification. Ismail (36) points out that the present lack of knowledge concerning the vast areas ecosystems is essentially due to the great expense involved in research in the Sunderbans.

The major species is <u>Heritiera</u> <u>fomes</u> which does not possess an adaptative mechanism towards increasing salinity. The distribution of the mangrove species in the region is related to the more or less permanent salinity patterns. Ismail (36) also argues for special urgency in expanding the knowledge of the Sunderbans mangrove ecosystems.

Burma

No recent information appears to be available. It is likely that any current biosaline research, if any is taking place, will be centered around mangroves.

Malaysia

Almost the entire biosaline work in Malaysia is related to mangroves. However, in Sabah the Fisheries Department is carrying out a project with a view to the introduction of <u>Eucheuma</u> aquaculture. So far, this has not been very successful because of heavy predation and lack of local interest (Fong, pers. comm.).

In Sabah the Forest Research Centre is carrying out studies on the national regeneration of mangroves following logging in Sandakan and Tawau. Trial plantings are also being used in order to see how effective the technique is in relation to natural regeneration (Hepburn, pers. comm.). At the Forest Department Headquarters in Sarawak (Kuching) a number of research projects are underway, some of which have been completed and will be written up (Chai, pers. comm.). The seven projects are as follows:

1. Typing of the forest and taxonomy of the flora;

2. Effects of tidal flooding and soils (physical and chemical) on the zonation, regeneration and succession of the tree species;

3. Natural regeneration and its problems;

4. Planting experiments and growth rates of commercial species;

5. Phenology of flowering and fruiting in the Rhizophoraceae;

6. Dispersal and establishment of propagules; and

7. Litter production and decomposition of leaves of Rhizophoraceae.

In the case of Item 3, it had been noted that natural

regeneration was more or less confined to low-lying areas along rivers and channels and to other inland areas subject to frequent tidal inundation. Chai and Lai (37) point out that there are four major factors affecting natural regeneration:

a) Thalassina lobster mounds elevates the land;
b) Acrostichum aureum fern preventing access of propagules;
c) Reduction of tidal inundation due to past accretion and the lobster mounds; and
d) Felling intensity which has failed to leave behind sufficient mature trees to provide a stock of seedlings.

Experiments with planting propagules of Rhizophora and Bruguiera, showed that mortality varies greatly from plot to plot. Heavy loss is caused by attacks from caterpillars, snails, crabs and borers (Poicilips spp.). In the case of Acrostichum eradication may be necessary over large areas and experiments have shown that it is best done manually rather than by the use of herbicides (17).

In Malaya at the University Sains Malaysia (Penang) a major research project is under way, dealing with the productivity of the mangrove ecosystem. This is subdivided into a number of sub-projects, as follows (Khoon, pers. comm.):

1. Preliminary studies on biomass, productivity and allo-metric relationships in a managed mangrove ecosystem;

2. Studies in organic production and mineral cycling in a mangrove forest;

3. Nutrient cycling, productivity and stream profile (mixing diagrams) of a tropical mangrove estuary;

4. An environmental and ecological study of a mangrove eco-system in relation to conversion of areas for aquaculture purposes;

5. Regeneration of certain tree species; and

6. Estimation of productivity in a tropical mangrove eco-system using gas exchange and micro-climatological methods.

In the Matang mangrove reserve it has been found that where there is a felling rotation a large scale mortality takes place between regeneration years 2-15 and this is an aspect that is in need of study (38). In this reserve there is also a continuing study to try and determine the reasons behind declining

productivity (39).   In Malaya the Nypa swamps are of considerable
importance and value.   Fong (40) has described the extent of the
swamps throughout the peninsula and also has pointed out the
serious threat to them because of increased coastal zone development.   It is clear that the time has arrived for research into a
proper management plan for this particular species.   At the recent
Asian Symposium on the mangrove environment at Kuala Lumpur there
were numerous papers dealing with the mangroves of this country.
Nixon et al (41) have commenced a study of the part played by mangroves in the carbon and nutrient dynamics of estuaries, and much
more sampling will be necessary before reaching a conclusion as to
whether mangroves do serve as a significant source of nutrients for
adjacent coastal areas. Gearing and co-workers (42) have commenced
a rather similar study using the carbon isotope ratios to determine
the supply and consumption of detrital material.

     Marine fungi are assuming more importance in different parts
of the world (Chapman, in press). In Malaya, Kuthabutheen (43) has
studied the leaf fungi associated with mangroves and except for
Pestalotia and Zygosporum the fungi are very similar to those found
on most non-mangroves of the regions.   There does not seem to be
any relationship between the tannin content of the leaves and the
total number of fungi.   Pestalotia is the major genus on leaves
with higher tannin content, whilst Fusarium is most abundant where
there is less tannin.

     A study of the salt glands of Acanthus ilicifolius (44) by
electron microscopy reveals the lack of any aperture for salt
excretion.   The same feature has been reported by Field (44a) for
glands of Aegiceras. In Acanthus the cuticle is apparently porous,
whereas in Aegiceras excretion takes place through the gap around
the gland. Mention has already been made of the use of mangrove to
treat sewage effluent (India) and it is significant that the heavy
metals, zinc and lead, do not appear to exert any serious effect on
either Rhizophora or Avincennia.   In the former it is suggested
that there may be an exclusion mechanism operating in the roots
whilst in the latter, excess metal may be excreted through the salt
glands together with the salt (45).   A study of litter production
and decomposition in the South Banjar Forest reserve, has shown
that   total litter   production in Avicennia,   Sonneratia and
Rhizophora zones to be 15.3, 13.9 and 15.7 tones/ha x yr respectively.   Leaves of Rhizophora species have the slowest rate of
decomposition (46).   The Malayan Forestry Department has set aside
three areas, one each in Kapar Forest, Banja and South Forest and
Pulan Klang Forest, for research over the next ten years.   The
Zoology Department of the University of Malaya (Kuala Lumpur) will
study the energy budget and transfer pathways in the mangrove
ecosystem (46).

Thailand

Considerable research has been and still is continuing on the
biology of the mangroves in Thailand. The management of these
forests has been the subject of two seminars in 1976 and 1977 at
the Phuket Marine station.   In 1978 Christensen (48) reported on
the amount of biomass and primary productivity of Rhizophora
apiculata and recorded for a 15 year old stand an annual increase
of 159 tonnes/ha with a leaf production of 6.7 tonnes/ha x yr. One
year earlier with Wium-Andersen (49) he had shown that leaf production for the species underwent two maxima in the year, one at the
start and the other at the end of the monsoon.   In the case of
Lumnitzera and Bruguiera there was but a single maximum during the
monsoon.   With two other genera, Ceriops and Avicennia, there were
two maxima in leaf productivity, but these maxima did not bear any
relationship to rainfall.

A number of contributions on a range of topics was also made
to the BIOTROP symposium on mangrove vegetation in South East Asia.
A study of the mangrove forest at Amphoe Khlung (50) showed that
the lowest stand volume occurred on the area where Rhizophora,
Avicennia and Bruguiera were predominant. The author argued that a
number of significant management questions remained unanswered and
that studies were still essential. A study of the genus Rhizophora
(51) showed that of the two species, R. mucronata is much more
tolerant of salt water than R. apiculata and this determines their
distribution. Although very little is known about the orchids that
occur in mangrove forests, a survey identified 18 species, none of
which are restricted to mangrove and because they are epiphytic
they are not affected by any salt (52).   Very little work anywhere
has been carried out on the bacteria found in mangroves despite the
fact that they are a group of micro-organisms that play an important role in the nutrient cycles of both soil and water.   They are
primarily responsible for the break-up of deposited organic matter.
In the mangrove of Amphoe Khlung some 25 bacteria have been isolated and further work is to continue (53).

According to Sornsomboon (54) in a report to the Mildura
Conference there are at least two biosaline research projects
under-way in Thailand.   One concerns the salt tolerance of rice,
corn, sorghum, sweet potatoes, cotton, soy beans, peanuts and mung-
beans at the Petchburi agricultural Research Centre, and the other
is involved with plant response to salinity at the Agricultural
Faculty of Kasetsart University.

At the most recent symposium at Kuala Lumpur, Chansang (55)
used the complexity index as an indicator of combined floral
characteristics and she also reported that zonation of the sea
grasses in S. Thailand was correlated with the level of inundation.

In a detailed study of nutrient cycling within a mangrove forest
Aksornkoae (56) found that the average rate of litter fall is 830
g. dry wt/m^2 x yr, being highest at the seaward edge.   The rate of
litter decomposition is about 55% of the total litter fall, again
being highest with Rhizophora, Bruguiera and Avicennia at the sea
margin.   It is here also that the highest amount of nutrients are
released. The average values of the nutrient contents are:

Material	N	P	K	Ca	Mg	Na
Leaves	1.53	0.09	0.84	0.67	0.19	4.58
Branches	0.77	0.07	0.54	0.40	0.08	2.81
Trunks	0.55	0.43	0.33	0.27	0.05	1.63

In the southern parts of Thailand much of the mangrove occurs
on areas rich in tin ore, so that extraction processes have damaged
much of it.   The rate of regeneration of such mangrove is very low
(based on dbh and total height) as compared with natural managed
forest (57).

Apart from mangroves Poorachiranon and Hansa (58) have studied
the heterotrophic and mixotrophic capacity of ten species of
benthic diatoms growing on mud flats within the mangrove.

Vietnam and Kampuchea

It has not proved possible to obtain any information from
these two countries, nor is it known whether any research is in
progress.

Indonesia

There are extensive mangrove swamps in this region, so that
one. may expect some active research to be in progress.   In 1979
Soekardjo and Kartawinata (59) showed that in mangrove of the Musi
River estuary there was no definite relationship between the
species distribution and soil characters.   There was, however, a
gradual increase in tree heights with decreasing soil salinity.   A
further study in this area (60) has provided information on zonation, stand volume and diversity of species.   Leaf litter is
clearly important as a source of food for the mangrove animals and
Brotonegoro and Abdulkadir (61) have recorded a range of fall of
1.5 g dry wt/m^2 x day to 4.06 g dry wt. in forest on the Isle of
Rambut.

Studies in this region have shown that there are differences

in regeneration after cutting depending on the locality.   Thus in
Kalimantan after clear felling in alternate strips the main species
regenerating is Rhizophora, whilst there was very little Bruguiera.
In Sulawesi, on the other hand, mangrove does not regenerate and is
replaced by Acanthus ilicifolius and the fern Acrostichum aureum
(62).   Similarly in Sarawak, Chai and Lai (37) have also reported
that natural regeneration in exploited forests is not satisfactory
due to a number of factors and research is being concentrated
towards finding possible solutions.

    At the International symposium in Papua New Guinea this year,
Soegiarto (63) discussed the mangrove system in Indonesia, its
problems and management.   He pointed out that studies in universities and research institutions are underway in order to understand
thoroughly how the mangrove ecosystem functions ecologically.
Arising from such research studies it is hoped that better management will be achieved.   Finally, at the New Delhi Wetlands
Conference, Polunin (64) put forward proposals for the study of the
fate of detrital matter in the Indonesian forests.

Papua New Guinea

    Most of the current biosaline work in this country is centered
around the mangroves.   Swift (University of Papua New Guinea) and
Cragg (Forest Products Laboratory) have commenced what is likely
to be a long-term study of the impact of fungi and marine borers to
wood decay of some of the mangrove species.   They have recently
(65) reported the decay of wood in a Rhizophora-dominated shoreline stand.   Of dead fallen wood, that most vulnerable to attack
are small branches and small prop roots.   In the case of the roots
the primary attack is from marine borers (Teredinidae and isopods)
and they can affect up to 30%.   This primary attack upon the wood
is followed by an invasion of marine fungi, termites and beetles as
decomposers.   These can be especially prevalent in timber above
high water mark.   In the case of dead roots buried in the mangrove
mud, decomposition takes place primarily through bacterial action.

    The distribution and taxonomy of the mangrove species in Papua
New Guinea is being studied by Frodin (66).   Apart from 31 woody
'obligate' specific and intra-specific dicotyledonous species,
there are two species of Acanthus, the palm Nypa fruticans, the
fern Acrostichum aureum and a number of epiphytes, parasites and a
few lianes.   On the land-ward fringe of the swamps one may find a
number of what Frodin terms "facultative" woody dicotyledons.
There is a distinct difference between the "obligate" north and
south coast floras with the former being the more peculiar.   As a
result of his studies Frodin argues (66) that the distribution and
establishment of mangrove propagules is not a random phenomenon.
Despite all the past work some of the genera still present

taxonomic problems, notably Avicennia, Rhizophora, Ceriops and Sonneratia. This is implicit also in the fact that Womersley (67) in an account of the mangal of Papua New Guinea only recognises 24 taxa and the palm Nypa. This could well be compounded by current speciation that may be taking place (68). Another current piece of research that may assist in this matter is that being prosecuted by Leach (69), who is carrying out a Chemosystematic revisionary study, mainly involving the leaf flavenoids, in mangrove species and in the sea-grasses.

The sea grass taxonomic problem probably originated with earlier work by Johnstone (70) on the sea grasses of Papua New Guinea. Eight species are widely distributed, three are restricted to one area (Daru) and one of unknown taxonomic status is confined to the Solomon sea coast. Johnstone (70) reported on the leaf growth rate of Enhalus acoroides and more recently (71) has given an account of the productivity of Thassodendron in Papua New Guinea and in northern Australia. He has also described phenotypic variation in Cymodocea serrulata.

Whilst at the University of Papua New Guinea, Johnstone (72) also studied the mangal and as a result has come to the conclusion that succession in mangrove is not, as is commonly supposed, from the seaward edge to the landward fringe, but that it takes place from both directions, so that the climax is in the heart of the mangal swamp. This is an extremely interesting and provocative concept and is clearly worthy of further investigation. Johnstone (73) has also studied herbivory of PNG mangroves for three reasons: (a) to determine the influence of forest composition on the degree of herbivory; (b) to determine the actual level of direct herbivory (many leaves are partially eaten before they drop); and (c) to determine if leaf herbivore population explosions occur in mangrove ecosystems, and if they are associated with forest defoliation. The results obtained were all negative so that the author concluded that attacks are largely on an individual plant or individual species basis. It could be determined by the presence or absence of specific antifeedant, e.g. there is generally a low attack on Excoecaria agallocha. Christensen (17) in his report on mangrove resources in S.E. Asia gives 222,624 hectares of different types of mangrove swamps in the country. He also reports that young radicles of Bruguiera, fruits of Sonneratia alba and fat from Cerbera seeds are used as food, and that an alcoholic drink is still made from Rhizophora fruits. Regeneration appears to take place readily where there has been well drilling or harvesting for cutch.

The Pacific Region

There appear to be two areas where biosaline research has been, and will be prosecuted: these are Fiji and the Philippines.

In the former country, Baines (74) has given a history of resource management of the mangroves over historical times. He has outlined a multiple use approach for the mangroves and identified the research which is regarded as necessary for this purpose. For the Samabula estuary in South western Viti Levu, Hassall (75) has prepared a vegetation map using units that could well be used to enumerate mangrove forests not only in the remainder of Fiji, but also in Western Samoa and Tonga. Sea grasses are also common around Fiji, the most abundant being Syringodium isoetifolium with Halophila minor and Halodule pinifolia locally abundant. In the Suva region the beds of these grasses can be totally destroyed by sand extraction. However, recolonization of the dredged areas does occur from fragments swept in from surrounding beds that have been left untouched. Establishment is a slow process bcause of induced substrate instability and modifications to the lagoonal hydrography caused by the dredging. Penn (76) reports that replenishment of the sediments is very slow, but it can be accelerated if the seagrasses recolonize and also bring about in situ carbonate production.

In the Philippines, where population pressure is considerable and productive and arable land are scarce, the International Rice Research Institute carries on intensive research into the use of saline lands (mangrove swamps) for potential rice crops. Soil studies, together with studies on the physiology of the many different strains of rice, encourage the researchers to believe that millions of hectares of mangrove can be converted to rice growing (77). In view of the fact that there is also pressure to convert mangrove to fish ponds, it would seem that unless great care is taken the importance of mangrove in maintenance of the offshore fisheries may be completely jeopardised. At the Mildura Conference, Ponnamperuma (78) outlined in detail the background to the rice research. He pointed out that Asia needed 5 million tons more of rice every year, either by increasing the yield per area or increasing the area. The latter virtually means the use of saline soils where salinity is the main stress on the plants. There are rice varieties which are pest resistant and these give a good yield when grown on saline soils. Of 14,000 rices screened between 1975 and 1977, 109 have an acceptable degree of salt tolerance. Breeding of other rice types also goes on at the Institute in order to develop lines that will suit the environmental conditions of other Asian countries. Experimental work at the Institute has also shown that salt injury in rice is related to water stress. There is also a good correlation between root salt content and salt tolerance of a strain. There may also be some relationship between the protein content of plants and their tolerance towards salinity, but this is an issue that requires further study.

Studies on the biology of mangrove-associated marine algae are also taking place in the Philippines. Cordero (79) reported 48 mangrove associated algae from Aklan province, of which thirty five are new records. One is a blue-green, 134 are Chlorophyceae, 10 are Phaeophyceae and 24 belong to the Rhodophyceae. Farming two species of Caulerpa in abandoned fish ponds has been described by Fortes (80). The changes in the standing crop of natural and 'farmed' populations could be correlated with the monthly variations in nitrate nitrogen and reactive phosphate of the water and degree of water movement.

Most of the research in the Philippines has centred around the mangroves and continues to do so. Lauricio (81) has examined the problems involved in mangrove management and has made recommendations for research. Bina (82) has used Landsat data in order to assess the remaining mangrove forests and determine which shall be preserved untouched, which will be managed for forestry, and which can be eliminated and replaced by fish ponds or rice fields. The current situation of mangrove species and ecology has been well described by Arroyo (83) who also provides figures for stand density. Growth data from samples in Pangasinan and Quezon showed that Bakauan seedlings of Rhizophora grew at the rate of 14.93% for the 5 cm dia. class, at 3.37% for the 10 cm dia. class and at 1.37% for the 20 cm dia. class. Because there could be problems of regeneration in healthy exploited areas, Sabala (84) has emphasised the importance of tree breeding, using seedlings from superior trees in order to hasten regeneration. Finally, Pahm and Aspiras (85) have looked at the micro-organisms involved in the breakdown of organic N to inorganic in mangrove soils. In the presence of Rhizophora more NH_4 nitrogen continued to be produced but under Nypa ammonification was greater in the lower soil layers. Soils with Avicennia also had high ammonia nitrogen.

In 1978 Chansang (86) prepared an ICLARM report and outlined a mangrove research programme for S.E. Asia, much of which would probably be carried out at ICLARM at Manila in the Philippines. She pointed out that basic ecological research is needed in planning any management of mangrove forests. In particular, one needs to determine the productivity of the different forest types. So far as silviculture is concerned the main issue needing research centers around regeneration problems, e.g. Rhizophora being succeeded by less valuable species, pest problems, insufficient seedlings and the unwanted growth of Acrostictum aureum.

CONCLUSION

It can be seen from this account that there is active bio-saline research in these countries, but that there is still scope for more to be done.

REFERENCES

1. P. A. Gillespie and A. L. McKenzie, Abst. Intern. Symp. Trop.
    Shallow Water Community, Port Moresby, p. 28 (1980).
2. V. J. Chapman, Department Lands & Survey, Auckland, 28 pp.
    (1976).
3. V. J. Chapman, Ibid, 32pp. (1978a).
4. V. J. Chapman, Ibid, 32 pp. (1978b).
5. V. J. Chapman, Ibid, pp. 35 (1978c).
6. V. J. Chapman, Abst. 2nd Inter. Symp. Biol. & Mangement
    Mangrove, Port Moresby, p. 20 (1980).
7. F. J. Taylor, Ibid, p. 51 (1980a).
8. F. J. Taylor, Ibid, p. 51 (1980b).
9. G. S. Maxwell,, M.Sc. Thesis, Auckland University, New Zealand
    (1978).
10. R. H. Quereshi, Abst. Plant Response to Salinity & Water Stress
    Conf., Mildura, p.24 (1977).
11. K. C. Basuchaudhary, G. N. Choudburi and N. K. Agarwal, Abst.
    2nd INTECOL Conf., Hebrew Univ., Jerusalem (1978).
12. G. N. Choudhuri and S. P. Varshney, Ibid, (1978).
13. D. N. Sen and K. S. Rajpurohit, Ibid (1978).
14. S. P. Varshney and G. N. Choudhuri, Ibid, (1978).
15. R. S. Rana, See ref. 10, p. D1, (1977).
16. A. J. Joshi, Ph.D. Thesis, Univ. Bharnagar (1979).
17. B. O. Christensen, Mangrove forest resources and their manage-
    ment in Asia and the Far East, FAO, Bangkok, Thailand (1979).
18. R. Herrest, See ref. 11 (1978).
19. K. Krishnamurthy and M. J. J. Prince, Abst. Asian Symp. on
    Mangrove Manage., Kuala Lumpur, pp. 37, 38 (1980).
20. S. C. Datta and S. Chanda, See ref. 11 (1978).
21. A. K. Mukherjee and K. K. Tiwari, See ref. 19, pp. 30, 31
    (1980).
21a.B. B. Mukherjee, Abst. Asian Symp. on Mangrove Environ., Kuala
    Lumpur, p. 4 (1980).
22. S. L. K. Bannerjee, Ph.D. Thesis, University of Kalyani (1979).
23. S. L. K. Untawale, S. Wafar and T. G. Jagtap, Abst. Wetlands
    Symp. New Delhi (1980).
24. A. G. Untawale, See ref. 19, pp. 33, 34 (1980).
25. S. N. Dwivedi and K. G. Padamakumar, See ref. 6, p. 24 (1980).
26. L. J. Bhosale and L. S. Shinde, Ibid, p. 18 (1980).
27. T. H. Rae, J. Bhattacharya and J. C. Das, Proc. Ind. Acad. Sci.
    87,B(8):191-195 (1978).
28. G. V. Joshi, S. D. Soutakke and L. J. Bhosale, See ref. 6,
    p. 33 (1980).
29. G. V. Joshi and A. P. Waghmode, Ibid, p. 33 (1980).
30. T. J. Andrews and B. F. Clough, Ibid, p. 15 (1980).
31. M. S. Balakrishnam and V. R. Gunde, See ref. 11, (1978).
32. B. D. Sharma and S. P. Varshney, Ibid (1978).

33. P. S. Rao, S. J. Jarwade and K. S. R. Sarma, Bot. Mar.
    23:599-601 (1980).
34. A. Karim, Z. Hossain and K. J. White, Abst. Asian Symp. on
    Mangrove Environm. Kuala Lumpur, p. 15 (1980).
35. N. Ahmad, Ibid, p. 38 (1980).
36. M. Ismail, Bangladesh Observer, July 12, Dacca, (1980).
37. P. P. K. Chai and K. K. Lai, See ref. 24, p. 36 (1980).
38. L. Srirastava and T. S. Poh, See ref. 6, p. 49 (1980).
39. W. K. Gong, J. E. Ong, C. H. Wong and G. Dharnarajan, See
    ref. 34, p. 23 (1980).
40. F. W. Fong, Ibid, p. 36 (1980).
41. S. W. Nixon, B. N. Farnas, V. Lee, N. Marshall, J. E. Ong,
    C. H. Wong and A. Sasekumar, Ibid, p. 12 (1980).
42. J. Gearing, P. Gearing, M. Rodelli, A. Sasekumar and N.
    Marshall, Ibid, p. 16 (1980).
43. A. J. Kuthaburtheeen, Ibid, p. 16 (1980).
44. C. H. Wong and J. E. Ong, Ibid, p. 17 (1980).
44a.C. D. Field, See ref. 6, p. 26 (1980).
45. T. Christy and J.E. Ong, See ref. 34, p. 26 (1980).
46. J. S. Loi and A. Sasekumar, See ref. 34, p. 42 (1980).
47. S. N. Poo, BIOTROP No. 10:85-91 (1979).
48. B. Christensen, Aquat. Bot. 4:43-52.
49. B. Christensen and S. Wium-Angersen, Ibid, 3:278-296 (1977).
50. S. Aksornkoae, BIOTROP No. 10:13-22 (1979).
51. W. Noonnitee, Ibid, 23-32 (1979).
52. O. Saharacharin, T. Bonkerd and P. Patanaponpaiboon, Ibid,
    51-60 (1979).
53. W. S. Daengshuba, Ibid, 121-124 (1979).
54. S. Sornsomboon, See ref. 10, D29 (1977).
55. Changsang, Hansa, See ref. 34, p. 4 (1980).
56. S. Aksornkoae, Ibid, p. 14 (1980).
57. J. Kongsangchai, Ibid, p. 26 (1980).
58. S. Poorachiranon and H. Changsang, Ibid, p. 24 (1980).
59. S. Sukardjo and K. Kartawinata, BIOTROP No. 10:61-80, (1979).
60. S. Sukardjo, K. Kartawinata and I. Yamada, See ref. 34, p. 6
    (1980).
61. S. Brotonegoro and S. Abdulkadir, Pros. Seminar, Jakarta, p. 81
    (1978).
62. P. Burbraidge and Koesoebiono, Ibid, p. 34 (1980).
63. A. Soegiarto, See ref. 6, p. 47 (1980).
64. N. Polunin, Abst. Wetlands Symp. New Delhi, India (1980).
65. S. M. Cragg and M. J. Swift, See ref. 6, p. 22 (1980).
66. D. G. Frodin, See ref. 6, p. 27 (1980).
67. J. S. Womersley, See ref. 6, p. 54. (1980).
68. V. J. Chapman, "Mangrove vegetation, 2nd Ed., Cramer (In
    press).
69. G. J. Leach, in: "Research in the Biology Department," UPNG, p.
    9, Ed. 17, J. Swift (1979).

70. I. M. Johnstone, Royle Aquat. Bot. 7:197-208 (1979).
71. I. M. Johnstone, See ref. 6, p. 32 (1980a).
72. I. M. Johnstone, Ibid, p. 32 (1980b).
73. I. M. Johnstone, INTECOL Conf., Jerusalem (1978).
74. G. Baines, See ref. 19, p. 41 (1980).
75. D. Hassa, Ibid, p. 7 (1980).
76. N. Penn, See ref. 6, p. 42 (1980).
77. F. N. Ponnamperuma, See ref. 19, p. 39 (1980).
78. F. N. Ponnamperuma, See ref. 10, D32-42 (1977).
79. P. A. Cordero, See ref. 6, p. 22 (1980).
80. M. D. Fortes, BIOTROP No. 10, pp. 111-120 (1979).
81. F. D. Lauricio, See ref. 19, p. 41 (1980).
82. R. T. Bina, Ibid, pp. 35,36 (1980).
83. C. A. Arroyo, BIOTROPP, No. 10, pp 33-44 (1979).
84. N. Q. Sabala, Ibid, pp. 93-100 (1979).
85. A. Pahm. Mirram and R.B. Aspiras, Ibid,, pp. 93-100 (1979).
86. H. Chansang, ICLARM, Manila, Philippines, 83pp. 1978).

BIOSALINE RESEARCH IN AUSTRALIA, CHINA,

JAPAN, SOUTH KOREA AND SRI LANKA

R. R. Walker

Division of Horticultural Research
C.S.I.R.O., Private Mail Bag
Merbein, Victoria, 3505, Australia

SUMMARY

The majority of plant-salinity research in Australia is either problem-oriented research which is directly associated with regional salinity problems or is of a more general nature aimed at understanding the reactions and adaptive strategies of halophytes and non-halophytes under saline conditions. A much smaller proportion of the total research effort is on crop response to irrigation with industrial or municipal wastewater, or on genetic methods of improving salt-tolerance in plants on which most research has only recently been initiated.

Plant-salinity research in China, Japan, South Korea and Sri Lanka is largely associated with regional salinity problems. Most research is on rice, particularly in Korea and Sri Lanka. Some research in Japan is associated with the problem of salt (residual fertilizer) accumulation in greenhouse soils. All four Asian countries have been involved to some extent in the reclamation and utilization of coastal (or tidal) land. Saline and saline-alkali soils are also being reclaimed in various parts of China. Consequently, procedures for reclamation and in particular the selection of salt-tolerant plant species most suitable for growth during the initial years of utilization have received much attention.

INTRODUCTION

The biosaline concept envisages the exploitation of deserts, inland brackish waters, oceans, etc. for production of food, fibre, chemicals and many other products. Biosaline research can be on any aspect of growth in a saline environment, including studies on the response of crop plants to salinity, studies on the suitability of plant species for reclamation and treatment of salt-affected areas, studies on halophytes and marine organisms, etc. Most aspects are being investigated to some extent in Australia and various aspects are being investigated in China, Japan, South Korea and Sri Lanka. Because of limitations on the amount of information which could be obtained on biosaline research in the Asian countries, the greater part of this paper will be on biosaline research in Australia. Emphasis is given to research on the plant-salinity interaction, although brief consideration will also be given to research on marine algae. The first part of this paper will consider biosaline research in Australia, while the second part will be concerned with salinity problems and associated plant-salinity research in each of the Asian countries.

BIOSALINE RESEARCH IN AUSTRALIA

The majority of projects can be grouped into four categories: (I) Problem-oriented research; (II) General biosaline research; (III) Studies on genetic methods of improving salt-tolerance; and (IV) Studies on the disposal of industrial and municipal wastewater. Categories I and II comprise the bulk of biosaline research in Australia. Most research is reported annually in the Australian Salinity Newsletter.

I. Problem-Oriented Research: A significant proportion of plant-salinity research in Australia has been stimulated by the decline in crop productivity due to secondary salinization of soils in some areas. Much of this research is conducted within the various State Departments of Agriculture, Forestry Commissions and the Common-wealth Scientific and Industrial Research Organization (CSIRO).

Secondary salinization of soils occurs as a consequence of irrigation, agricultural practices or the clearing of native vegetation. These soils are distinct from soils which are primarily saline (36). There are two forms of secondary salinity. One occurs in irrigated soils, while the other occurs in non-irrigated soils. Both problems are described in turn.

Secondary salinity in irrigated regions. More than 70% of the irrigated land in Australia lies within the Murray River Basin. The major stream is the Murray River (2590 km) and its principal tributaries are the Darling, Murrumbidgee and Goulburn rivers. Two distinct physiographic zones exist within the Basin, namely the Riverine Plains and the Mallee zone (Fig. 1). Natural groundwater salinity is high in both zones.

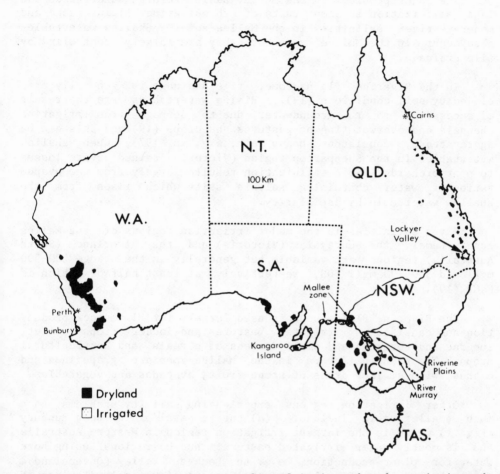

FIGURE 1. The location in Australia of irrigated and non-irrigated
(dryland) regions in which occurrences of secondary salinization of
soils have been reported (from refs. 21,22,29,31,33,36). The areas
affected by dryland salinity are scattered within the regions shown
and represent only a small proportion (mostly <3%) of the cleared
land in each region (29). The estimated total area of land affected
by dryland salinity in each State is: Western Australia, 264,000 ha
(C.J. Henschke, in preparation); Victoria, 85,000 ha (33); South
Australia, 14,000 ha (31); New South Wales, 8,000 ha (G.J. Hamilton,
pers. comm.); and Queensland, 7,300 ha (22). Estimates of the total
area of land affected in irrigated regions are not available, with
the exception of the Riverine Plains zone, where approximately
287,000 ha was affected by shallow watertables (<2m from surface)
and soil salinization in 1976-78 (32).

The major problems in the Murray Basin are high watertables and land salinization in some parts of the Riverine Plains zone and moderate river salinities in the Mallee zone. Shallow watertables also occur in the Mallee zone but they are largely controlled by tile drainage.

In the Riverine Plains zone, tile drainage is generally precluded by soil conditions (19). Rising watertables are the result of accessions to the groundwater due to seepage from irrigation channels and overwatering of pastures and crops (19) and this can be aggravated by consistent heavy rain, e.g. in 1973, when shallow watertables in the Shepparton region (Victoria) caused severe losses to pear orchards. Soil salinization results largely from the evaporation of water containing soluble salts which rises from the shallow watertable by capillarity.

River salinities in the major irrigation regions of the Mallee zone, namely the Sunraysia (Victoria) and the Riverland (South Australia) regions, are variable but generally in the range 200-500 mg/l and 300-700 mg/l TDS, respectively, at least half of which is NaCl (19).

The Riverine Plains zone is used largely for livestock production and consists of irrigated pastures and fodder crops (millet, lucerne and sorghum) and smaller areas of cereals and horticultural crops. The principal crops in the Mallee zone are grapevines and citrus, with smaller areas of stone fruit, avocados and vegetables.

Other irrigation regions experiencing salinity problems on a much smaller scale include: (a) the coastal plain near Bunbury (Fig. 1) which is the largest irrigation region in Western Australia and is mostly under irrigated pasture; and b) regions using bore water for plant production, e.g. the Lockyer Valley (Queensland). The Lockyer Valley (Fig. 1) accounts for 36% of Queensland's area of irrigated vegetables. One third of this area is serviced by groundwater containing 350-700 mg/l chloride, the prolonged use of which is believed to be responsible for rising soil salinities (40). Major crop losses have been associated with increased usage of irrigation water during dry years, e.g. spring 1977, when yield losses in beans were in the vicinity of 30-100% (40).

Dryland salinity. Problems of this type have been reported in most Australian states (Fig. 1).

Secondary salinization in non-irrigated areas is attributed to the removal (clearing) of perennial deep-rooting native trees and shrubs and their replacement by annual crops and pastures with shallower root systems and a lower total water consumption than the native plants they replaced. This leaves an excess of water for dissipation, leading to increased surface and sub-surface water movement, and re-mobilization of soluble salts into streams and low-lying areas and also increased re-charge of saline groundwater tables (37 and refs. therein).

Where rising watertables intersect the soil surface, e.g. on hillsides or in gullies, areas of saline seepage develop which can often be concentrated by evaporation to give salinities much greater than that of the underlying groundwater. Where watertables intersect drainage lines in catchment areas, the result is increased salinity in streams and eventually reservoirs and dams for irrigation or for stock or human consumption. Where watertables approach but do not intersect the surface, e.g. in valley floors, water and soluble salts rise to the surface by capillarity and salts are concentrated by evaporation. This type of saltland is sometimes called watertable saltland to distinguish it from saltland caused by saline seepage (30). A third type of saltland has also been distinguished and given various names (30, 31, 33). It is not formed as a result of saline watertables or seepage - instead, it occurs on soils kept bare by overgrazing and overcropping and is due to movement to the surface of soluble salt which is naturally present in the subsoil.

In Western Australia, the areas most affected are the agricultural areas in the south west, particularly in the wheatbelt region where watertable saltland comprises the largest proportion of the area affected. This is also the main problem in South Australia being common on Southern Eyre Peninsula, Yorke Peninsula and Kangaroo Island (31). In Victoria, the areas affected are principally the valleys and lower slopes in the hill country of west-central Victoria, the extensive alluvial plains in the north and the dune and dune corridor terrain in the north-west (33). Watertable saltland in Queensland occurs in scattered locations along a wide belt in the east of the State, extending from the southern border to as far north as Cairns (25). Seepage and watertable saltland also occurs in both the eastern tablelands and slopes and in the western Mallee of New South Wales (G.H. Hamilton, pers. comm.) and in scattered locations in Tasmania (8).

Research on problems in irrigated regions

(a) The effects of salinity on crop production. Assessment of yield reductions due to salinity is often difficult, especially for horticultural crops where salinity effects must be distinguished from other factors which also influence yields, e.g. climate, soil and management. Few studies have critically examined the effects of salinity on irrigated crops under Australian conditions. Most assessments of yield reduction are based largely on data obtained in other countries. This gap in knowledge has been especially recognized in citrus and is currently being investigated.

The major approaches are to compare yields in plots which are:
(a) irrigated with de-ionized water or River Murray water (South
Australian Department of Agriculture); or (b) irrigated with water
artificially salinized to varying extents (New South Wales Department of Agriculture). Other current approaches are surveys of
citrus yields in areas with different soil and salinity regimes and
the statistical evaluation of past climate, river salinity and
citrus yield data. Similar approaches to assess the effect of
salinity on grapevine yields have not been initiated.

A proposed solution to the problem of shallow watertables in
parts of the Riverine Plains zone is to pump groundwater from the
shallow aquifers. A major objective in disposing of the drainage
water is to minimize further accessions of salt into the Murray
River and one of the options, besides diversion to evaporation
basins, is to re-use some of the water for irrigation of pastures
and crops on either irrigated or formerly non-irrigated land. This
has stimulated research to evaluate the productivity of varius crops
under irrigation with drainage water of varying salinity, up to
3000 mg/1 TDS. Crops presently being evaluated (by the Victorian
Department of Agriculture) include perennial pasture, annual pasture, various row crops (sugar beet, sunflower, sweet sorghum,
tomato, millet) and horticultural crops (pears).

(b) The interaction between salinity and waterlogging. Plants
grown under conditions of low oxygen tension (waterlogging) and high
salinity accumlulate greater concentrations of foliar salt than
plants grown in saline, but well aerated soils (39 and refs. therein). The implication of this finding is that waterlogging should be
prevented, but this may not always be possible. Some amelioration
of the effect of anaerobiosis can be achieved by altered irrigation
management, e.g. irrigation with waterlogging at night, and results
in lower levels of foliar salt than irrigation with waterlogging
during the day (39).

Although some species of agricultural value are reasonably well
adpated to either salinity or waterlogging alone, there are few
reports of agricultural species being well adapted to both conditions together. While it may be possible to breed varieties which
are tolerant of both factors, present research is concerned with the
further evaluation of salinity and waterlogging on existing pasture
and horticultural species (Victorian Department of Agriculture).

(c) Irrigation technique and management. The method of irrigation and the management of water and fertilization regimes to maintain adequate leaching fraction and crop nutrition are important
factors when water is appreciably or even marginally saline. In

this respect irrigated horticulture in the Mallee zone of the Murray Basin has been a matter of concern for some years. The majority of citrus orchards in this region are irrigated using overhead sprinklers. Foliar damage has been severe in years of high river salinities, e.g. during 1966-69, when river salinities in the South Australian region reached 700 mg/l (TDS). Reductions in foliar salt levels can be achieved by irrigation at night. The preliminary data of Grieve (17) suggests that foliar penetration of salt into citrus leaves is due to cuticular rather than stomatal penetration and that reduced penetration at night may be related to lower night temperatures. Alternative methods of irrigation which avoid the wetting of foliage also reduce foliar penetration of salt. For example, a 30-50% reduction in leaf chloride levels can be achieved with low level, full ground cover sprinklers compared with night-time irrigation by overhead sprinklers (18). Further evaluation of under tree microsprinklers, drag-hose sprinklers and low level impact sprinklers is being made (New South Wales Department of Agriculture). Similar work has not yet been initiated with grapevines.

Where shallow watertables are a problem one of the principal objectives is to minimize accessions to the watertable, e.g. by irrigation methods that allow more accurate and even application of water. Research on alternative methods of irrigation and improved management systems is being conducted concurrently with research on the potential use of saline drainage water for irrigation of various crops, e.g. trickle irrigation of tomatoes, and irrigation of pear ochards using under-tree sprinklers compared to the normally-used but less-efficient flood method (Victorian Department of Agriculture).

In respect to reducing accession to watertables the water use efficiency of various crops is also important. With the exception of a current study on lucerne (Victorian Department of Agriculture) little attention has been given to this aspect of field crop production in areas subject to shallow watertables and soil salinization.

(d) Selection and testing of salt-excluding rootstocks. Salt-excluding rootstocks enable the maintenance of low foliar salt concentrations in horticultural species grown in saline areas. A large range of varieties and hybrids of grapevine, citrus, avocado, pome and stone fruit rootstocks are being screened for their potential to exclude Na^+ and Cl^- (CSIRO, New South Wales and Victorian Departments of Agriculture).

Most work has been on grapevine rootstocks. It has been shown that grapevines on certain rootstocks, e.g. Salt Creek, generally contain less chloride but more potassium in petioles and berries than own-rooted grapevines (10 and refs. therein). The rootstock-induced changes in berry ion composition are also reflected in wine

composition.   However, the higher K^+ content of berries from grapevines on rootstocks may be an undersirable factor in wine production
(10, 20)

Obviously, while certain rootstocks may be valuable as 'salt-
excluders' they may not be commercially desirable in other respects.
Other factors that must be considered are resistance to pathogens,
effects on fruit size and quality, etc.   Complete evaluation can
take many years.   Nevertheless, rootstocks do offer a means of
regulating Cl^- concentrations in scion varieties and a means of
studying the effects of elevated tissue Cl^- levels on scion development.   For example, it has been shown that budburst is delayed on
grapevines grafted to certain rootstocks and that this is related to
cane Cl^- concentrations (11).

Research on problems in dryland areas

Considerable attention has been given in various Australian
States to the selection of plant species suitable for the treatment
of salt-affected land in non-irrigated areas.   The major approach
has been to re-vegetate the affected areas with the most useful
plants that can survive, e.g. species which are both salt-tolerant
and suitable for use as stock feed (30).   Ultimately, however, the
value of any plant species resides in its ability to combat the
process of secondary soil salinization.   In areas affected by
shallow watertables and saline seepage it has been assumed that useful species are those with relatively high rates of water consumption such that watertables in areas revegetated with these species
would be lowered and secondary salinization of soils reduced.   Until
recently, this assumption had not been critically examined (15).
However, it has since been demonstrated that Atriplex spp. do modify
the hydrological regime.   Evaporation of water from ungrazed plantations of A. vesicaria increased with closeness of plant spacing
leading to a 10-20 mm reduction in the level of the watertable under
the closest, compared to the widest plant spacing (16).

Deep-rooting, fast growing tree species might also modify the
hydrological regime through relatively high rates of water consumption and a greater capacity (of foliage) to intercept and evaporate
rain (15).   Again, there is little information on either transpiration or evaporation of water from wet leaves of tree species (15).
The only available data on transpiration is from juvenile Eucalyptus
species planted upslope of saline groundwater seeps.   Although
transpiration rates varied widely between species, considerable
amounts of water were transpired vis. 26-37 l/day per tree (15).
Experiments of this type are continuing with Eucalyptus and Pinus
spp. The effect on groundwater levels, groundwater salinity and on
agricultural production downslope of the plantation will eventually
be determined (CSIRO, West Australian Forests Department).

Desalination of land re-vegetated with salt-tolerant species has also been reported (33). Comparisons of soil salinity have been made between bare salt-affected patches of land before, and several years after re-vegetation with <u>Puccinellia</u> <u>ciliata</u> and <u>Agropyrum</u> <u>elongatum</u>. During this period there was a marked reduction in the salt content of the top 30 cm of soil in all areas (varying in rainfall) which had been re-vegetated with the two species. In contrast, soil profiles in bare areas outside the re-vegetated areas were found to contain salt-concentrations similar to those in profiles before the land was re-vegetated, which indicates that soil desalination was specific to the re-vegetated sites. This is probably a result of increased ground cover reducing the evaporation and concentration of salts in the surface layer combined with periodic leaching of salts by rainfall.

A large range of plant species have been tested for establishment in salt-affected areas in southern Australian States. Between 1968 and 1975 over 700 local and introduced shrubs and grasses were tested over a range of climatic conditions and soil types in Western Australia alone (27). Many more species have been evaluated since that time and many are currently being evaluated. Many have been assessed for their suitability as forage and their ability to recover from heavy grazing. The most promising species have been listed in various publications (e.g. 28, 33). However, further appraisal of these and other species may be requird for the reclamation of salt-affected land in northern areas, e.g. Queensland.

In contrast, little is known about the tree species most suitable for treatment of salt-affected areas. Several investigations are in progress. The general approaches are the evaluation of tree species for growth rates and water use in field situations and screening and selection for tolerance to salinity using artificially salinized media. <u>Eucalyptus</u> spp. represent the major species under evaluation because some species show apparent ability for deep rooting, significant water use in summer, withstanding salinity and waterlogging and resisting <u>Phytophthora</u> and leaf miner attack (5).

II. General Biosaline Research: This includes studies on the physiological responses of non-halophytes to salinity and studies on the ecology and physiology of various halophytes ranging from arid and semi-arid halophytes, e.g. <u>Atriplex</u> spp., to semi-aquatic and aquatic halophytes, e.g. mangroves, salt-marsh plants, seagrasses and marine algae. Much of this work is conducted within Universities, the Australian Institute of Marine Science (AIMS) and CSIRO.

Over the years there have been many studies on the response to salinity of various non-halophytic crop plants and some halophytic spp., particularly in terms of growth and tissue ion composition.

Several studies have also considered plant water relations and
osmoregulation. There have been fewer studies on the effects of
salinity on in-vivo metabolism, with some exceptions (see Walker and
Downton, this volume). Much of the available information on these
and other aspects relevant to the plant-salinity interaction, e.g.
processes of ion uptake and the regulation of ion transport, have
been discussed elsewhere (e.g. 12, 14). Brief outlines of current
research on physiological responses to salinity of various non-
halophytic and halophytic species are given in Table 1.

Atriplex biology has been given much attention over the years.
However, apart from the studies listed in Table 1 and a continuing
program on the evaluation of the grazing potential of Atriplex and
Maireana spp., there seems to be little work on arid or semi-arid
halophytes currently underway in Australia.

On the other hand, there is a great deal of work on mangroves,
salt-marsh plants and seagrasses (Table 2). Interest in mangroves
at AIMS stems from the potential for coastal mangrove systems to
contribute to offshore reefal systems. For the distribution of
various mangrove and salt-marsh species within Australia the reader
is referred to Saenger et al. (38). Current knowledge on mangrove
physiology and mechanisms of adaptation to salinity have recently
been outlined by Clough et al. (6). The range of plant species
present in saline lakes in inland Australia has been briefly
described by Bayly and Williams (4) and by Saenger et al. (38) but a
lack of salinity data in many investigations was noted (4). Few
studies have followed the changes in vegetation pattern with changes
in salinity. One exception is a recent study made in the Gippsland
Lakes in South Eastern Victoria, where a decline in the reed
Phragmites australis has been attributed to a gradual increase in
salinity (7).

A considerable amount of money has been made available for
research in marine biology in Australia in recent years. Some of
this has been used for research on marine algae which is in addition
to research by commercial organizations with an interest in exploiting marine algae for food and other purposes. Most aspects of
marine algal biology are being investigated – taxonomy and distribution, nutrition, ultrastructure, physiology, biochemistry, etc.
Studies on osmoregulation are being conducted using various species
including Dunaliella and some Characean algae, e.g. comparisons
between euryhaline and fresh water Characean algae in terms of turgor regulation, osmotic control and membrane responses (University
of Sydney). These and similar studies with other algae, e.g.
Chlorella emersonii (University of Western Australia) which lives in
both fresh and brackish water, may give some insight into how
various algae can successfully adapt to highly saline environments.

TABLE 1

CURRENT RESEARCH IN AUSTRALIA ON PHYSIOLOGICAL
RESPONSES TO SALINITY IN VARIOUS NON-HALOPHYTES
AND HALOPHYTES

Research Topic	Species Involved
Salinity effects on plant growth and ionic relationships	Hordeum spp., Disphyma australe, Trifolium fragiferum, Eucalyptus spp., tomatoes, mung beans, red beet, Lolium spp., lupins, Cakile maritima, Atriplex spp.
Salinity and metabolism	
Protein synthesis	Various non-halophytes and halophytes.
Photosynthesis	Horticultural spp., tropical grasses, lucerne, Disphyma australe.
Nitrogen metabolism	Disphyma australe.
Carbohydrate metabolism	Beans, cassava.
Salinity and ion localization in cells	Atriplex spp., Wedelia biflora, mung beans, lupins, citrus.
Salinity and compatible solutes	Atriplex spp., Wedelia biflora, Hordeum spp.
Salinity and regulation of ion uptake	
Transpiration effects	Atriplex spp., citrus, wheat
Hormonal regulation	Lupins.
Root membrane lipids	Citrus.
Foliar uptake of salt	
Sea Spray-cyclic salt	Various species.
Irrigation water	Citrus, grapevines.

TABLE 2

CURRENT RESEARCH IN AUSTRALIA ON MANAGROVES,
SALT MARSH PLANTS AND SEAGRASSES

Plants	Aspects under Investigation
Mangroves	•Distribution, taxonomy and ecology of Qld., N.T., W.A. and NSW mangroves •Sediment chemistry •Nutrient and organic fluxes in mangrove communities •Geomorphology of mangrove swamps •Growth and primary productivity •Photosynthesis and gas exchange in relation to salinity, root and leaf temperature and nutrition •Salt balance, salt excretion, water relations and osmoregulation •Phenology.
Salt-Marsh Plants	•Distribution, taxonomy and ecology of spp. in southern Australia •Distribution in salt lake ecosystems and associated salt marshes in relation to salinity (particularly Ruppia spp) •Nutrient and organic fluxes •Growth and salt-balance of various spp. •Ultrastructure of salt glands in Samolus repens.
Seagrasses	•Distribution, taxonomy and ecology of seagrasses in N.S.W., W.A. and Vic. •Growth and primary productivity •Photosynthesis in relation to light and temperature •Biochemical aspects of evolution in seagrasses •Flowering and pollination •Morphology and ultrastructure •Turgor regulation

III. Research on Genetic Methods of Improving Salt-Tolerance:
Increasing awareness of the potential offered by the genetic
approach to improving the tolerance of plants to salinity has led to
the initation of several projects. Various short growth cycle crops
e.g. lucerne and barley, and some horticultural crops, e.g. grapevines and citrus, are being screened for the identification of
tolerant and susceptible lines for studies on the heritability of
salt-tolerance (CSIRO and the Universities of Melbourne and
Adelaide). However, there are extremely few projects specifically
involving breeding for salt-tolerance. The few exceptions are a
program involving citrus rootstocks (CSIRO) and a new program
involving wheat (Victorian Department of Agriculture).

IV. Research on the Disposal of Industrial and Municipal Wastewater:
Most research is on crop response to sewage effluent. However,
other effluents, e.g. winery, cheese factory, piggery and pulp mill
effluents, are also being evaluated for irrigation of crop, pasture
and tree spp. on a small scale.

   In the Northern Adelaide Plains of South Australia, sewage
effluent has been used commercially as an alternative to bore water
for the irrigation of various crops because of diminishing supplies
of bore water. The average salinity of the effluent is 1500 mg/l
(TDS) with relatively high concentrations of Na^+ (440 mg/l) and Cl^-
(550 mg/l). With suitable management procedures, satisfactory
yields have been obtained from various vegetable and pasture crops
(9). Winegrape production in a large commercial vineyeard trickle-
irrigated with effluent has also been satisfactory, although the
concentrations of NaCl in wines produced form the vineyard are
higher than in normal wines. The implications of this finding for
wine quality are not yet clear (McCarthy and Downton, pers. comm.).

   The salinity tolerance of some aquatic macrophytes, e.g. Typha
spp. and Phragmites australis, is being evaluated as part of a
program which aims to investigate the suitability of aquatic plants
for the treatment of various nutrient-rich wastewaters, e.g. piggery
effluent which is rich in KCl. It is assumed that salt-tolerant
aquatic plants might 'extract' considerable amount of K^+ and Cl^-
from the wastewater. The general principles have been outlined by
Mitchell (34).

                SALINITY PROBLEMS AND PLANT-SALINITY RESEARCH
                        IN SOME ASIAN COUNTRIES

   Much of the available information on biosaline research in
China, Japan, South Korea and Sri Lanka concerns plants and salinity, particularly the effects of salinity on rice. The work on
salt-tolerance and breeding for salt-tolerance in rice at the Inter-,

national Rice Research Institute (Philippines) is in addition to work conducted within China, Japan, South Korea and Sri Lanka. Other aspects of biosaline research have been given some attention, particularly in Japan, where there have been many studies on marine algae, including research on algal products for use in pharmaceutical science (35). Macro-algal farming is practiced in coastal waters of China, Japan and Korea. However, only local salinity problems and associated plant-salinity research will be considered in this section.

China

Saline soils occur in many parts of China. A recent survey by Liu et al. (26) has distinguished six regions of salt-affected soils suitable for reclamation and utilization. These are: (1) the coastal sea-soaked region; (2) the north-eastern sodic-alkaline region; (3) the patch-like salt-affected region in the plain of the Yellow (Huaiho) river; (4) the sheet-like salt-affection region in Mongolia and Ninghsia; (5) the salt-affected region in the Kansu-Sinkiang endorheic basin and; (6) the alpine-arctic salt-affected region in Chinghai-Tibet.

The procedure for reclamation of saline-alkaline soils as practiced in the Shantung province in north-eastern China involves digging ditches 1m deep, addition of new soil, followed by leaching with fresh water and then growing (in succession) salt-tolerant grasses, cotton and rice after which the land is ready for normal cropping (C.T. Gates, pers. comm.). An afforestation technique has also been reported for reclamation of saline-alkaline soils in Tientsin.

China has in the past been involved in the reclamation of land in coastal areas, often requiring special attention to occur to some extent in both old-established and new irrigation schemes. For example, some problems with salinity have occurred in the Sinkiang region where 36,000 ha of former desert has been opened up by utilizing the waters of the Tarim river (24).

Few physiological studies on salt-tolerance have apparently been made in China. However, some studies on rates and dates of application of water and fertilizer to give high yields in wheat grown on saline soil have been carried out at the North China Agricultural College.

Japan

Few soils in Japan are primarily saline. However, there are soils which are affected by secondary salinity, although the problems are not considered to be serious when compared with those occurring in other countries of the world (41). The existing problems occur either in greenhouses for vegetable production or in regions of lowland rice production.

The problem in greenhouses is due to the accumulation of residual fertilizer, with the affected soils containing particularly high levels of nitrate (41). Much research has been devoted to this problem, e.g. studies on specific effects of high levels of nutrients on plants, numerous studies on the salt-tolerance of various vegetable and ornamental spp. (41) and an evaluation of various forage grasses as a remedial measure against excessive salt-accumulation in greenhouse soils (Chiba University).

The regions of lowland rice production which have experienced salinity problems are the deltaic plains of the coastal region where seawater intrusion is often a problem and in the polders or areas of reclaimed tidal land, of which more than 30,000 ha has been reclaimed in the last 30 years (41). Research on rice has included the establishment of salt-tolerance limits, studies on the physiology of tolerance and breeding for salt-tolerance (2).

South Korea

Present salinity problems in Korea are largely associated with rice production on reclaimed tidal land. Nearly 150,000 ha of tidal land has been reclaimed since the 1920's (25).

Fresh water reservoirs are an integral part of most reclamation schemes, as adequate supplies of good quality water are necessary for desalination (25). Rice yields increase as soil salinity is progressively reduced but the time taken to reach normal productivity depends on the method of drainage which is further hindered by the cost involved. For example, normal productivity is achieved in approx. 5 years on land serviced by underground drainage systems compared with approx. 14 years on land drained by cheaper surface methods (25).

A further limitation during the early years of reclamation is the lack of high yielding, salt-tolerant rice varieties. The japonica varieties, Mankyeong, Jinheung and Gancheok 9, have been traditionally sown in reclaimed areas but there has been an increase in the use of improved semi-dwarf varieties, Tongil, Yushi and Milyuang 3 (25). Introduced salt-tolerant varieties, e.g. Pokkali, CK 32, Getu and IR 24, are also being crossed with local varieties and the progeny tested. Other aspects of rice production that have received attention are fertilizer requirements, seeding methods, physiology of salt-tolerance and the control of perennial weeds,

e.g. Scripus maritimus L. in reclaimed paddy fields.   There is
little information on crops other than rice except for some research
on the effect of salinity on Phragmites longivalvis grassland.

    The lack of fresh water in some smaller schemes for rice
production has led to research into the use of mixed fresh and sea
water on the growth and yield of rice.   However, the varieties
tested so far, Mankyeung and Tongil, do not have a high degree of
salt-tolerance and major benefits in this respect await the release
of improved varieties.

Sri Lanka

    There are three major climatic zones in Sri Lanka, namely wet,
intermediate and dry zones in which the   rainfall is respectively
>250, 190-250 and 100-190 cm annually.   The main irrigation regions
are in the intermediate and dry zones where rice is the principal
crop.

    Soil salinization and intrusion of seawater into cultivated
areas occurs to varying degrees in Sri Lanka but does not represent
an immediate hazard to agriculture (1).   Abeygunawardene (1) has
listed the main areas where problems occur and these will now be
considered.

    Most problems with plant productivity concern rice and have
been encountered in the older irrigation schemes where a steady
increase in soil salinization has occurred in some areas.   Reduced
yields and some crop damage have been reported in most of these
schemes, particularly during the dry season when irrigation water is
often limiting.   There are also areas where secondary salinization
of soil has occurred due to shallow watertables caused by overwatering and lack of adequate drainage.   Poor quality irrigation water
(relatively high salinity and medium to high sodicity) has also
caused concern in a tube well irrigation project in the dry north-
west.

    Other problems have been reported in low-lying coastal areas
particularly in the south and south-west and in the northern parts
of the country, due to salt-water intrusion into rice land.   Germination and growth of rice is affected by high soil salinities
particularly during the dry season.   More tolerant varieties are
presently being evaluated.   The variety Pokkali, which grows
naturally along the coast of Sri Lanka in deep, brackish water, has
been traditionally grown in the salinity affected regions of Sri
Lanka, but its yield is relatively poor and it does not respond to
fertilizer like the high yielding varieties.   The salt-tolerance of
Pokkali has been found to be inherited quantitatively and more than

2000 pedigree lines from 20 crosses involving Pokkali have been tested in the greenhouse at the International Rice Research Institute (Philippines) while a further 3000 pedigree lines have been tested in saline coastal areas of the Philippines. So far, two advanced breeding lines have consistently shown salt-tolerance comparable with Pokkali and may be of promise in future plantings (3). However, some benefits have already been obtained in Sri Lanka with new improved varieties such as IR8, BG34-8 and BG 11-11 (1).

ACKNOWEDGEMENTS

The author wishes to thank Dr. W.J.S. Downton (CSIRO, Horticultural Research) and Prof. M.J. Pitman (University of Sydney) for helpful discussions during the prparation of this manuscript and the many people who contributed details of current research projects, with particular thanks to Dr. B.F. Clough (Aust. Institute of Marine Science, Townsville), Dr. J. Gibbs-Clema (University of Adelaide), Dr. E.A.N. Greenwood (CSIRO, Land Resource Management), Dr. A.M. Grieve (Department of Agriculture, New South Wales), Dr. C.V. Malcolm (Department of Agriculture, Western Australia) and Dr. R.A. Wildes (Department of Agriculture, Victoria).

REFERENCES

1. D. V. W. Abeygunawardene, in: "Plant Response to Salinity and Water Stress," W. J. S. Downton and M. G. Pitman, eds., pp. 25-28, Australian Dept. of Science, Canberra (1977).
2. M. Akbar, T. Yabumo and S. Nakao, Jap. J. Breed. 22:277-284 (1972).
3. Anon., IRRI Reporter, No. 3, 2pp. (1978).
4. I. A. E. Bayly and W.D. Williams, "Inland Waters and their Ecology," pp. 207-208, Longmans, Melbourne (1973).
5. E. F. Biddiscombe and A. L. Rogers, Aust. Salinity Newsl. 4:93 (1976).
6. B. F. Clough, T. J. Andrews and I. R. Cowan, in: "Structure, Function and Management of Mangrove Ecosystems in Australia," B. F. Clough, ed., ANU Press, Canberra (1980).
7. R. D. Clucas and P. Y. Ladiges, Ministry for Conservation, Victoria, Environmental Studies Series, Publication No. 32 (1980).
8. J. D. Colclough, Tasm. J. Agric. 44: 171-180 (1973).
9. P. G. Cooper, L. Goss and W. E. Matheson, Proc. Aust. Water and Wastewater Assoc., Sixth Federal Convention, pp. 67-119, Melbourne (1974).
10. W. J. S. Downton, Aust. J. Agric. Res. 28: 879-889 (1977).
11. W. J. S. Downton and A. W. Crompton, Aust. J. Exp. Agric. Anim. Husb. 19:749-752 (1979).

12. T. J. Flowers, P. F. Troke and A. R. Yeo, Ann. Rev. Plant
    Physiol. 28:89–121 (1977).
13. L. E. Francois and R. A. Clark, J. Amer. Soc. Hort. Sci.
    104:11–13 (1979).
14. H. Greenway and R. Munns, Ann. Rev. Plant Physiol. 31:149–190
    (1980).
15. E. A. N. Greenwood and J. D. Beresford, J. Hydrol. 42:369–382
    (1979).
16. E. A. N. Greenwood and J. D. Beresford, J. Hydrol. 45:313–319
    (1980).
17. A. M. Grieve, Aust. Hort. Res. Newsl. 50:33 (1980).
18. A. M. Grieve, Aust. Hort. Res. Newsl. 50:32–33 (1980).
19. Haskins Gutteridge and Davey Pty. Ltd., Murray Valley Salinity
    Investigations, River Murray Commission, Canberra, Vol. 1
    (1970).
20. C. R. Hale and C. J. Brian, Vitis 17:139–146 (1978).
21. R. E. R. Hartley, in: "Soil Conservation Branch Report S3/80,"
    Department of Agriculture, South Australia (1980).
22. K. K. Hughes, in: "Assessment of dryland salinity in Queensland,"
    Queensland Dept. of Primary Industries, Division of Land
    Utilisation, Report No. 7, 21pp. (1979).
23. V. A. Kovda, C. Berg and R. M. Hagen, in: "Irrigation, drainage
    and salinity: An International source book, FAO/UNESCO,
    Camelot Press, London, 510pp. (1973).
24. P. Kung, Wld. Crops 27:54–66 (1975).
25. M. Lim and E. Lee, See ref. 1, pp. 15–23 (1977).
26. W. S. Liu, T. C. Wang and Y. Hseung, Acta Pedol. Sin. 15:101–112,
    (In Chinese with English summary) (1977).
27. C. V. Malcolm, West Aust. Dept. Agric. Techn. Bull. No. 21,
    34 pp. (1973).
28. C. V. Malcolm, J. Agric. West Aust. 15:68–74 (1974).
29. C. V. Malcolm and T. C. Stoneman, J. Agric. West Aust 17:42–49
    (1976).
30. C. V. Malcolm, J. Agric. West Aust. 18:127–133 (1977).
31. W. E. Matheson, J. Agric. South Aust. 71:266–272 (1968).
32. Maunsell and Partners, "Murray Valley Salinity and Drainage
    Report," Murray Valley Study Steering Committee, Canberra,
    94pp. (1978).
33. A. Mitchell, S. Zallar, J. J. Jenkin and F. R. Gibbons, in:
    "Dryland – Saline – Seep Control," H.S.A. Vander Pluym, ed.,
    11th International Soil Science Society Congress, Edmonton,
    Canada (1978).
34. D. S. Mitchell, Water (Aust. Water and Wastewater Assoc.),
    5:15–17 (1978).
35. K. Nisizawa, in: "Marine Algae in Pharmaceutical Science," H.A.
    Hoppe, T. Levring and Y. Tanaka, eds., Walter De Grutyer and
    Co., Berlin and New York (1979).
36. K. H. Northcote and J. K. M. Skene, Aust. Commonw. Sci. Ind. Res.
    Organ. Soil. Publ. No. 27, 62pp. (1972).

37. A. J. Peck, Aust. J. Soil Res. 16:157-168 (1978).
38. P. Saenger, M. M. Specht, R. L.Specht and V. J. Chapman, in: "Wet
    Coastal Ecosystems (Ecosystems of the World)," V. J. Chapman,
    ed., Elsevier, Amsterdam, pp. 293-345 (1977).
39. D. W. West and J. A. Taylor, Plant and Soil 56:113-121 (1980).
40. B. J. White, in: Salinity in the Lockyer Valley: "A preliminary
    evaluation," Queensland Dept. of Primary Industries, Division
    of Land Utilisation, Report No. 4, 31 pp. (1980).
41. H. Yokoi, See ref. 1, pp. 2-14 (1977).

BIOSALINE RESEARCH ACTIVITIES IN KUWAIT

A. J. Salman, I. Y. Hamdan and N. M. Hussain

Kuwait Institute for Scientific Research (KISR)
Food Resources Division
P. O. Box 24885, Kuwait

SUMMARY

Toward its goal to increase production of feed and food in hot arid regions, the Food Resources Division (KISR) has embarked on a research program with the aim of exploring new avenues in food producing systems through conventional and non-conventional technologies; while exploiting its natural resources.

Various programs have been initated and are ongoing at present, among which are: Development of truffles cultivation in the field and laboratory; Rangeland evaluation, its forage production and desert combating in Kuwait; Optimization of poultry production systems under hot arid zone climates; Efficient use of water in irrigation; Utilization of brackish treated sewage water as well as industrial waste water in agricultural crop production.

Controlled Environmental Agriculture is being looked into very actively at present at KISR, as an alternative to conventional agriculture for food and feed production. The Food Resources Division is also involved in intensive programs in Mariculture and Fisheries Management. The goal of the first program is to augment supplies of fish and shrimp while that of the second program is to elucidate the overall status of the fishery and fish stocks to provide sound management advice to the fishery and government.

INTRODUCTION

Kuwait lies in the hot arid region of the Arabian Peninsula. It is located at the northwestern end of the Arabian Gulf and is bordered by Iraq to the north and northwest, and Saudi Arabia to the south and southwest.

The climatic conditions are characterized by long hot summers where daytime temperatures seldom fall below $40^{\circ}C$ and daily maximum temperatures in excess of 45° are frequently recorded. The relative humidity during the summer usually remains low (20–30%) leading to an increase in evaporation which reaches 24.3 mm/day in July and August.

During the summer which extends from early May to late September, the incidence of solar radiation is high and averages 740 cal/cm^2 per day during June. This level of radiation is equal to three times that received in most northern hemisphere countries. Most of the annual rain of 110 to 120 mm falls in the winter months of December and January; however, there is great temporal and spatial variation (30.3–242.4 mm). Often the rain falls in one or two heavy showers spread over a few days.

The soils are not well developed due to the influence of limited soil forming factors. This is attributed to climatic aridity, sandstorm parent material, a meager influence of vegetation and insufficient chemical and biological weathering. Soils are generally sandy with occasional occurences of hardpan which outcrops on the surface. In addition, soils suffer from a high salt content, low phosphorus, low organic matter, low cationic exchange capacity and a clay silt portion that does not exceed seven percent. It is evident that several physical environmental factors have contributed to the stressed agricultural productivity.

Desalinated sea water provides the main source of drinking water and over sixty million gallons per day are produced in Kuwait. The Rawdatain and Um-Alish underground fresh water acquifers produce six million gallons per day; the salinity of this water does not exceed 650 ppm. Other sources of underground water vary in salinity from 3,000 to 10,000 ppm, making a total production of 100 to 120 million gallons per day.

In view of the climatic and water resource constraints, production of food and feed by direct conventional agricultural systems has been restricted, resulting in a meager agricultural contribution of less than one percent of the country's total output. Recently, extensive efforts have been made to increase local food production to about 13% of the national food needs.

The Food Resource Division (KISR) has been exploring methods to increase food production by focusing attention on maximizing production in arid areas with special emphasis on biosaline research activities. This has been achieved by adopting and/or developing new technologies that are appropriate to Kuwait and other countries of similar background. In addition, at KISR we are investigating the potential development of novel food industries. In this regard, the desert, the sea, the climate, saline and sewage water, as well as solid wastes, are considered as resources rather than deterrents. Although other institutions in Kuwait have been involved in some agricultural research activities, the present text will touch upon KISR efforts related to food resource development.

Agro-production

The main objective of this department is to increase food and feed production in a desert environment, by combating stress effects that result in reduced productivity of plants and animal species.

Range Mangement Project

Rangelands form 90% of the land base and can provide renewable products on a continuous basis; e.g., sheep production systems through appropriate utilization and management.

KISR, in cooperation with the Dept. of Agriculture, Ministry of Public Works, has initated a program with the following specific objectives: (1) To record vegetation structure and herbage production data within major vegetation types found in rangelands; (2) To investigate means of increasing forage production on rangelands; and (3) To evaluate and direct range use to obtain sustained maximum animal production without rangelands degradation (desertification).

Data on vegetation structure forms the basis for resource planning use. To study this ecological parameter an investigation was initated on the prime grazing regions (Fig. 1). Only Rhanterium and Cyperus were selected since they provide the bulk of forage for livestock.

Suitable study sites were selected within these two regions using restricted random procedures (Fig 1). Investigation was commenced with the Rhanterium stand. The vegetation is mainly pure and mixed stands of Rhanterium epapposum, a deciduous shrub growing up to about 60 cm in height and with a deep tap root often reaching 1.8-2.0 m. Major species associated with the mixed stands are Convolvulus oxyphyllus, Stipagrostis plumosa and Molkiopsis ciliata.

Phenological development and growth curves for the four

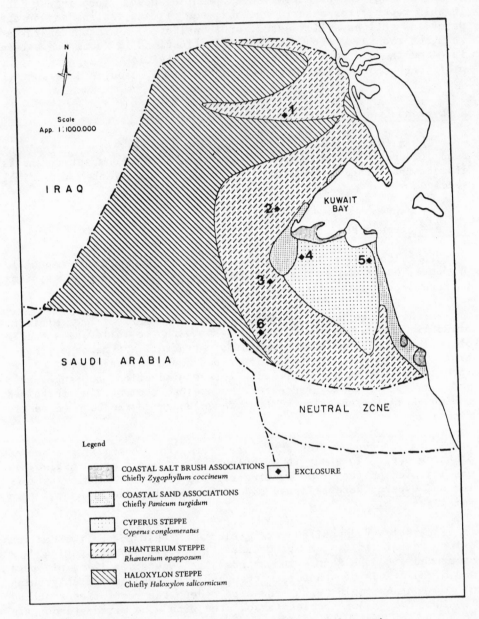

Figure 1. Major Grazing Regions of Kuwait.

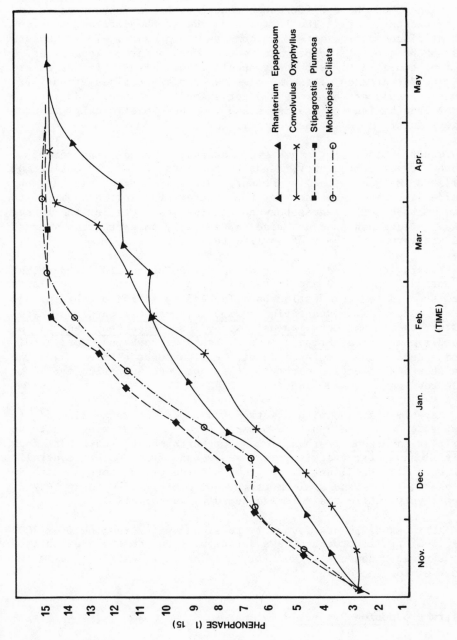

Figure 2.   One year average growth for prime species of Rhanterium steppes.

primary species of the Rhanterium zone are shown in Figure 2. It is evident that the four species differed in their growth pattern as well as phenological development. This was particularly evident in marked phenophase differences between the shrubby species and the herbaceous ones. The former proceeds in phenological development at a much slower rate when compared to grasses. This suggests that shrubs are able to provide "green" forage for a longer period of time compared to herbaceous species. It also indicates possibilities for grazing these species at different times to utilize each, when they are best suited for sheep production, and in this manner lengthening the grazing season. These preliminary results suggest future more intensive field work.

Other ecological parameters measured included measurement of percent cover, frequency, density and production by utilizing published methods (1-6). Measurements were taken inside and outside the exclosures set at the different desert sites. Data showed marked differences between inside and outside the fenced areas. Decreaser species reacted favorably to protection whereas invaders were more dominant outside the fence.

In studies aimed at increasing forage production for animal production from native and introduced species, eight native plants were studied for germination behavior and seedling establishment. The eight species: Rhanterium epapposum, Panicum turgidum Zygophyllum coccineum, Convolvulus oxyphyllus, Aeluropus lagopoides, Cutandia memphitica, Stipa capensis, Schismus barbatus, were germinated under three simulated Kuwait environments (Table 1). The data showed that environment [1] was the most suitable for large scale seeding of mixed species.

Experimentation with introduced species was limited to xerophytic and halophytic plants (7-8). Four-wing saltbush (Atriplex canescens) were selected for performance study under four irrigation waters; brackish, sea, sewage and fresh water (control). Four-week old seedlings were grown inside green houses with an average maximum and minimum temperature of $34.7^{\circ}C$ and $26.2^{\circ}C$, respectively, while the relative humidity was 63.81%.

Atriplex responded favorably to saline water ranging from 3000 to 10,000 ppm for twenty weeks duration. Due to the availability of these water sources in Kuwait, it seems that desert shrubs might be useful for increasing the forage supply.

Poultry Production Project

Poultry production is affected by various factors in hot arid regions, the most important of which is thermal stress where

TABLE 1:   Percent germination of eight native species at three
different environments. Day and night temperatures
were 35,25;   25,14   and 35,20;   whereas the Relative
Humidity (%) was   60,40;   74,34   and   65,35   for the
environments 1, 2 and 3, respectively.

| Treatments | Germination (%) | | |
Species	Environment [1]	Environment [2]	Environment [3]
Rhanterium Expapposum	79.6	0	0
Pancium turgidum	90.0	–	–
Zygophllum coccineum	52.4	45.3	95.3
Convolvus Oxyphyllus	0	1.6	23.4
Aeluropus lagopoides	96.0	–	–
Cutandia memphitica	77.1	40.6	76.6
Stipa capensis	52.6	98.4	96.9
Schismus barbatus	33.3	53.1	50.0

TABLE 2:   Shrimp (Panaeus semisulcatus)   production rates from
the rearing tanks at Salmiya during the period   1974
to 1979.

Year	Total Number Produced x 10^6	Volume (m^3)	Production (number/m^3)
1972[a]	11.10	1,200	9,250
1973	3.75	1,875	2,000
1974	9.13	1,875	4,869
1975	20.6	2,200	9,364
1976	7.68[b]	2,200	3,491
1977[a]	24.91	2,800	8,896
1978[a]	17.40	2,800	6,214
1979[a]	ca. 26[c]	1,400	18,571
TOTAL	120.57		

[a]Includes production of Penaeus japonicus and/or Metapenaeus
affinis.
[b]Reduced production due to inclement weather during breeding time.
[c]Estimated, as juvenile shrimp not harvested at P_{20} (20 days after
metamorphosis).

temperatures exceeding body temperatures are prevalent for both a large portion of the day and extended periods. This results in reduced feed and energy intakes and subsequently, lower growth and productivity. Thermal stress induced by high environmental temperature leads to reduced survival and large losses due to heat stroke, especially if the birds are not adequately sheltered.

Mixed poultry feeds and feed ingredients are virtually all imported and stored until utilized. The quality of the imported materials is of importance to formulate balanced poultry rations. Studies conducted at KISR revealed variability in nutrient composition of the same variety of feedstuffs. Forty-seven percent of the feeds and feed ingredients contained aflo-toxins, while layer and broiler rations were deficient in lysine levels for supporting optimal performance. Amino acid content of various soybean meals also differed. The differences were attributed to processing conditions and protein contents which were as high as 60% in soybean meal of Chinese origin.

Performances of broilers grown at constant temperatures of 35° C, or fluctuating ambient temperature ($29-39^\circ$C), were compared with controls that were started at a brooder temperature of 35°C and reduced by 2.8°C per week to 27°C. Mortality rates at the constant temperature of 35°C and the fluctuating ambient temperature $29-39^\circ$C were as high as 42 and 51.5%, respectively; while it did not exceed 3.3% in the control treatment. Feed consumption and growth rates were significantly depressed at high constant and fluctuating temperatures.

In a comparative study under a controlled constant ambient temperature of 27°C, and a fluctuating ambient temperature range of $29-39^\circ$C in a naturally ventialated, insulated unit in Kuwait, egg production rates, egg size, feed consumption and body weight were adversely affected. A comparison of data obtained under the above mentioned fluctuating temperatures for energetic efficiencies indicated close agreement with those reported by Smith (9).

Better performance and lower mortality rates could be achieved at high ambient temperatures if the animals are acclimatized. Laying hens that are reared in the fall and come into lay in February do not suffer high mortality rates although production rates did fall down to 52% in midsummer.

Poultry production in hot arid regions is also limited by feed source supplies. Recycling of nutrients from the established poultry industry indicated that hatchery by-product meal and dried poultry waste were capable of supporting good performance. Further, dates can replace a large extent of the energy usually supplied by corn in poultry rations. The dates supplied energy without adversely affecting growth or feed efficiency.

Brackish saline water was adequate in growing broilers and replacement pullets but with incurring adverse effects. The brackish water affected the poultry equipment rather than the animal itself.

The concept of constructing poultry houses partially sunken to benefit from the reduced temperature of the soil during the summer has been attempted; while insulation costs were reduced for the houses, additional evaporative cooling was needed during the day.

Future studies will identify optimal nutrient levels during each season and compare different breeds and strains for broilers and layers to study their capability for acclimatizing to the hot arid climate. Management – nutrient interaction studies will identify the most economical systems for production in hot environments which maximize use of the available feed resources.

Truffles (Kame)

Four types of desert truffles "Kame" are found in the desert: Terfezia boudieri, Terfezia claveryi, Tirminia nivia and Tirminia pinoyi. These truffle species are also found in Saudi Arabia, Syria, Iraq and North Africa. Truffles are considered a delicacy in Kuwait. They contain a relatively high nutritive value and are of good taste and aroma. They are in season for a short period from March to April. Kame (desert truffles) are hypozygeous mycorrhizal fungi. Their production is affected by environmental factors.

KISR initiated activities in the truffle area in 1977, the achievements to date can be summarized as follows: Successful germination of spores (Kame seeds) of the four types of local Kame species was possible using fresh and dry Kame ascocarps. Isolation between Kame mycelium and the host plants H. edifolium and H. salicifolium were completed. Field data indicates a natural relationship between Kame and the host plant H. edifolium and H. salicifolium.

In the future, the research at KISR will focus on the following objectives: (1) To continue field work on truffles and to determine the environmental factors associated with Kame fruit production. In this study special attention will be given to soil characteristics and the natural relationship between Kame and adjacent plant communities; (2) To develop method(s) for innoculation of desert land with Kame spores and/or planting pre-innoculated hosts (mycorrhizal plants); (3) To study host specificity of Kame mycelium with other native plants (especially H. Lippie and others); and (4) To study the morphological and physiological characteristics of Kame mycelium and its effect in the development of mycorhiza and fructification.

The outcome of the research outlined above will help in the generation of food in an arid environment, from soils suffering from water stress in addition to a deficiency in nitrogen, phosphorus and potassium.

Miscellaneous activities

Controlled Environmental Agriculture (CEA)   Initial studies will concentrate on the selection of candidate cultivars of tomatoes that are tolerant to saline water and high temperatures. In the following studies, germplasm of wild relatives of tomato will be used to cross with the desert tolerant varieties and to test the progeny for temperature and salinity tolerance.

Water management techniques will be studied and losses reduced to a minimum. This will be achieved by careful monitoring of application rates, soil moisture levels and evaporative losses. Methods using various levels of fresh water at critical plant growth stages, and brackish water for other growth stages, will be evaluated.

Other aspects of CEA which will be examined include structure design and environment control, nutrient studies and studies concerned with photosynthetic and respiration processes with particular reference to CO_2 enrichment of the atmosphere within the production units. The objective of this research is to demonstrate the feasibility of commercially producing vegetables in protected environments using brackish water.

Intensive Animal Production: Studies on sheep are also being considered by KISR. The adaptation of imported breeds to a desert climate and the effects of crossing them with local breeds will be investigated. Brackish water will be used as the main water supply in such production systems.

Efficient Use of Irrigation water for Cash Crop Production: Alternate irrigation methods (trickle drip and sub-surface) will be introduced to demonstrate their feasibility under Kuwait's conditions. Municipal sewage water, produced at the rate of 9.5 million gallons per day, will be utilized in the studies.

The irrigation systems will be used either alone or in combination with petroleum mulch or latex film application on the surface to minimize and/or prevent evaporation and surface salt accumulation, in addition to ensuring even fertilizer distribution.

Mariculture and Fisheries

The objectives of the research conducted at the Mariculture and Fisheries Department are: (1) To aid the State of Kuwait in achieving a greater measure of self-reliance as a safeguard against possible future decrease in fisheries resources for human consumption; (2) To develop viable commercial alternatives for the present and future fishing industry and related technology in Kuwait; and (3) To provide economically sound choices for the protection and enhancement of Kuwait's natural fish resources.

In the early stage of this work, fish and shrimps were raised and released into the sea to augment Kuwait's natural stocks, based on results obtained in the Seto Inland Sea of Japan. However, since no data were available at that time on the movement and habitat of Kuwait's fish and shrimps, it was decided to concentrate the research activities mainly on the artificial cultivation of fish and shrimp in tanks or ponds until such time as fisheries management could be conducted and certain necessary parameters established.

The following are some of the achievments and objectives of specific projects initiated at the Mariculture and Fisheries Department.

Shrimp and Culture Project

The decline in shrimp landings since the 1966/67 season was probably directly due to overfishing. However, reduction of the fishing effort (from ca. 120 vessels to 40) has not resulted in the recovery of the fishery. The large scale releasing of artificially cultured juvenile shrimps into Kuwait waters at selected sites could help to increase recruitment to former levels and as a result promote the recovery of the shrimp fishery. Studies to investigate the feasibility of such an operation in Kuwait were initiated in 1971 with subsequent development of the shrimp hatchery techniques at Salmiya.

From 1974 onwards efforts to increase production levels of the hatchery and to improve reliability and efficiency were made. The production levels and tank volumes utilized since 1972 are given in Table 2. The total number of shrimp produced each year has steadily increased and production rates have been significantly improved recently.

The reasons for the increased production are primarily better
management and tank design, and improved feeding methods. Problems
of cannibalism, uneven distribution of larvae and feed, and
accumulation of detritus in the tanks observed at the start of
these studies were alleviated by the introduction of mechanical
agitators in the culture tanks and reducing the amount of minced
trash fish supplied to the early post-larval stages. More recently, renewed emphasis has been placed on the food supplied to the
larval and juvenile shrimp. Marine yeast (Candia Flavus) supply
50% of the requirements for the protozorel stages. It is hoped
that dependence on phytoplankton produced in the tanks will be
reduced in future studies.

The mysis and early post-larval stages of the larval shrimp
were fed in the past on brine shrimp nauplii (Artemia saline)
hatched in special tanks after incubation for 24-30 h in seawater.
The dependence on this food supply was reduced through increased
use of rotifers (Brachionus plicatilis); currently about 50% of the
total daily food requirement is provided by rotifers. Studies are
continuing to determine whether it is possible to eliminate
completely the use of brine shrimp nauplii. Compounded feed
especially formulated for shrimps has been used to replace it.
Bulk storage is much more convenient and it is easier to handle.
Furthermore, work is in progress to develop locally prepared
compounded diets to avoid the necessity of importing compound food
from Japan or elsewhere. These modifications resulted in significant improvement in the production rates of shrimp. The increased
production rates provided the basis for a large number of trials
from 1974 onwards for the releasing of juvenile shrimp into Kuwait
waters. To estimate how many of the shrimp released into the sea
were actually caught by the shrimp fishermen, it was necessary to
develop an effective and safe method of marking the shrimp before
release.

Preliminary marking studies on adult shrimp (1974-76) using
anchor tags have provided basic information on the migration of
shrimp in Kuwait waters. It appears that shrimp migrate out of
Kuwait Bay and there is a net movement southwards along the coast
of Kuwait. It is interesting to note that a cost-benefit analysis
showed that if only 20% of the released shrimp are finally caught
by the fishery, then the operation breaks even. Similar studies in
Japan (10) have confirmed 7% recovery of released shrinmp and these
contribute to 50% of the total landings of the gillnet fishery.

The commercial culture of shrimp to market size requires the
availability of large numbers of juveniles for stocking purposes.
Since reliable shrimp hatchery techniques have already been developed, it was decided to investigate the feasibility of commercial
culture to market size.

The shrimp naturalization pond (1.2 ha) at Al-Khiran originally built for holding shrimp for a few weeks before release into the sea is now being used for commercial culture studies and these are already in use.

Preliminary studies have been promising and particularly good growth rates are currently being achieved using the Shigueno-type tanks, P. semisulcatus, stocked at a mean body weight of 0.0187 g, have reached 2.86 g in only eight weeks. Similarly, a polyculture experiment in 1978 using shrimp and maid (Liza macrolepis) together was very successful. More extensive studies in this field are planned for the new ponds at Al-Khiran. Polyculture is of particular interest as two crops are being produced simultaneously from the same facilities using approximately half the quantity of food required if the two species were grown separately.

Control environment raceways for high density culture of shrimp, similar to those used at Puerto Penasco, are planned for future construction at Al-Khiran.

Fish Culture Project

An initial survey showed that the most suitable species for culture studies were: hamoor (Epinephelus tauvina), sheim (Acanthopagrus latus) and maid (Liza macrolepis).

In 1975, broodstock of these species was collected from the sea using RV Asmak 4. Natural spawning of hamoor in 90 m^3 spawning tanks was attempted and the newly-hatched larvae were reared until 26 days old.

In 1976, the culture season emphasized the development of an effective egg collection system from the spawning tanks. Biological studies of the cultured species were conducted and their spawning seasons identified. Fundamental fingerling production techniques for maid and sheim were developed. Hamoor fingerlings were raised at the end of the year and their life history was identified.

Since 1977, large-scale fingerling production techniques of the above species have been developed by using circular tanks (60 m^3 capacity) and net cages suspended in 200 m^3 capacity retangular tanks. About 100,000 maid and 50,000 sheim reached the fingerling stage in 1978. Hamoor were grown from newly hatched larvae to market size of 1.5 kg in less than two years, with a relatively good food conversion ratio of around 3:1.

Studies on young sobaity (Acanthopagrus cuvieri) began in 1976

in open circulation tanks. Fed on a pelleted food imported from
Japan, they reached 37 cm total length and 900 g body weight in two
years. The fish spawned in capacity for the first time during the
1979 breeding season. Preliminary large-scale fingerling production techniques for sobaity was investigated with good results and
preliminary commercial-size culture of sobaity has shown that its
growth is relatively fast (it grows from 23 g to 192 g within six
months). It accepts pelleted feed with a conversion ratio of 1.81
-3.3:1 depending on water temperature. Tilapia spp., safy (Siganus
oramin) and bouri (Mugil cephalus) have been selected as additional
species for study. Induced spawning of safy was attempted and the
early life history of the fish was identified. Tilapia aurea
fingerlings inmported from Syria and the USA were successfully
acclimatized to full-strength seawater (salinity, 38-40%).

The main emphasis of the current Fish Culture Program will be
on the commercial culture of Sheia sobaity on Tilapia fish to
marketable size in Al-Khiran ponds as well as the release of fish
fingerlings of one to two of these species into the sea to improve
the commercial catch. Commercial culture of the selected fish
species using floating net-cages in the coastal waters of Kuwait
will be investigated.

Plans are being made at present to undertake the scale up of
the shrimp and fish culture projects to the pilot commercial scale
level of production in Al-Khiran with the aim of eventually
establishing a commercial fish farming industry in Kuwait.

Fisheries Mangement Project

The fisheries management project was established to provide
sound management advice to the fishery and to the government. The
objective was to determine optimal management strategy so that
natural fish stocks be conserved and to exploit new stocks
rationally. To attain these aims, the following tasks have to be
undertaken: (1) To study the population of wild fish at sea; (2)
To determine the landings of Kuwait's principal species and catch
by the fishing fleets; and (3) To investigate the oceanography of
Kuwait waters with a view to determine its effects on fish
population and their habitats.

The quantity and seasonal distribution of Kuwaits fish stocks,
variety of species, size frequencies, sex composition, stomach
contents, length/weight relationships and estimates of growth,
mortality, and recruitment will be determined. Thereafter, a basic
picture of the population dynamics of most of the species taken
will be provided from which their maximum sustainable yield can be
determined.

The fisheries oceanography research will evaluate how the physical and chemical properties of the waters in which the fish live influences the distribution and life cycle of the fish population. For this purpose water from various depths is sampled at three fixed stations once every one or two months. Preliminary results from all three research areas are interesting, especially those indicating that major inputs of freshwater from the Shatt Al-Arab occur in the Gulf between March and May. Variation in the flow of the Shatt Al-Arab appears to be a major environmental influence on fish populations.

The major incursions of less saline water, composed of mixed sea water and Shatt Al-Arab water, may have crucial effects on the fish and shrimp fisheries since the nutrient content of Shatt Al-Arab waters may provide a major stimulus for plankton growth at a time when temperatures may become important in tripenning spawning. These combined effects may affect larval survival.

Dissolved oxygen levels as high as 7 to 8 mg/l is characteristic of Kuwait waters, but levels as low as 4.8 mg/l were observed in May 1979 between Ras-Al-Ard and Ras-Al-Ajuza. This could be attributed to the municipal sewage which is commonly allowed into the sea along the southern coast-line in Kuwait bay. The low dissolved oxygen may result in a real problem to fish and shrimp nursery areas in the area.

Ichthyoplankton Abundance and Diversity

A cooperative project between the Mariculture and Fisheries Department and the University of Miami has been established in order to determine the distribution, kinds and abundance of eggs and larvae of fishes in the spawning areas and during the spawning seasons of important commercial fishes, as well as estimates of biomass for some species based on egg and larval numbers. The potential yield to the commercial fisheries can be estimated if the adult biomass is known. Relationships between the ichthyoplankton and other planktonic organisms are being investigated in cooperation with the Zooplankton Project. The role of oceanographic factors and their effects on spawnings is also being considered. In addition to providing baseline data on the early life stages of fishes, an important objective of the project is to provide train - ing in ichthyoplankton methodology.

Five cruises in Kuwait waters and a single Gulf cruise have been completed. Eggs and larvae are being sorted from the samples and preliminary identification of some larvae has been made. The early analyses indicate that Arabian Gulf ichthyoplankton is abundant and moderately diverse. Collections made in September and

October 1979 contained a high proportion of pelagic species, with larvae and eggs of anchovies predominating, but those of sardines and jacks were also common. A preliminary examination of some samples from November and December indicated that demersal species may be more common during the cooler months.

Diversity, Distribution and Biomass of Zooplankton

Zooplankton samples are being collected by using 333 μm mesh paired Bongo nets and 110 μm mesh plankton nets towed for ten minutes at a 45° angle from 5m above the sea bed to the surface. Initial processing of samples as well as dry weight measurements are being carried out in Kuwait while identification of zooplankton species is continuing in Miami. Our scientists will participate in zooplankton identification and data analyses at the University of Miami during the second year of the Project, from June 1980 onwards.

References

1. Dorothy Brown, "Methods of Surveying and Measuring Vegetation," Commonwealth Agriculture Bureau, Farham Royal, Bucks, England (1954).
2. R. E. Winkworth and C. O. Rossetti, J. Range Management 15: 194-196 (1962).
3. L. A. Stoddart, A. D. Smith and Thadis Box, "Range Management," New York: McGraw-Hill Book Co. (1976).
4. H. G. Fisser, J. Cox and F. K. Taha, Herbage Structure, Production and Phenology Monitoring on the Black Thunder, Wyo. Agr. Exp. Sta., Scientific Rpt. No. 779, p. 162 (1976).
5. H. G. Fisser, J. Cox, F. K. Taha and Mike Mecke, Ibid, No. 856, p. 157 (1977).
6. F. K. Taha, H. G. Fisser and R. E. Ries, J. Range Management, (In Press).
7. J. R. Goodin and C. M. McKell, in: "Proceedings XI Int'l. Grassland Congress, Brisbane, Australia," pp. 158-161 (1970).
8. Nasser Nemati, J. Range Management, 30:268-269 (1977).
9. A. J. Smith, Trop. Ani. Health Prod. 5:250-271 (1973).
10. H. Kutuba, in: "Proceedings of the Int'l. Shrimp Releasing, Marking and Recruiting Workshop," Kuwait Bulletin of Marine Science, pp. 25-29, (In Press).

REVIEW OF BIOSALINE RESEARCH IN SAUDI ARABIA

Y. M. Makki

Dept. of Crops and Forage
College of Agricultural Sciences and Food
King Faisal University, P. O. Box 380
Al-Hassa, Saudi Arabia

SUMMARY

Saudi Arabia is one of the most arid countries in the world, and until recently its economy was based on nomadic pastoralism and oasis agriculture. Recently, the rapid rise in oil revenues has enabled us to embark on an ambitious development program for land water resources development with the main emphasis on attainiing self sufficiency in food production. This objective can be achieved only if an optimum level of diverse soil, water and crop management factors is developed under severe agroclimatic conditions to obtain a maximum harvest from a given unit of land and water resources. Among different research areas, biosaline research is likely to play an important role in the development of the desired agriculture achievements.

Pioneer Crops for Reclamation of Salt Affected Soils

Rhodes and Bermuda grass yield over 8 tons dry matter/ha for summer reclamation and over 7 tons dry matter/ha for mid-summer reclamation of saline soil in Al-Hassa oasis in the Eastern Province. Hence Rhodes grass is a good pioneer crop for summer reclamation (10-12). On the other hand, Barley (winter) followed by Sorghum was established as the best cropping system for reclamation, yielding a total of 32 tons dry matter/ha (12). However, the yield of forage barley followed by Sorghum was positively correlated with gypsum and negatively correlated with the calcium carbonate content of soils (12). Oats was established as the pioneer crop for winter reclamation of sandy calcerous and heavy saline desert soil with a yield of 10.6 tons dry matter/ha compared to alfalfa with a yield of 5.1 tons dry matter/ha (13,14). A more recent trial proved that Rhodes grass is suitable as a reclamation crop. It was further suggested that management oriented trials are required to determine optimum management practices for its successful cultivation under reclamation (15).

The application of farm manure during reclamation experiments contributed significantly towards crop stand establishment on saline soil but was a major source of weed introduction. It was also observed that Rhodes grass will considerably outyield alfalfa. Overall it was recommended that farm manure application is beneficial in increasing crop yield under local conditions (5, 6).

Reuse of Drainage Water

Due to limited water supply, the reuse of drainage water for irrigating salt tolerant crops is now under investigation at Hofuf Agriculture Research Center (HARC) in cooperation with AlHassa Irrigation and Drainage Authority (HAIDA) in the Eastern Province.

Saline water of EC 8, 6, and 4 mmhos/cm leads to reductions in okra-fruit yield of 37, 24, and 17%, respectively, as compared to the fruit yield obtained by irrigation with water of EC 2.3 mmhos/cm (19). For cauliflower the fruit yield reductions were 30, 25, and 15%, respectively, and for Rhodes grass the average dry matter yield reductions were 20, 14, and 17%, respectively (20,21). It is recommended that if saline water (EC 2.3 mmhos/cm) is used for irrigation purposes, considerable leaching and adequate drainage are required to control soil salinity.

Quality of Irrigation Water

Utilizing the comprehensive chemical equilibrium program GEOCHEM, a relationship between the sodium absorption ratio based on free-ion concentrations, and the practical sodium absorption ratio based on total-ion concentrations, was determined for the Eastern Province water by Elprince (8).

According to Elprince et al. (7) the clay material, palygorskite, precipitates under Al-hassa oasis conditions. This process decreases the hazard from salinity (ca 1500 ppm) of the irrigation water and the Mg^{2+} level in the soil solution. Presently Elprince (pers. comm.) is working on a chemical model to explain, among other things, the genesis of soil salinization in relation to the chemistry of irrigation water.

Soil Microorganisms - Salinity Research

Halophylic bacterial strains were isolated from soil and rhizosphere of six salt marsh plants, Aeluropus massauensis, Sporobolus spicatus, Suaeda monico, Halopeplis perfoliata, Anabasis setifera and Arthrocnemum glaucum, for biochemical studies of acid production from dextrose, mannitol and sucrose and of hydrolysis of starch, pectin, cellulose, gelatin and fat (3). Although rhizospheric isolates were similar qualitatively, they showed a marked quantitative difference in their activities and in terms of unit incidence of particular characteristics for each isolate, indicating a degree of biochemical specialization. The halophiles were inactive in these studies. Root exudates probably play an important role in this connection. Salt marsh halophiles are well adapted to salty environments and their closeness to plant roots in a particular community is undoubtedly related to organic matter mineralization in the rhizospheres of these plants.

Range - Salinity Research

Abo-Hassen et al (2) studied the effect of salinity on seed generation and growth of the Orache plant (Atriplex spp.) The results showed that seed germination decreased with increase of salinity. The highest concentration (5000 ppm) inhibited completely the germination of A. visicaria. Kochia indica, which grows naturally, is promising under salinity stress and can be considered as a potential non-conventional forage plant in Saudi Arabia (31).

Ecology of an Inland Salt Marsh

The distribution and biomass of emergent plants, macrodetritus, and the substrate organic content was investigated in an inland salt marsh (north of Al-Hassa Oasis) in relation to salinity, chloride content and conductivity of the water (29).

The dominant emergent plant was the reed, Phragmites australis covering 18% of the total area of the marsh, more abundant near the inflow of water and absent from the farthest point. Its biomass ranged from 96 to 1721 g dry weight/m^2. Typha domingensis, another emergent plant, was found in small clumps and covered 1% of the area of the marsh. Macrodetrius, formed by a green filamentous alga and allochthanous organic matter (derooted Ceratophylum from the canal), was more abundant in open areas of the marsh near the point of inflow of water and decreased as the distance from the source of water increased. Biomass of macrodetrius ranged from 32 to 306 g dry weight/m^2; organic content of the substrate showed an almost similar pattern of distribution.

The salinity of the water, together with the chloride content
and conductivity, increased as the distance from the point of
inflow of water in the marsh increased. It appears that the
establishment and success of P. australis was mainly influenced by
the salinity of the water, being more abundant in less saline water
(4-6 ppt) and absent from more saline water (6-9 ppt).

Marine Biology and Fisheries

There are a few reports on the salinity of the Gulf (17, 18,
23), the surface temperature (9) and distribution of nutrients and
dissolved oxygen (18, 27). Ecology of a high saline environment
was studied by Jones et al (22) and temporal variation in abundance
of Penaeid shrimp larvae by Price (26). Basson et al (4) published
a book on the biotopes of the Western Arabian Gulf.

Preliminary observations on some aspects of breeding the
African cichlid, Sarotherodon spilures, were reported by Osborne
(25). Fecundity, egg size, brood size, and number of eggs per
brood were reported.

Some aspects of salinity tolerance and subsequent growth of
three Tilapia species Sarotherodon aureus, S. Spilurus and T.
zillii were investigated by Osborne (25). Egg hatchability was
reduced as salinity increased. No yolk sac larvae survived beyond
day 5 at salinities higher than 20%.

William (30) presented the results of survey carried out along
the Red Sea Coast of Saudi Arabia between May 1973 and December
1974. It was reported that a fish farm with an annual output of
830 tons would give a return on investment of nearly 20 percent per
annum (16).

An ecological survey of the main Al-Hassa oasis, drainage
canal and marsh was conducted to determine the suitability of
drainage water for fish culture. The physio-chemical and biological properties of the water were suitable and a fish farm could be
established with minimum financial investment (28).

Current Biosaline Research in Saudi Arabia

Mashady and Sayed (24) carried out an experiment at the
College of Agriculture, University of Riyadh, to study salinity
tolerance of wheat and triticale. They tested three wheat and one
triticale cultivars at four levels of salinity ranging from EC 3.7

to 11.0 mmhos/cm. All cultivars were tolerant at germination. However, grain yield was severely reduced under high levels of salinity.

The Saudi Arabian National Center for Science and Technology (SANCST) is sponsoring an ambitious project on the use of saline water irrigation under an arid environment. Preliminary results of a screening program to test local and introduced groups of wheat and barley against salinity levels up to 15,000 ppm indicated that several local strains are more salt tolerant than introduced ones. Presently, more plant groups are being tested to evaluate their physiological responses to salinity. Priority is given to those plants which proved to be more salt tolerant in the first screening program.

Recently, a French group was appointed by the Ministry of Agriculture and Water to conduct an investigation on the reuse of drainage water at the Irrigation and Drainage Project at Al-Hassa. They designed a system to test several agronomic and vegetable crops under various saline levels of drainage water.

Water use efficiency is a very important concept in arid land agriculture. Presently invstigators at the University of Riyadh, King Faisal University, and King-Aziz University are independently conducting experiments to evaluate water use efficiency and to determine water requirements for various crops grown in different regions of Saudi Arabia.

Desertification

Another cause of problems in areas of variable rainfall is overgrazing or in some cases over-cultivation of the land which accelerates the process of desertification. When subnormal rainfall is combined with such overuse, the results are particularly disastrous. This leads to another biosaline objective in which fodder or grain are not the key but merely stabilization of the soil.

The best estimate as to the earth covered by sand dunes is about 13 million hectares. The shifting of these sand dunes has been a problem for centuries in different parts of the world. Where sand is moved by wind in deserts, three types of movements can be distinguished: (i) surface creep; (ii) saltation; and (iii) suspension.

We should note that there are alternates to the plant techniques which draw on biosaline research and include straight-forward engineering with mechanical and physical efforts to

control sand movement.   Among agricultural methods tried,   sowing
grass was effective but slow (1).   Afforestation, both with and
without irrigation, was also tried.   Species which are drought
resistant, saline tolerant and can withstand a wide range of air
and extremes were used by Abu-Hassan (1) at Al-Hassa Oasis, who
developed a new method of planting long cuttings (1 to 2 cm) of
Tamarix aphylla without irrigation.   This method showed a great
promise of establishing inexpensive and successful plantations on
sand dunes.   In Al-Hassa, mobile sand dunes 2 to 12 miles high were
shifting from the north every year.   Thus, the inhabitants of three
areas were forced to move south to new locations, which in many
cases are less productive.   Nearly half the oasis was lost to
desert and the old capital of Al-Hassa, once in the middle of the
oasis, was covered by sand dunes.

The Ministry of Agriculture and Water has taken strong
measures to bring the situation under control.   The project on sand
stabilization is now under direct supervision of the Ministry of
Agriculture and Water.   We hope to: (i) stop sand movement; (ii)
protect the cultivated land from being lost to desert; (iii)
increase the area of land under cultivation and the area of grazing
lands; (iv) conserve forest and timber resources; and (v) improve
the climatic conditions through all possible conservation
practices.

Agronomic measures used to stabilize the sand include the
erection of fences and the establishment of forest.   For the use of
fences the lands to be stabilized were divided into squares of 5m x
5m or 4m x 4m and fenced with wire and palm leaves.   Then the
squares were sown with quick growing grass seeds and trees and
irrigated with water.   Stabilization by afforestation proved to be
the most suitable and successful method.   The basins were planted
with cuttings and trees adaptable to the region that could resist
drought, salt, wind, high temperature and sand movement, such as:
Tamarix gallica, T. aphylla, Accacia cyanophylla, Parkinsoia,
aculeata, Prosopic juliflora and Eucalyptus camandulansis.

The project described above was started in 1969 and its total
area is some 4000 ha.   The planted area is more than 1800 ha with
about 6 million timber trees, of which 90% are the local Tamarix
trees.   Some 500 ha are under irrigation; more than 1300 ha are
without any irrigation (dry farming system).

References

1. A. Abu-Hassan and M. Habib, College of Agric. Riyadh Research
      Centre, Res. Bulletin 3, (1980).
2. A. Abo-Hassan, H. Tawfic and M. Habib, Saudi Biol. Soc.
      Fourth Symposium on the Biol. Aspects of Saudi Arabia,

Riyadh (1980).

3. N. A. Bafshin, A. S. Hamed, M. J. Sejiny and M. M. Zaki,
    Ibid. (1980).

4. P. W. Basson, J. E. Burchard, J. T. Hardy and A. R. G. Price,
    "Biotypes of the Western Arabian Gulf: Marine Life
    and Environments of Saudi Arabia." 284 pp. Aramco,
    Dhahran, Saudi Arabia (1979).

5. J. R. G. Bloomfield and I. B. Ruxton, Min. of Agric. & Water,
    Pub. 58, Saudi Arabia (1975).

6. J. R. G. Bloomfield and I. B. Ruxton, Ibid. (1977).

7. A. M. Elprince, A. S. Mashady and M. M. Aba-Husayn, Soil Sci.
    128:211-218 (1979).

8. A. M. Elprince, The search for suitable land for cultivation in
    the Eastern Province - Saudi Arabia. Progress Rpt. (March
    1980-June 1980). (SANCST) (1980).

9. Y. Enomota, Bull. Tokai Reg. Fish Res. Lab. 66:1-74 (1971).

10. J. Farnworth, Res. Dev. Pro. Min. of Agric. & Water,
    Saudi Arabia, Pub. 39 (1974).

11. J. Farnworth and I. B. Ruxton, Ibid, Pub. 19 (1973).

12. J. Farnworth and I. B. Ruxton, Ibid, Pub. 47, (1974).

13. J. Farnworth and I. B. Ruxton, Ibid, Pub. 41, (1974).

14. J. Farnworth, I. B. Ruxton and D. Younie, Ibid, Pub. 49,
    (1974).

15. J. Farnworth and R. J. Williams, Ibid, Pub. 113, (1977).

16. Fisheries Devel. Project, Tech. Rpt. Min. of Agric. & Water,
    Saudi Arabia.

17. K. Grasshoff, Meteor Forsch. Ergebn. A 6:1-76 (1969).

18. K. Grasshoff, in: Rep. Consultation meeting of Marine Sci.
    Res. for Gulf States, UNESCO, Paris (1975).

19. Z. Hussain, Hofuf Agricultural Res. Ctr., Saudi Arabia,
    Pub. No. 35 (1978).

20. Z. Hussain, Ibid, Pub. No. 33 (1978).

21. Z. Hussain, Ibid, Pub. No. 36 (1978).

22. D. A. Jones, A. R. G. Price and R. N. Hughes, Estuarine and
    Coastal Marine Sci. 6:253-262 (1978).

23. P. Koske, Meteor Forsch. Ergebn, A 11:58-73 (1972).

24. A. S. Mashhady and I Sayed, Saudi Biol. Soc. 4th Symp. on the
    Biological Aspects of Saudi Arabia (1980).

25. R. S. Osborne, Fish Dev. Pro. Tech Rpt. 48, Min. of Agric. &
    Water, Saudi Arabia (1979).

26. A. R. G. Price and D. A. Jones, Bull. Mar. Res. Ctr.,
    Saudi Arabia No. 6 (1973).

27. U. Rabsch, Meteor Forsch. Ergebn. Reihe A 11:74-88 (1972).

28. A. Q. Siddiqui, Proc. Saudi Biol. Soc. 3:329-335 ((1979).

29. A. Q. Siddiqui, Proc. First Conf. on Marine Environments and
    Fisheries Resources in the Gulf, Univ. of Basrah, Basrah,
    Iraq, (1980).

30. T. M. William, Marine Research Ctr., Jeddah, Tech. Rpt.
    Min. of Agric. & Water, Saudi Arabia (1974).

31. M. A. Zahir and Y. Maghrabi, Saudi Biol. Soc. Fourth Symp. in
    the Biol. Aspects of Saudi Arabia, Riyadh (1980).

FOOD AND ECONOMIC PLANTS

CHAIRMAN'S REMARKS

Emanuel Epstein

Department of Land, Air and Water Resources
University of California
Davis, CA 95616 USA

Ladies and gentlemen:

This meeting we are attending is a workshop on biosaline research, and the present session is devoted to Food and Economic Plants. What is this session all about? There are two aspects to it, one quite specific, the other more general or philosophical.

As for specifics, the topic, I repeat, is Food and Economic Plants, meaning, in the context of this biosaline meeting, such plants under saline regimes. Plant scientists are increasingly becoming aware of both the need and the possibility of raising economically useful plants under more saline conditions than has so far been conventional. The key to doing this is an understanding of genetic diversity in plants. This specific aspect of our session, then, deals with the building of bridges between plant scientists interested in the salt relations of plants, on the one hand, and geneticist-breeders, on the other.

But beyond that lies a second and broader consideration. It is not just in the matter of salinity that plant scientists working on problems of mineral metabolism need to introduce a genetic dimension into their thinking and doing. Salinity is just one such aspect of mineral metabolism; there are at least two more.

First, acid soils often have excessive concentrations of aluminum, manganese, and other heavy metals. Alkali soils may have potentially harmful concentrations of boron, and so may seawater. Selecting and breeding crops for these conditions will be increasingly important as time goes on. Like salinity, these toxicities are problems of "too much."

There are also problems of "too little," viz. soils with too little in the way of fertility. Nitrogen and phosphorus deficiencies are common. Depending, however, upon where you are, you may encounter deficiencies of every element we know to be essential to plants, except only chlorine. Fertilizers will not, in future, be as readily available as in the past, their cost in terms of dollars and energy will skyrocket, and so will the cost of applying them. Breeding crops for efficiency of nutrient absorption and metabolic utilization therefore is yet another field for collaboration between mineral nutritional plant physiologists and geneticist-breeders.

I suggest, then, that our particular concern here with salt tolerant crops, with halophytes, with marine algae, is part and parcel of a fresh departure in plant science in which all manner of problems of mineral metabolism will be tackled jointly by plant physiologists and geneticist-breeders. This combined endeavor will deepen our understanding of the mineral nutrition and salt relations of plants, and will make possible the biological utilization of mineral media that have biologically counterproductive properties. It is in this broad, novel context that we should see our particular concern with food and economic plants under saline regimes.

I mention this broader scope for two reasons. The first is that we ought to be aware of the intellectual context in which our specialized researches belong. And secondly, some of these other problems of mineral metabolism may show up in our biosaline sphere. For example, I already mentioned the excessive boron concentrations of many alkali soils. High-pH soils may pose problems of the availability of iron, manganese, zinc and other micronutrients. It will often be desirable to breed into crops the physiological competence needed to make them thrive where these conditions go hand in hand with salinity

To conclude, the kind of biosaline research we are about to discuss is ultimately based on the enlarged awareness of the role that a genetic dimension can play in the science of mineral plant metabolism. We are fortunate in being able to assist in the initiation of a novel joint venture of basic and applied plant science.

FOOD AND ECONOMIC PLANTS: GENERAL REVIEW

G. Fred Somers

University of Delaware
Newark, DE   19711   USA

INTRODUCTION

This summary of recent and current biosaline research on food and economic plants will draw upon information from two sources: the published literature and information on current plans and recent progress provided by Biosaline Newsletters and by the summary prepared by the Smithsonian Science Information Exchange this past summer for the National Science Foundation and the Kuwait Institute for Science Research (1).   For a comprehensive bibliography to research in this field from 1900 to 1977 see Francois and Maas (2).   Attention is called to recent, comprehensive reviews on the mechanism of salt tolerance in halophytes (3) and nonhalophytes (4).

Interest in this field has been growing so rapidly that no review can claim to be all inclusive.   No doubt some significant research will be missed inadvertently.   But the main purpose of this review is not to provide an exhaustive summary of current literature in the field, rather to provide an assessment of the current status of this field and to provide a basis for discussion in this Workshop. I will focus upon what is being done and comment on what I think needs to be done--the problems to be solved.   I will draw upon research with which I am most familiar.

Basically there have been two approaches to the development of crops which might be grown using highly saline water--water equal to, or approaching the oceans in salinity.   One of these proposes to select and improve salt-tolerant wild plants to meet the food needs of man or domesticated animals. The other proposes to select for salt tolerance among conventional crops or to introduce such germplasm from wild sources by breeding and selection.   The two

127

approaches share a common goal: to provide new options for
agricultural production using presently unused, or underutilized
resources. Arguments can be made for both. At this juncture it
seems prudent to pursue vigorously both approaches.

Various reviews have made arguments for the desirability of
undertaking this kind of research. Classical in this regard are
two books edited by Boyko (5, 6). In 1975 Somers (7) argued in
support of this concept. In the proceedings of the first Workshop
in this series, Epstein et al. (8) and Somers (9) presented further
evidence to support the validity of these approaches. These were
reinforced in later symposia (10-14). This thesis was developed
further in a recent article by Epstein and his colleagues (15).
They point out that there are a variety of technical approaches
that might be, or have been used in addressing this problem. Some
of these will be mentioned more specifically in what follows.

More limited in the scope of the plants involved has been the
long-standing research in Australia to develop salt bushes, especially species of Atriplex, for forage use in saline areas. In 1970
the Biology of Atriplex was published (16). Much of the information on this species was summarized, including anatomy, physiology
and technological uses. In 1974 Malcolm (17) summarized additional
information, particularly his extensive experience in adapting this
species for use as a forage by sheep. A very recent publication on
physiological plant ecology features Atriplex (18). Obviously this
genus has received much attention and merits serious consideration
in future work, especially that aimed at growing forages in saline
areas.

Let's now turn to a more specific review of publications,
particularly those of the last few years.

Culture of Crop Plants with Saline Water

Many experiments have had as their main aim the culture of
crops using saline water for irrigation. Phills et al (19) grew
Lycopersicon esculentum, L. peruvianum, Solanum penellii, S. lycopersoides and an F_1 (L. esculentum x S. lycopersoides) in sand
culture outdoors at Geneva, New York, U.S.A. using a nutrient
solution containing 0-294 meq. NaCl per liter. Among other things
they were testing the hypothesis that salt-sensitive plants
excluded salt from their leaves and salt-tolerant ones do not (see
20). They found that the salt sensitive taxa were more effective
at excluding Na^+ uptake to the leaves, but their uptake of K^+ was
not influenced by the salinity level. There was no real difference
in Cl^- content.

Ahmad and Z. Abdullah (21) grew several varieties of potatoes on soils salinized with a salt mixture to salinities comparable to Pakistan soils.   They judged varieties to be "tolerant" when the yeild was decreased only 20-50% at the highest salinities (0.9-1.0%) and to be "susceptible" when there was a 50-85% decrease in yield.   Even so, one variety considered to be "susceptible" by this criterion yielded more than one classed as "tolerant."   Maximum protein content was found at low salinity levels.   More recently, Z. Abdullah and Ahmad (22) reported that amendments with Ca^{2+} and N resulted in increased tuber initiation in salinities up to 0.8% Also, the mean weight of the tubers was greater than that of the control.   They report also growing eggplant in a sandy soil using seawater dilutions (TDS to 37,000 ppm).   Height and dry matter were reduced as the concentration increased--a common observation. However, at 15,000 ppm the yield of fruit was reportedly increased, but the size of the fruit was not impoved.   They conclude that salt tolerance in this case depended mainly upon the exclusion of Na^+ from the roots.

Epstein and his colleagues continue to report advances in the selection of salt-tolerant strains of barley and wheat from existing collections of germplasm.   The incidence of salt-tolerant lines is low, but useful results have been obtained.   They have also introduced a high degree of salt tolerance into tomato by hybridizing with salt-tolerant biotype of Lycopersicon cheesmanii collected by Rick from the Galapagos Islands.   The selections were made under controlled conditions in Davis, California and promising lines were field tested using seawater for irrigation at Bodega, California (8, 10, 11, 15, 23).   They argue the desirability of selecting under controlled conditions using salinized solution culture.   Their success using such an approach is meritorious.

However, Hoffman and Jobes (24) state: "Crop tolerance is not absolute but depends on environmental factors such as relative humidity (RH), as well as management of irrigation and fertility." They studied, under controlled climatic conditions, the influence of RH on plant growth and how it interacts with salinity to influence salt tolerance and water relations of barley, wheat and sweet corn.   They found that high RH increased the salt tolerance of barley and corn, but did not affect the tolerance of wheat. They conclude that there is a relatively narrow range of leaf total water potential in which barley and wheat will grow.   They found that the leaf total water potential was influenced by salinity.   In light of this, it is interesting to note that Fontes and Glen (pers. comm.) have successfully germinated selected lines of salt-tolerant barley using very high salinity, but the seedlings did not grow to maturity in Sonora, Mexico, where they conducted the experiment.

Siegel and Siegel in Hawaii (25) reported the successful culture of pepper, tomato and Wong Bok in the greenhouse using 17,000 ppm diluted seawater, and corn in the field using 17,000 ppm diluted seawater and a very porous substrate. Shomer-Ilan et al. (26) grew Rhodes grass (Chloris gayana), a perennial, salt-excreting, C_4 plant, at various salinities. A high level of NaCl plus high N-fertilization had a negative effect on growth and tillering, leading to decreased yield with age. The major effect of salinity, they concluded, seemed to be upon protein synthesis. Plant growth was improved by a salinity treatment of up to 100 mM, at higher atmospheric humidity. M. Abdullah and Qureshi (27) report studies in which they grew "Kaller grass" (Diplachne fusca) in sand culture using Hoagland's solution amended with NaCl, $CaCl_2$ $MgCl_2$ and Na_2SO_4 to produce osmotic pressure levels of 5, 7.5 and 10. Mannitol was used also to produce osmotic stress. They conclude that Na^+ is less toxic than Ca^{2+} and Mg^{2+} and Cl^- is less toxic than SO_4^{2-}.

Rai (28) grew 13 varieties of rice in saline water at EC 2.1, 4.0, 6.0, 8.0 and 10.0 mmho/cm produced by a mixture of $CaCl_2$, Na_2SO_4, and $NaHCO_3$. He was interested in a relationship between salinity and "dead heart damage" (stem borer). The first year showed no difference among salinites, but during the second year damage was positively correlated (r= 0.92) with salinity. He suggests that this may result from an increased succulence of the stem resulting from the increased salinity.

Species of Atriplex continue to attract attention. Goodin has summarized several experiments (29, 30): For A. canescens, earlier work has shown that it was superior in production of dry matter to alfalfa under dry land or low irrigation regimes. "Considering all tests over the past several years, it would appear that sustained yields of over 1000 kg/ha/yr are possible with various saltbrushes, including A. canescens." "Under conditions of restricted rainfall, water use efficiency in Atriplex far exceeds any of the traditional agricultural crops that we have studied." He reports successful growth using blowdown water with a salinity of 8,000 ppm from an electric generating plant for irrigation when needed. He addressed the question of the balancing of Na+ uptake by oxalate synthesis in this plant and says, "After several years of study, we have not found oxalate levels to increase dramatically with increasing levels of salinity, especially in A. canescens. Instead the charge balance is maintained by chloride." "There is no evidence that accumulation of either Na^+ or Cl^- is great enough to cause metabolic disorders." He refers to the tremendous phenotypic variability in this species as documented by Stutz (31). "Ecotypes of Atriplex canescens which show great genetic variability, including selectivity for low oxalate accumulations, have been selected" (30).

Most forage species of <u>Atriplex</u> tolerate exceptional levels of substrate salinity and are considered to be halophytes and (see 30) Greenway and Osmond (32) class them as "salt accumulators." Experiments in progress at Puerto Penasco, Sonora, Mexico by Fontes and his colleagues have demonstrated a high salt tolerance for several species (M. Fontes et al, pers. comm.) They are growing <u>A</u>. <u>lentiformis</u>, <u>A</u>. <u>nummularia</u>, <u>A</u>. <u>barclayana</u> and <u>A</u>. <u>canescens</u> using effluent from a shrimp culture facility. The water is pumped from shallow wells near the shore of the Gulf of California, has a salinity of 40,000 ppm and is flooded onto outdoor plots on very sandy soil. Some of the yield for the first growing season were equal to those reported in the literature for conventional forage crops grown using fresh water. Feeding trials are under way and the preliminary results are promising. In some cases the NaCl content of the forage was relatively high, but techniques developed by Katzen appear promising as a means of leaching this out.

Omar et al in Kuwait (33) grew seedlings of <u>A</u>. canescens using brackish water (TDS 3.48°/oo), seawater diluted 1:3 (TDS 11.6°/oo) and sewage water (TDS 2.28°/oo). Seedlings grew well with all sources of water with little difference among the treatments.

An annual <u>Atriplex</u> variously referred to as <u>A</u>. <u>hastata</u>, <u>A</u>. <u>patula</u>, <u>A</u>. <u>patula</u> var. <u>hastata</u>, <u>A</u>. <u>triangularis</u> appears to be a highly variable taxon which has been widely used for a variety of researches. As a young plant it has been used casually as a potherb or salad green. We have a number of selections of this taxon and have been growing it for the past several years in plots flooded 3 x weekly with estuarine water, the salinity of which ranges from about 10,000 to 32,000 ppm, mostly 25,000 - 32,000 during the summer growing season. A yield of 7000 kg/ha was obtained in Aughust 1977 when the plants appeared to be suitable for forage. The yield of seed at maturity for this crop was 1200 kg/ha. The seed has a lipid content of 10% with an iodine value of 90 (maize oil = 140; soybean oil = 130) and a protein (N x 6.25) content of 15%. The spectrum of essential amino acids indicate a good quality protein (13). But we have encountered problems. In 1978 the growth was very disappointing on the same plots using water from the same source. Soil tests showed no salt accumulation in excess of that in the irrigation water; pH was normal for this soil. However, the water infiltration rate had decreased appreciably, so much so that the soils were waterlogged at least part of the time. We have found that this plant does not tolerate this condition. However, if the soil is shaped into furrows (about 20-25 cm deep) and seed is sown midway along the slopes, the resulting plants will grow satisfactorily when flooded to the planting line 3 x weekly. Yields are now being obtained. In other studies we have found deposits of ferric oxide on the surface and in the outer cells of roots of this plant grown in soils repeatedly saturated with water from our estuarine sources (34).

    Selection for Salt Tolerance – Whole Plants.   Salt tolerance
is an attribute that is not easily defined; it is not readily
measured without ambiguity.   The most extreme response to salinity
is death, commonly observed in plants which are thereby judged to
be salt sensitive or intolerant.   Short of this the limits are much
less clear.   Commonly some parameter associated with growth is
measured and its resistance to change with increasing levels of
salinity is judged to be a measure of tolerance.   Often the
criterion is simply survival, an all-or-none response.   We will use
the term "salt tolerance" ambiguously to refer to either survival
or the resistance to change of some parameter of growth as the
salinity is increased.   Resistance in most cases is only partial-
-growth, for example, commonly decreases as salinity is increased.
Some species, however, exhibit maximal growth at a low level of
salinity.   Such plants are frequently referred to as halphytes.
Even a casual acquaintance with the growth and distribution of
plants is sufficient to establish a premise that variations in salt
tolerance exist.   Some species, e.g. Salicornia spp., Spartina
alterniflora, Distichlis spicata, etc. are found growing naturally
only in saline habitats.

    Apparently, man's early experience with plants established
that salt was inimical to his crops.   As a result there is a large
body of literature dealing with salinity, saline soil properties,
and plant growth.   From the simple observation that some plant
species grow only in saline areas and that soil or water salinity,
even at rather low levels, is lethal to other plant species, one
can conclude that there must be an inheritable attribute for salt
tolerance.   But the nature of such an inheritance and how it is
effected phenotypically are not at all clear. Presumably the
inheritance is determined genetically, but the connection between
genetic information and effected response is not at all clear.
Undoubtedly the mechanism for effecting salt tolerance is not the
same in all species.   As a consequence, those who would select for
salt tolerance are bound to use largely empirical approaches; to
ask simply, will a plant grow satisfactorily in the presence of
high levels of salt.   But what levels of salt? How applied? When?
These are some difficult questions.   They are, however, now getting
some attention.

    One of the problems encountered in developing salt tolerant
crops is the technology required to evaluate them, especially under
field conditions.   It would be desirable to grow plants under
conditions which are realistic in terms of agricultural applications but this usually means a greater expense than can be afforded
with the large numbers needed for effective selection.   The

general approach has been to more or less simulate an applied
situation and to minimize the influence of variables thought to be
less important. For example, an excess of saline water might be
applied to provide sufficient leaching and hence avoid an accumulation of salts in the "soil." The limitations of this approach are
not unique to this area of research and are not uniformly
encountered, but they should be recognized.

As noted above, Epstein and his colleagues have had considerable success in selecting barley and wheat for salt tolerance and
in transferring this trait from a related wild species into the
tomato. They make their selections in defined, controlled conditions using salinized solution media (8, 15, 35). Successful
selections have been field tested using natural seawater to grow
the plants in coastal sand plots. Their results show clearly that
salt tolerance is genetically determined. In a series of barley
cultivars the more times the genome of California Mariout was
included in the pedigrees of the cultivars Numar and Briggs, the
higher the level of salinity required to reduce the growth. This
suggests complex genetic control, a view supported by comparisons
of several progeny lines (36). Kirkham and Faden (37) report
studies with wheat which they are growing in soil or sand in pots
salinized with NaCl.

Staples, et al (38) introduced the salt tolerance of Solanum
penellii into the tomato. S. penellii, among the most salt
tolerant of the wild relatives of tomato they tested, was crossed
with New Yorker, a cultivar with desirable horticultural traits.
The F_2 generation showed extreme segregation, but none displayed
fully acceptable horticultural attributes. Backcrossing the F_1 to
New Yorker resulted in many plants with good fruit. In sand
culture, using Hoagland's solution salinized with NaCl, six lines
produced 70% or more of the dry weight of the control at 5,000 ppm
and three lines produced 40% of the control at 11,600 ppm. They
reported that these lines were under field test in early 1980 (see
Sacher et al., this volume).

Kelly, et al (39) report an effort to associate various
morphological attributes of the tomato with tolerance to salinity.
They measured such characters as trichome density, ontogeny and
size; leaf succulence, expansion and abscission; distribution of
ions; localization of Na^{22} in various organs.

Flowers and Yeo (40) report experiences in selecting rice for
salt tolerance. Some of their observations merit special attention
because they point up the problem of assessing salt tolerance
quantitatively. They say: "We have subsequently established that
within the more salt sensitive varieties that we have tested there
was a negative correlation between the sodium concentration in a

leaf sample taken three days after imposing a 50 mM NaCl treatment and the survival of plants. Thus, plants with high initial salt concentrations in their leaves do not survive for very long. However, for more tolerant varieties this correlation did not hold good. The reason for this is not entirely clear but it may in part be due to the ability of these more tolerant varieties to dump ions into older leaves thus maintaining the younger leaves at a lower salt concentration." In a further discussion of this response they add: "Current results indicate that although relatively simple correlations can be demonstrated for the salt sensitive varieties, more factors are involved when a range of varieties come to be considered. . . ."

Datta (41) evaluated rice varieties for salt tolerance, using "tube culture" under laboratory conditons and NaCl levels up to 17.5 mmho/cm. He observed a number of morphological changes at high levels of salinity. He says: "From the studies made on the tolerance of salinity under laboratory and glass house conditions and later corroborated by field trials, two varieties, Hamilton and Matla, were developed as suitable for growing under saline areas with a reasonably good yield of grain (2.5-3.0 t/ha)." Datta and Pradhan (42) describe a rapid technique for mass screening of rice at the seeling stage for salt tolerance. They used 0 and 17.5 mmho/cm NaCl solutions and consider this approach preferable to selection in the field. They report considerable difference among varieties based upon size and fresh and dry weight.

Jana and Slinkard (43) screened lentils (Lens culinaris) for salt tolerance using Hoagland's solution salinized to 6.0 mmho with NaCl. Considerable genetic variability for salt tolerance was found in the USDA World Collection. Salt-tolerant and salt-susceptible lines were identified. Field trials confirmed the classification based upon solution culture tests.

Various groups report that selections from wild populations yeild differences in salt tolerance. Reference has been made to selections of Lycopersicon cheesmanii, some of which was reportedly more salt tolerant than others (see later). Silander and Antonovics (44) selected Spartina patens, sometimes used as forage, from "dune," "swale," and "marsh" habitats. When grown in the presence of salt at 8.8, 17.5 and 35⁰/oo, differences in salt tolerance were observed at the two higher levels (P ≤ 0.05). Wrona and Epstein (45) collected Distichlis spicata, a plant grazed by animals under some circumstances, from a variety of habitats in northern California. When grown in solution cultures in the greenhouse with salinity increased at the rate of 10% seawater/week until a level twice that of seawater was reached, inland ecotypes died; coastal ecotypes still appeared healthy.

Glenn and Fontes (46) are growing plants collected from coastal salt marshes on all three coasts within the Sonoran Desert and from seed obtained elsewhere. They are being evaluated for growth in the greenhouse and in open field trials at Puerto Penasco, Mexico. In the field the salinity of the irrigation water is 40°/oo. The trials include Atriplex glauca, A. polycarpa, A. repanda, A. canescens, A. patula, A. nummularia, A. barclayana. A. lentiformis, Salicornia europaea, Distichlis palmeri and Cressa truxillensis. These field trials continued through the summer of 1980. Plants which appear superior are selected for further propagation by cutting and/or seeds. Best growth is being obtained with Atriplex species which are native to the Sonoran Desert (pers. comm.).

My colleagues and I (J. L. Gallagher, D. M. Grant, M. Siegel and R. W. Smith) have for the past several years been evaluating plants collected from a variety of saline habitats, as well as some conventional crop plants. We've used two approaches: Germination and growth in growth chambers and growth in field plots flooded, usually three times weekly, with water from one of two estuarine streams which differ somewhat in salinity (47, 48). Some germination results are given in Table 1. These are species which for one reason or another might be considered for development as food. Chenopodium quinoa is, of course, already used as a food. Zizania aquatica is used as a food and considerable progress has been made in domesticating it for freshwater habitats in Minnesota, U.S.A. (49). Atriplex patula and Chenopodium album have been used casually as foods and the latter proved satisfactory as a forage in an animal feeding trial (50). In a subsequent test, Spartina alterniflora, selected for large inflorescences, was germinated in full strength seawater, grown as seedlings using artificial seawater, and transplanted into field plots where it was grown using estuarine water of a salinity up to that of seawater. It grows well in Delaware in plots which are waterlogged and at a salinity (interstitial water) equal to seawater. This selection and a few other selections from northeastern United States were transplanted into the plots of Fontes et al in Puerto Penasco, Mexico. The Delaware selection grew best. None of them grew well; most of them died. This is illustrative. There were two differences between the Delaware and the Sonoran sites. Salinity was higher at the latter (40,000 ppm vs. $\leq$ 32,000) and the climate and daylength were very different. Daylength influences the flowering of this species (51, 52).

Despite the fact that germination of Chenopodium quinoa was obtained in the presence of 33% seawater, subsequent attempts to grow these plants at this salinity were not successful, either in the laboratory or in the field. The experience with Kosteletzkya virginica (seashore mallow), a perennial, has been similar.

Seedlings do not grow at the salinity at which germination was obtained. On the other hand, if the plants are seeded in the field and allowed to become established for several weeks they then will tolerate high salinity. In established plots of plants 1-2 years old, flooded three times weekly with estuarine water, the salinity of water extracted from the root zones was 25-30 °/oo. These plants grew well and showed no specific sysmptoms of salt damage. Thus with some perennial plants it apparently will be feasible to start them using whatever water of low salinity is available and to use very saline water later for crop production. This raises a question of the most efficacious manner of screening for salt tolerance.

Shannon (53) addressed the problem of finding adequate rapid screening techniques for plant salt tolerance. He points out that heretofore such techniques are "untried, untested or unproven." "....selections made for tolerance at early growth may not prove salt tolerant during later growth, or even at other salinity concentrations or compositions," or, one might add, under different environmental situations. He continues: "The development of rapid screening techniques based upon physiological factors that confer salt tolerance could circumvent many of the problems." "Our lack of knowledge about physiological mechanisms of salt tolerance has

TABLE 1. Germination of some species in the presence of artificial seawater mixtures between filter paper sheets in Petri dishes. Only the maximum salinity at which germination was obtained is given.[a]

Species	% Germination	Salinity
Atriplex patual var. hastata	90	50% seawater
Bromus tectorum	5	50% seawater
Chenopodium album	17	33% seawater
Chenopodium quinoa	100	33% seawater
Distichlis spicata	70	66% seawater
Kosteletzkya virginica	20	66% seawater
Salicornia europaea	100	100% seawater
Spartina alterniflora	74	66% seawater
Strophostyles helvola	100	33% seawater
Zizania aquatica	5	33% seawater

[a]Selected from Pihl et al (47).

hampered the development of reliable and rapid screening procedures." "The tolerance of a plant to salinity is a measure of
its ability to withstand the effects of soluble salt concentrated
in its root zone. The simplicity of this definition is deceiving;
however, soil salinity includes a wide variety of ions in a dynamic
function dependent upon soil composition and structure, and its
equilibrium with the ever-fluctuating soil moisture content.
Variation among these parameters within a plant's root zone may be
quite extensive and difficult to duplicate or quantify in the
laboratory. Additionally, little is known about how the plant
integrates the heterogeneity within its root zone with its shoot
environment and its own genetic potential."

   Some Interactions Between Salinity and Plant Physiological
Processes. Various aspects of ion uptake continue to attract the
attention of researchers in this field. Salinity is, above all
else, an expression of ion activities. The manner in which various
plants deal with the ions present in excess and at the same time
obtain those needed in metabolism poses intriguing questions (3,
4). Answers to them might lead to better techniques for selecting
for salt tolerance and better management of agricultural production
in saline habitats.

   It is reasonable to suppose that selective ion uptake by roots
is membrane mediated. A relatioship between salt stress and
phosphate transport in barley roots is indicated by studies of Maas
et al (54) who found that treating excised barley roots with cold
NaCl solutions resulted in the release of proteins. Associated
with this was a reversible reduction in the capacity of the roots
to absorb inorganic phosphate. Following treatment with 0.25 M
NaCl the roots lost phosphate to the surrounding solution. After
treatment with 0.16 M NaCl they resumed active uptake equal to
about 60% of that of the controls in 1-2 hours. Inhibitors of
protein synthesis inhibited this recovery. The authors suggest
that the proteins eluted by NaCl are required for active phosphate
transport. Another membrane-centered response was reported by
Muller and Santarius (55) who found in barley adapted to 400 mM
NaCl that the content of galactolipids of chloroplast membranes had
decreased. Santarius et al (56) studied the senstitivity and
resistance of biomembranes toward extreme temperature and high
salinity. Various ATPases associated with salinity ahve been
noted. Erdei et al (57) studied the effect of salinity on the
lipid composition and activity of Ca^{2+}-stimulated and Mg^{2+}-
stimulated ATPases in salt-sensitive and salt-tolerant species of
Plantago. Mishustina et al (58) isolated from Halocnemum strobila-
ceum membranes that exhibited $(N^{+}+K^{+})$-stimulated ATPase activity.

   Differential ion uptake under saline conditions has been
documented repeatedly. Tal (59) points out that plants of Solanum

penellii suffer less under salt stress than do plants of their cultivated relative, tomato. Associated with this, the wild plants accumulate more Na^+ and Cl^-. He is examining this using also isolated tissues and cells of these species. Rush and Epstein (60) compared Na^+ accumulation by Lycopersicon esculentum with that of L. cheesemanii and hybrid of the two (see also 61, 62). L. cheesemanii under salt stress accumulates Na^+ in its leaf and petiole tissue to an excess of 20% of the total dry weight. L. esculentum, on the other hand, tends to exclude Na^+ from its leaves under similar conditions. In this species a Na^+ concentration in excess of 5% of the dry weight is lethal. The accumulation of Cl^- by these species under these conditions does not differ significantly. A hybrid of the two species accumulated Na^+ in the manner of L. Cheesemannii. Wienecke and Lauchli (63, 64) compared the soybean varieties Lee ("salt tolerant") and Jackson ("salt sensitive"). When exposed to 66.5 mM NaCl Jackson contained more NaCl in its leaves than Lee. "Both soybean varieties retained Na^+ in the proximal root and stem." This suggests that the varieties differed in transport rather than in uptake by the roots.

A number of workers have used various other wild species that at present have little or no recognized potential as crops to study interactions such as these, and others. No doubt these studies will be helpful in interpreting the responses of present or putative crops under saline conditions. A study of salt secretion by the salt gland of Glaux maritima in response to salinity in the root medium, time and relative humidity showed that: "At increasing salinities the amount of secreted ions showed a five-fold increase, whereas the osmotic potential was raised only twofold." A low level of secretion of K^+ was found (65).

Distichlis spicata and Spartina alterniflora exhibit selective uptake of potassium and exclusion of sodium (66). It must be recognized that such discrimination is not absolute. Both of these species obviously take up appreciable quantities of NaCl, as is evident from the secretion of this salt by the salt glands of the leaves. Species of Plantago differ in responses to salt (67). P. media was killed at 75 mM NaCl; P. maritima, maintained growth at 300 mM; and P. coronopus, was of intermediate salt tolerance. The roots of these species apparently are more salt sensitive than the shoots. When the Na level was greater than 2 meq/g dry weight and the Na/K ratio exceeded 4, the roots became damaged irrespective of the Na content of the shoot. The authors suggest that halophytic Plantago species have develped an efficient translocation system between xylem parenchyma cells and the xylem vessels to avoid excessive Na accumulation in the roots. It is not clear why the shoots of tolerant species will tolerate the higher Na content. It is consistent, however, with other studies. Lauchli (20) reviewed comparative studies concerning the regulation of salt transport in

relation to salt   tolerance of glycophytes and halophytes.   "It is
demonstrated that among the glycophytes salt exclusion from the
leaves (that is the exclusion of Na^+ and/or Cl^-) occurs in very
salt-sensitive and moderately salt-sensitive species.   On the other
hand, the majority of halophytes belong to salt includers." An
exception, he points out, is Puccinellia peisonis in which the
exclusion of Na by the roots is very pronounced (68).   Kramer et al
(69) observed transfer cells in the root of Atriplex hastata in
response to salinity and studied their function with x-ray microprobe.

Morphological compartmentalization of salts has been reported
for Salicornia pacifica var. utahensis (70).   In young shoots the
palisade cells (outer portion of the stem) had low contents of Na^+,
K^+ and Cl^-, whereas the spongy cells (interior portion of the stem)
had much higher contents of Na^+ and Cl^-.   The difference in K^+
content between the two regions was much less marked, though the
inner portion contained more K^+ than the outer.   As the stems
matured, contents of these elements increase in both regions of the
stem.   In Suaeda monoica the chlorenchyma occurs in two layers, an
outer one with small chloroplasts and an inner one with large
chloroplasts rich in starch.   "Sodium was found to be selectively
accumulated in the outer layer of chlorenchyma, whereas $Cl^-$was
concentrated mostly in the cells of the inner chlorencyma" (71).
The Nernst criterion suggests that $Cl^-$is actively transported into
root cells of Salicornia bigelovii, but that active transport need
not be involved to explain the accumulation of Na^+ at all
salinities investigated (up to the concentration of full seawater)
nor for K^+ at moderate to high salinities (72).

There continues to be considerable interest in enzymes, their
activity as influenced by salinity, and various aspects of metabolism in halophytes.   Boucard and Billard (73) grew Saueda maritima
var. macrocarpa and Phaseolus vulgaris in culture solutions containing 0 - 500 mM NaCl.   When enzyme activities were measured in
extracts in the presence of NaCl, glutamine synthetase activity
from neither plant was influenced by concentrations up to 500 mM.
However, the activity of glutamic dehydrogenase from beans markedly
decreased when the concentrations of NaCl exceeded 100 mM.   This
enzyme from Suaeda, on the other hand, was influenced little by
NaCl in the test solution.   Addition of NaCl to the culture
solution stimulated the glutamine synthetase activity and lowered
the glutamic dehydrogenase activity of shoots and roots of Suaeda.
The reverse was true for beans.

Poljakoff-Mayber (74) observed that the relative increase of
pentose phosphate pathway of pea roots seems to parallel higher
sensitivity to salinity.   But the whole pattern of changes in
activity of enzymes active in these pathways in response to salin-

ity does not permit a clear-cut conclusion.   With enzymes isolated
from Halimione portulacoides, malic dehydrogenase and catalase
activities were reduced in plants exposed to salinity while
peroxidase and superoxide dismutase were not affected. Based upon
PEP carboxylase activity and the presence of Krantz anatomy,
Kuramato and Brest (75) concluded that Spartina foliosa and
Distichlis spicata are C_4 plants and that Salicornia europa and
Batis maritima are C_3, not CAM, plants.

Jeffries et al (76) studied the response of the halophytes
Plantago maritima, Limonium vulgare, Triglochin maritima and
Halimione portulacoides to stress induced by salts or polyethylene
glycol (PEG 6000). They responded by accumulating organic solutes
regardless of the stress agent.   Depending on the species,
sorbitol, proline, reducing sugars, quaternary ammonium compounds
and α-amino nitrogen accumulate in tissues as the water potential
of the tissue falls.   Jeffries (77) has recently reviewed in depth
the role of organic solutes in osmoregulation in halophytic higher
plants.   There is considerable evidence that such substances serve
as osmotica in the cells of such plants, possibly being most
effective in the cytoplasm.

Rozema, et al (78) using Glaux maritima as a model system
found that anaerobiosis resulted in higher levels of malate,
citrate, and malic dehydrogenase activity in the roots.   Adding
mannitol (to 600 mM) or NaCl (to 300 mM) to the nutrient solution
resulted in an increase of proline from about 5 to 25-30 mM in the
plant.   Some of cell sap constituents inhibit malic dehydrogenase
activity and others have little or no effect.   On this basis they
suggest that proline and potassium are largely in the cytoplasm and
that the major portion of Na^+, Cl^- and soluble sugars are in the
vacuole.   Similarly, Storey and Wyn Jones (79) observed, with
Atriplex spongiosa and Suada monoica, that the proline and glycinebetaine increased greatly in response to NaCl (to 800 mM).
However, the amount of glycinebetaine was about 10x that of proline
and was highly correlated with the osmality of the shoot sap. They
suggest that this compound is the major cytoplasmic osmoticum (with
K^+ salts) in these species at high salinities and that Na^+ salts
may be preferentially utilized as vacuolar osmoticum.

A few additional notes indicate other interest in plant
physiological responses to salinity.   Ungar (80) has reviewed
papers dealing with the germination of seeds and halophytes.   Some
generalizations (see his review for references) are: Seed from
plants growing in saline environments are more salt tolerant at the
germination stage than those from plants growing in less saline
environments.   Furthermore, comparisons of various taxa indicate
that natural selection for salt tolerance is taking place in
nature.   Studies with the germination of seed in Atriplex canescens

indicate a strong interaction between temperature and salinity. Numerous investigations indicate that halophytes are generally more salt tolerant than glycophytes at the germination stage. "There appears to be a continuum of change in salt tolerance from the least tolerant glycophyte to obligate salt marsh species. However, all vascular plant species investigated display both a delay in the time of germination and a reduction in the total number of seeds germinated as salt stress is increased beyond the optimal levels for seed germination." "An important characteristic of seeds of halophytes, distinguishing them from glycophytes, is their ability to remain dormant at high salinities and then germinate at a later date when salinites are reduced." "Generally, NaCl has been found to be the least toxic in tests with isotonic solutions and significantly it is one of the more important salts influencing the distribution of halophytes."

Longstreth and Nobel (81) measured changes in characteristics of leaves of Phaseolus vulgaris, Gossypium hirsutum and Atriplex patula in response to salinity. An increase in the latter resulted in substantially higher ratios of mesophyll surfaces to leaf area (A^{mes}/A) in the case of beans and cotton and a smaller increase in this ratio for Atriplex. CO_2 resistance on the basis of mesophyll cell wall area (r_{cell}) increased even more in beans and cotton with increasing salinity. There was only a small increase in r_{cell} for Atriplex under similar conditions. They conclude: "The constancy of r_{cell} for Atriplex over a wide NaCl range as compared to the variation for bean and cotton presumably indicates differences at the chemical level."

A similar difference between species was reported by Rush and Epstein (82) who found that Lycopersicon esculentum and L. cheesemanii differed in sensitivity to high concentrations of KCl and NaCl. Epstein (62) in summarizing interactions between Na^+ and K^+ says: "Taken together, evidence from short-term experiments with excised tissues and long-term experiments with growing plants suggests that resistance to sodium salts of necessity involve potassium transport mechanisms that are highly indifferent to or even stimulated by sodium."

Callus and cell suspension cultures. All possibilities for selecting desirable attributes in salt-tolerant plants should be considered. Recent advances in generating whole plants from callus and cell suspension cultures offer new prospects. Nabors, et al (83) obtained NaCl-tolerant cell lines by exposing tobacco cell suspensions to increasing concentrations of NaCl (to 8.8 °/oo). Plants regenerated from resistant cell lines transmitted salt tolerance to two subsequent generations and tolerated salinities in excess of that tolerated by the parent plants from which the cell lines were derived.

Rains and his colleagues exposed cultured alfalfa cells to an agar-solidified medium containing 1% (w/v) NaCl (84, 85) and obtained a cell line with increased resistance to NaCl toxicity. This line grew better than the unselected culture at high levels of NaCl and performed poorly in the absence of this salt, an observation that suggests a requirement for NaCl for optimal growth.

Callus cultures have been used in an effort to elucidate the mechanism of salt tolerance and ion accumulation.   Orton (86) compared the NaCl tolerance of whole plants and callus cultures of barley (Hordeum vulgare) and H. jubatum. The plants and cultures of barley were less tolerant than those of the wild relative.   They suggest: ". . .that this tolerance is expressed at the tissue level and [the results] are compatible with a model where Na$^+$ and Cl$^-$ are compartmentalized in [H. jubatum] root tissue thus preventing transmission to the shoot and interference with endogenous K$^+$."

Diaz et al (87) report studies using callus and protoplast suspensions of tomato, mangrove, and salicornia. Tal and his colleagues (59, 88, 89) have studied salt tolerance and osmotic stress in the tomato and its wild relatives, L. peruvianum and S. pennellii, using callus and cell suspensions.   Calli derived from the wild species grew better in the presence of NaCl than those derived from the cultivated tomato and accumulated more Cl$^-$ and Na$^+$ and less K$^+$ than the latter.   The growth of callus from L. peruvianum was more inhibited by 3,4-dehydroproline and by mannitol and less inhibited by NaCl than callus from the tomato.

This approach has attracted much attention, especially since Carlson et al (90) described the production of interspecific hybrids of species of Nicotinia by regenerating plants following fusion of protoplasts. That the approach is fraught with difficulties is evident from what appears to be the slow progress.   Few successes have been reported with cultivars.   The complexity is underscored by the work of Shepard et al (91) with potato protoplasts.   They use, sequentially, five culture media to regenerate plants from single-leaf cells isolated from plants of the cultivar Russett Burbank, which had to be grown under precisely controlled conditions of nutrition, temperature and photoperiod. They obtained phenotypic variations in the regenerated plants without applying overt selection pressure.

Despite the difficulties, this is an approach that merits serious attention, though the rewards in terms of useful selections may come slower than with classical selection techniques, there is greater promise of eventually making greater strides in providing new phenotypes.

Current Research. The Smithsonian Science Information Exchange (1) has provided a useful "International Directory of Current Biosaline Research Projects." Frequent updating of such information would be of great assistance in facilitating exchange of information among workers in this field. (Possibly subsets of this Directory dealing with more limited topics would be useful.) Information in this Directory is germane to this Workshop. However, the various abstracts are so uneven in various ways that it is difficult, if not impossible, to provide a critical summary. Some abstracts represent research that has made substantial progress; others give little more than the objectives of planned research, simply, one assumes, because that is all that is available at present. Despite these limitations an attempt to provide an overview is included below in Summary and Conclusions.

Summary and Conclusions

1.   Interest in developing salt-tolerant crops appears to be growing markedly. Two approaches are being used: (a) Improving conventional crops by selection from the available gene pool or by introducing this trait from wild, naturally tolerant species; and (b) Improving naturally salt-tolerant wild species to suit them for agricultural use. Both approaches show promise of providing additional options for producing food and feed. The former may find greatest use in areas now being used for crop production but facing salinity problems. Because of other intrinsic traits of the species involved, the latter may prove more effective in extending agriculture into areas not used, or used minimally, for this purpose.

2.   The crops currently receiving most successful attention in selection programs are the cereals: rice, barley and wheat. Success is being reported with tomato in breeding and selection. A variety of other crops are being studied for salt tolerance, including potatoes, onions, cantaloupes, asparagus, cucurbits, lettuce, eggplant, oats, maize, sugar beets, sugar cane, cotton, sorghum and millet. Various legumes are being evaluated, including alfalfa, soybean, lentils, peanut, cowpea, and mung. Interactions between salinity and N_2 fixation is of concern with these crops. Rootstocks of walnut, pecan, grape, and citrus are being studied for salt tolerance. The length of this list is evidence of the pervasiveness of the concern for salinity problems in agriculture.

3.   Various wild plants currently under evaluation as potential crop plants for saline areas include various species of Atriplex, being studied by a number of investigators. Others receiving less attention as potential crop plants for saline areas include: Chenopodium spp., Cressa truxellensis, Simmondsia

chinensis (jojoba), Prosopis spp., Salicornia spp. and Kosteletzkya
virginica. Some grasses, mostly wild species, are being examined:
Rhodes grass (Chloris gayana), Kallar grass (Diplachne fusca),
Spartina alterniflora, S. patens, Distichlis spicata, D. palmeri,
Panicum virgatum and Sporobolus virginicus. Mangrove is being
evaluated for potential commercial use. (Earlier literature, of
course, includes many other species; see for example Mudie, 92).

4. Various physiological responses to saline stress and their
interaction with plant hormones continue to attract many investigators. It is clear that a variety of responses are exhibited, all
appear to be aimed at avoiding Na^+ and/or Cl^- or by compartmental-
izing of ions at the tissue or cellular level. Salt may be avoided
by selective ion uptake or excretion processes. Compartmentalization may be between adjacent tissues of a stem (Salicornia sp.),
contiguous organs, or within the cell. Exclusion is commonly at
the root level and secretion at the surface of the leaf, including
salt-laden vesicles in Atriplex spp.

Beyond this there appear to be few generalizations. An
understanding of the mechanism(s) in satisfactory basic biochemical
and physiological terms is wanting. Presumably membranes are
involved, but the study of these in plant tissues is intrinsically
very difficult. Yet intuition suggests that a thorough understanding of such phenomena would make applied research in this field
more effective. In its absence, empirical approaches must be
used.

5. Recent developments using tissue culture, cell suspensions
and protoplasts appear to offer new prospects for developing
salt-tolerant lines and might contribute to a better understanding
of the nature of the genetic control. Manipulation of the DNA
might prove possible and protoplast fusions could open up new
possibilities. Exciting as these developments are, they are as yet
mostly promises for the future. Attendant technical difficulties
remain to be resolved.

6. Success in improved crops for a variety of attributes has
been greatly enhanced by an understanding of the genetics of the
species involved and by the development of techniques needed to
apply such knowledge. Such information with regard to salt
tolerance is sadly lacking. It is encouraging to note that efforts
to correct this situation have been initiated.

REFERENCES

1. Smithsonian Science Information Exchange. International
      Directory of Current Biosaline Research Projects. National

Technical Information Service, U.S. Dept of Commerce, 5285.
Post Royal Road, Springfield, VA 22151 USA, 269 pp plus
indexes (1980).

2. L. E. Francois and E. V. Mass, "Plant Responses to Salinity:
An Indexed bibliography." Agr. Reviews and Manual, Western
Series, No. 6. Ofc. of Regional Adm. Agr. Res. (Western
Region), Science and Education Adm., US Dept. of Agr.,
Berkley, CA 94705 (1978).

3. T. J. Flowers, J. F. Trokes and A. R. yeo, Ann. Rev. Plant
Physiol. 28:89-121 (1977).

4. H. Greenway and R. Munns, Ibid. 31:149-190 (1980).

5. H. Boyko, (ed.) "Salinity and Aridity. New Approaches to Old
Problems." W. Junk, The Hague, 408 pp. (1966).

6. H. Boyko, (ed.), "Saline Irrigation for Agriculture and
Forestry," W. Junk, The Hague, 325 pp. (1968).

7. G. F. Somers, "Seed-Bearing Halophytes as Food Plants."
DEL-SG-75, College of Marine Studies, Univ. of Delaware,
Newark, DE, 156 pp. (1975).

8. E. R. Epstein, W. Kingsbury, J. D. Norlyn and D. W. Rush, in:
"The Biosaline Concept," (A. Hollaender, ed.), pp. 77-79,
Plenum Press, New York (1979).

9. G. F. Somers, Ibid., pp. 101-115, Plenum Press, N. Y. (1979).

10. E. Epstein, in: "Proceedings of a workshop on plant adaptation
to mineral stress. Cornell Univ. Agr. Expt. Sta., Special
Bulletin (1977).

11. D. B. Kelley, J. D. Norlyn and M. Epstein, in: "Arid Land
Conference on Plant Resources," (J. R. Goodin and D. K.
Northington, eds.) pp. 326-334, Texas Tech Univ., Lubbock,
TX (1979).

12. G. F. Somers, in: "Stress Physiology in Crop Plants," (H.
Mussell and R. C. Staples eds.), Wiley-Interscience, N. Y.
(1979).

13. G. F. Somers, M. Fontes, and D. M. Grant, in: "Arid Land Plant
Resources," (J. R. Goodin and D. K. Northington, eds,) pp.
402-417, Texas Tech Univ. Lubbock, TX (1979).

14. D. W. Rains, R. C. Valentine and A. Hollaender, (eds.) "Genetic
Engineering of Osmoregulation. Impact on plant producivity
for food, chemicals and energy." Basic Life Sci., Vol. 14.
381 pp., Plenum Press, N. Y. (1980).

15. E. Epstein, J. D. Norlyn, D. W. Rush, R. W. Kingsbury,
D. B. Kelly, G. A. Cunningham and A. F. Wrona, Science
210:399-404 (1980).

16. R. Jones, (ed.) "The Biology of Atriplex." Division of Plant
Industry, CSIRO, Canberra, Australia (1970).

17. C. V. Malcolm, J. Agric. W. Austr. 15:68-73 (1974).

18. C. B. Osmond, O. Bjorkman and D. Anderson, "Physiological
Processes in Plant Ecology. Toward a synthesis with
Atriplex." Ecological Studies, Vol. 36, 500 pp.
Springer-Verlag, N. Y. (1980).

19. B. R. Phills, N. H. Peck, G. E. MacDonald and R. W. Robinson, J. Amer. Soc. Hort. Sci. 104:349-352 (1979).
20. A. Lauchli, Ber. Deutsch. Botan Gesell, 92:87-94 (1979).
21. R. Ahmad, and Z.-N. Abdullah, Pakistan J. Botany 11:103-112 (1979).
22. Z.-N. Abdullah and R. Ahmad, Biosaline Newsletter 2:10, (A. San Pietro, ed.), Indiana Univ. (1980).
23. E. Epstein, Ibid. 1:4, (1979).
24. G. J. Hoffman and J. A. Jobes, Agron. Jour. 70:765-768 (1978).
25. S. M. Siegel and B. Z. Siegel, Biosaline Newsletter 1:23, (A. San Pietro, ed.) Indiana Univ. (1979).
26. A. Shomer-Ilan, B. Samish, T. Kipnis, E. Elmer and Y. Waisel, Plant and Soil 53:477-486. (1979).
27. M. Abdullah and R. H. Qureshi, Biosaline Newsletter 2:17, (A. San Pietro, ed.), Indiana Univ. (1980).
28. S. K. Datta and S. S. Pradhan, Ibid 2:12-13 (1980).
29. M. Rai, Ibid., 2:15-16 (1980).
30. J. R. Goodin, in: "Arid Land Plant Resources," pp. 418-424, Texas Tech Univ. Lubbock, TX (1979).
31. J. R. Goodin, in: "New Agricultural Crops," AAAS Selected Symposium 38, pp. 133-148, Westview Press, Boulder, CO (1979).
32. H. C. Stutz, C. L. Poper and S. C. Sanderson, Amer. J. Botany 66:1181-1193 (1979).
33. H. Greenway and C. B. Osmond, in: "The Biology of Atriplex," (R. Jones, ed.), pp. 49-56, Div. of Plant Industry, CSIRO Canberra, Australia, (1970).
34. S. A. Omar, F. K. Taha and A. Nassef, Kuwait Inst. Scie. Res. Ann. Report 1979, 7-8 (1979).
35. G. F. Somers, D. M. Grant and M. Siegel, Bot. Soc. Amer. Misc. Series 158:107 (1980).
36. E. Epstein and J. D. Norlyn, Science 197:249-251 (1977).
37. J. D. Norlyn, in: "Genetic Engineering of Osmoregulation. Impact on plant productivity for food, chemicals and energy," pp. 293-310, Basic Life Sci. Vol. 14, Plenum Press, N. Y. (1980).
38. R. C. Staples, R. W. Robinson and N. E. Oebker, Biosaline Newsletter 2:18-19, (A. San Pietro, ed.), Indiana Univ. (1980).
39. D. B. Kelley, E. Epstein and A. Lauchli, Plant Physiol. 65: (Suppl.) 83 (1980).
40. M. B. kirkham and A. O. Faden, Biosaline Newsleter 1:16, (A. San Pietro, ed.) Indiana Univ. (1979).
41. T. J. Flowers, and A. R. yeo, Ibid., 1:10-11 (1979).
42. S. K. Datta, Ibid.., 1:5-7 (1979).
43. M. K. Jana and A. E. Slinkard, Ibid. 1:15 (1979).
44. J. A. Silander and J. Antonvics, Evolution 33:1114-1127, (1979).

45. A. F. Wrona, and E. Epstein, Plant Physiol, 65: (Suppl.), 133, (1980).
46. E. P. Glenn and M. Fontes, Biosaline Newsletter 1:9, (A. San Pietro, ed.), Indiana Univ. (1979).
47. K. B. Pihl, D. M. Grant and G. F. Somers, DEL-SG-11-78, College of Marine Studies, Univ. of Delaware, Newark, DE 38 pp. (1978).
48. G. F. Somers, D. M. Grant and R. W. Smith, DEL-SG-12-78, College of Marine Studies, Univ. of Delaware, Newark, DE 35 pp., (1978).
49. E. A. Oelke, in: "Seed Bearing Halophytes as Food Plants," DEL-SG-3-75, College of Marine Studies, Univ. of Delaware, Newark, DE (1975).
50. G. C. Marten and R. N. Anderson, Crop Sci. 15:821-827 (1975).
51. E. D. Seneca, Amer. J. Botany 61:947-956 (1974).
52. G. F. Somers and D. M. Grant, Ibid., 68:6-9 (1981).
53. M. C. Shannon, Hort. Sci., 14:587-589 (1979).
54. E. V. Maas, G. Ogata and M. H. Finkel, Plant Physiol, 64:139-143 (1979).
55. M. Mueller and K. A. Santarius, Ibid. 62:326-329 (1978).
56. K. A. Santarius, U. Heber and G. H. Krause, Ber. Deutsch Botan. Gesell. 92:209-224 (1979).
57. L. Erdei, C. E. E. Stuiver and P. J. C. Kuiper, Physiol Plant. 49:315-319 (1980).
58. N. E. Mishustina, N. I. Tikhaya and N. S. Chaplygiva, Soviet Plant Physiol. 26:432-437 (translation from Fiziologiya Rastenii) (1979).
59. M. Tal, Biosaline Newsletter 1:26, (A. San Pietro, ed.) Indiana Univ. (1979).
60. D. W. Rush and E. Epstein, Plant Physiol, 65 (Suppl) 83 (1980)
61. D. W. Rush and M. Epstein, Ibid., 57:162-166 (1976).
62. E. Epstein, in: "Genetic Engineering of Osmoregulation. Impact on plant productivity for food, chemicals and energy," (D. W. Rains, R. C. Valentine and A. Hollaender, eds.) pp. 7-22, Basic Life Sciences, Vol. 14, Plenum Press, N. Y. (1980).
63. J. Wienecke and A. Lauchli, Z. Pflauzennahr, Bodenkd. 142: 799-814 (1979).
64. J. Wienecke and A. Lauchli, Ibid., 143:55-67 (1980).
65. J. Rozema and I. Riphagen, Oecologia 29:349-357 (1977).
66. R. M. Smart and J. W. Barko, Ecology 61:630-638 (1980).
67. L. Erdei and P. J. C. Kuiper, Physiol. Plant 47:95-99 (1979).
68. R. Stelzer and A. Lauchli, Zeit. Pflanzenphysiol 88:437-448 (1978).
69. D. Kramer, W. P. Anderson and J. Preston, Austral. J. Plant Physiol. 5:739-748 (1978).
70. D. J. Weber, H. P. Rasmussen and W. M. Hess, Canad. J. Botany, 55:1516-1523 (1977).

71. A. Eshel and Y. Waisel, Physiol. Plant. 46:151-154 (1979).
72. A. L'Roy and D. L. Hendrix, Plant Physiol. 65:544-549 (1980).
73. J. Boucard and J. P. Billard, Compt. rend. Acad. Sci.,
    D. Sci. Natur., 289:599-602 (1979).
74. A. Poljakoff-Mayber, Biosaline Newsletter 1:17, (A. San Pietro,
    ed.), Indiana Univ.. (1979).
75. R. T. Kuramato and D. E. Brest, Bot. Gaz. 140:295-298 (1979).
76. R. L. Jeffries, T. Rudmik and E. M. Dillon, Plant Physiol.
    64:989-994 (1979).
77. R. L. Jeffries, in: "Genetic Engineering of Osmoregulation.
    Impact on plant productivity for food, chemicals and energy"
    (D. W. Rains, R. C. Valentine and A. Hollaender eds.)
    pp. 135-154, Basic Life Sci., Vol. 14, Plenum Press, N. Y.
    (1980).
78. J. Rosema, D. A. G. Buizer and H. E. Fabritius, Oikos 30:
    539-548 (1978).
79. R. Storey and R. G. Wyn-Jones, Plant Physiol. 63:156-162 (1979)
80. I. A. Unger, Botan Review 44:233-264 (1978).
81. D. J. Longstreth and P. S. Nobel, Ibid. 63:700-703 (1979).
82. D. W. Rush and E. Epstein, Plant Physiol. 63 (Suppl.), 163.
    (1979).
83. M. W. Nabors, S. E. Gibbs, C. S. Bernstein and M. E. Meis,
    Zeit Pflanzenphysiol 97:13-18 (1980).
84. T. P. Croughan, S. J. Stavarek and D. W. Rains, Crop Sci.
    18(6), 959-963 (1978).
85. D. W. Rains, T. P. Croughan and S. J. Stavarek in: "Genetic
    Engineerig of Osmoregulation. Impact on plant productivity
    for food, chemicals and energy," (D. W. Rains, R. C.
    Valentine and A. Hollaender, eds.), pp. 279-292, Basic Life
    Sci., Vol. 14, Plenum Press, N. Y. (1980).
86. T. J. Orton, Z. Pflanzenphysiol 98:105-118 (1980).
87. J. L. Diaz, F. Lopez and R. Sanchez, Inform. Gen. Labor
    pp. 89-101, Cent. Invest. Biol. de Baja Calif., A.C.
    (1979).
88. M. Tal and A. Katz, Z. Pflanzenphysiol. 98:283-288 (1980).
89. M. Tal, H. Heikin and K. Dehan, Ibid. 98:231-240 (1980).
90. P. S. Carlson, H. H. Smith and R. D. Dearing, Proc. Nat. Acad.
    Sci. 69:2292-2294 (1972).
91. J. F. Shepard, D. Bidney and E. Shahin, Science 208:17-24
    (1980).
92. P. J. Mudie, in: "Ecology of Halophytes," (R. J. Reimold and W.
    II. Queen, eds.) pp. 565-597, Academic Press, N. Y. (1974).

BIOMASS PRODUCTION OF FOOD AND FIBER CROPS USING HIGHLY SALINE

WATER UNDER DESERT CONDITIONS

Rafiq Ahmad and Zaib-un-Nisa Abdullah

Brackish Water Irrigation Research Project
University of Karachi
Pakistan

SUMMARY

Beet root, desi and upland cotton were irrigated on coastal
sand with various dilutions of seawater supplemented with chemical
amendments. There was an increase in fresh weight of beet root
tuber up to 12,000 ppm salts and in their sugar content up to
16,000 ppm salts, respectively, in irrigation water. Though the
number of bolls per plant in both species of cotton were reduced
with the increase of salinity of irrigation water, the weight of
lint per plant was increased at lower salinities in upland cotton.
In comparison with upland cotton there was more chlorophyll, sugar
and proline in desi cotton and less reduction in the yield at
higher salinities. No adverse effect was found on the fiber quality in either species of cotton plant due to brackish water irrigation.

INTRODUCTION

Seepage of water from the river beds and canal system raised
the level of subsoil water to the extent that it came above the
ground level in some parts of Pakistan, creating an acute waterlogged condition. Salt dissolved in this water was precipitated
and left behind on the soil surface when the water level receded.
This created a great problem and out of 37.4 million acres of irrigated land about 10.1 million acres were affected by salinities of
various concentrations.

Some of the research undertaken in Pakistan to deal with this problem is as follows:

Soil Engineering and Management: Reclamation through leaching; construction of horizontal and vertical drainage; installation of tube wells; lining of canals and other distributaries; studies on carbon and nitrogen tranfsormation; chemical amendments and irrigation management, etc.

Biosaline Approach: Decomposition of organic matter by cellulolytic fungi on saline soils; physiology and biochemistry of salt tolerance in sunflower, wheat, maize and cotton, etc., with reference to mineral uptake, enzymology, hormonal imbalances; cultivation of salt tolerant grasses, e.g. Diplachne fusca for fodder; breeding for salt tolerance with special reference to sugar cane and rice; ecological studies on mangroves; growth and development of marine algae at Karachi coast and irrigation with brackish water or different dilutions of seawater on coastal sand for afforestation and agriculture, etc.

Since the work mentioned is being carried out by the author and his coworkers at Karachi University, it will be discussed with some detail.

Shortage of good quality water for irrigation has forced mankind to look into the possibility of using waters which heretofore have been considered unsuitable for irrigation. Recent work in the field of plant physiology has pointed out the utility of underground saline water for growing salt-tolerant plants on sandy deserts. If an area of coarse textured sand is selected for afforestation or agriculture, the highly saline sub-soil water, after being supplemented by chemical amendments, becomes a special kind of nutrient solution and we have an area of large scale artificial cultivation. The saline water percolates down into the sand, staying in the root zone just long enough for the root hairs to absorb mineral nutrients. The latent quality of the salt tolerant plants enables the roots to regulate ion fluxes to a certain extent, thus adjusting osmotic and ionic balances. Biochemical changes at sub-cellular levels in such plants prove advantageous for their survival and growth under saline environments. Forestry trails using saline water at Kuwait, irrigation with salty water at Morocco and Tunesia and similar experiments of afforestation and agriculture in Spain, Egypt, Israel and India have indicated that sub-soil saline water could render the sandy, arid regions capable of production.

It has now been made possible to work out water salt balance equations for a given climate, soil and crop, with the provision of efficient drainage systems, so that the positive balance (accumulation) turns into a negative balance (desalinization). Kovda's equation, as discussed by White (29), for inflow and outflow of salts is very helpful in this respect. Accumulation of exchangeable sodium in irrigation soil which may gradually lead to alkalization does not occur if water having even 60-70% of sodium cations is mineralized with 2-5 g/l dissolved gypsum (29). Chemical amendments used with various dilutions of sea water in our earlier experiments (2) have further reduced the probability of alkalization of sand. Limits of excessive salinity in irrigation water depends upon soil type and degree of salt tolerance of the plants being grown at a particular area. Under favorable drainage conditions, water containing 2-7 g salts per litre can be utilized for irrigation of most salt tolerant crops. Water having salinity equivalent to about one-fourth of oceanic strength has been used for irrigation of salt tolerant field and forage crops on sandy soils (24).

While working on a sandy beach of the Arabian Sea, off the Karachi coast, we observed that concentration of salt in sub-soil water is inversely proportional to the distance from the shore line. Hence, it was considered that at a suitable distance from the sea coast the salt concentration of sub-soil water may dilute to such an extent that it could be directly used, after chemical amendments, for irrigation of sandy soil. We have grown maize on coastal sand by irrigating with different dilutions of seawater supplemented with chemical amendments and found that a concentration of 6800 ppm salts in irrigation water would not reduce the vegetative yield, although the reproductive yield (grains) was reduced by about 47.72%. The present work was undertaken to find out the maximum permissible concentration of salts in irrigation waters which would still result in economically feasible produce. Beet root and cotton were grown on coastal sand and irrigated with various dilutions of seawater supplemented with chemical amendments.

Materials and Methods

Plant Material (i) Beet Root: Beet root (Beta vulgaris) is derived from Beta maritima which is found growing wild on the seashore of the Mediterranean region and some other countries. It includes different varieties used for table beet, fodder or sugar producing beet roots. This crop is considered to be highly salt tolerant, capable of growing on saline soils having EC of solution to an extent of 12 m.mohs, ESP = 40-60 (16). Irrigation water, having a salt concentration as high as 14 gr/l has been used for growing beet root in Holland, where, as in America, irrigation water having EC up to 10 m.mohs has been used for this purpose. Boyko (7) has grown beet root by irrigating them with sea water of Caspian type (10,000-13,000 mg/l T.D.S.) and North Sea Type (20,000-27,000 mg/l T.D.S.) on coastal sand. Hence it was decided to use beet root as one of the crops for our experiment. This crop is widely grown for food and fodder all over the world. Seeds of the beet root (Beta vulgaris) Ver. Crimson Red, supplied by Ayoub Agricultural Research Station, Faisalabad (Pakistan) were used in the present investigation.

(ii) Cotton: Cotton is the main fiber producing crop of the world. Its different cultivars are widely distributed in the tropical and subtropical regions of Africa, Asia, Australia and America. A mean annual temperature of over $60°F$ and average rainfall of even less than 6 inches is sufficient to give a good yield on irrigated land. It is capable of growing on marginally fertile soil and can tolerate salinity to a certain extent. Some of the old world cotton of herbacea series are better suited for growth on saline soil or can stand saline irrigation better than new world cotton of hirsuta series. Local cultivars, after passing several generations under saline environment, have developed salt tolerant strains which are better suited for saline agriculture. These plants do not produce good quality fiber, but are capable of producing sufficient lint and seeds which could be used in many industries.

Cotton also has been classified as a highly salt tolerant plant like that of beet root, capable of being irrigated with water having EC 10-16 m.mohs or 0.5-1.0% salts on dry soil bases (10). Cotton plants, which frequently are found growing on solonchak patches show a tendency of producing progeny of salt tolerant plants at a later stage. One of our local cotton variety, Gossypium arborium Var. D-9, has been grown for quite a few years on moderately saline soils of Pakistan and has developed some salt tolerance in the course of repeated selection. Hence, the above mentioned variety was selected for present investigation and another variety of new world cotton Gossypium hirsutum Var. M-100, was taken for comparison. Seeds of Gossypium arborium Var. D-9 (known as Desi Cotton) were collected from the stock of Ayoub Agriculture Research Station (Pakistan). These seeds were produced by the crops growing for many years on moderately saline soils. Seeds of Gossypium herbarium Var. M-100 (known as upland Cotton) were also supplied by the the research station.

Culture Technique

Cultivation practices of drum pot culture as recommended by Boyko (7) were adopted with some modification. Drum pots having an outlet for drainage near bottom were filled with 300 kg sand taken from the sea coast. They were capable of retaining 44 litres of water at field capacity and any additional amount of water, if used in irrigation, was leached out from the containers. At certain other localities these drum pots were actually sunk in the sandy strata after removing the basal plate. Seeds were sown directly in the drum pots and irrigated with normal water till they reached the three leaf stage. Five seedlings were left in each drum pot and irrigated with various dilutions of sea water with a difference of 4000 ppm salinity so that there were five treatments of 4000, 8000, 12000, 16000 and 20000 ppm salinities, respectively. Four replicates were kept for each treatment. Control plants were irrigated with half strength Hoagland solution. The composition of chemical amendments supplemented with each dilution of sea water has been described in our earlier paper (2). Height, weight, and leaf area of the plants (18) were taken as parameters for vegetative growth and number of bolls per plant and seeds per boll (in cotton) were taken as criteria for reproductive growth. The drum pots were flooded three times a week with irrigation water during the seedling stage but later the frequency was reduced to once a week. Chlorophyll was extracted as described by Maclachlam and Zalik (12). Total sugars were estimated by Nelson's method (17). Na^+ and K^+ were analyzed using Flame Photometry (27). Proline was estimated as described by Bate (4).

Results and Discussion

Different growth and biochemical parameters of beet root as affected by various concentrations of salts in the irrigation medium are presented in Table I. Critical analysis of the data shows a general increase in number of leaves, photosynthetic area, fresh weight, chlorophyll and sugar content at lower levels of salinites. This trend persists in the first two parameters, even up to the higher salt concentrations of the irrigation medium, though there appears a decrease in the remaining three parameters at this stage. It is interesting to note that in spite of the above mentioned reduction, the values still remain more than that of control. Concentrations of proline and sodium keep on increasing both in the leaves and root tubers with the increase of salts in the irrigation water. Potassium accumulation, after showing an initial enhancement at lower salinities, retarded at higher salinities, though the amounts accumulated are still greater than that of control.

Response of beet root towards salinity as shown in our investigation is not much different from results found by other workers (13, 28). It might be mentioned that sodium chloride occupies about 60% of total salt concentration in sea water. Better growth of beet root at lower salinities may be due to its essentiality for Chenopods as shown by Brownell (5) and Brownell and Crossland (6). However, higher concentrations in irrigation water still proved to be toxic. Selective transport of sodium in beet root (2)

TABLE 1. Growth Responses of Beet Root Under Sea Water Irrigation.

Parameters	Control (-sea water)	Concentration of salts in sea water (ppm)				
		4,000	8,000	12,000	16,000	20,000
Avg. No. of leaves/plant	13.7	25.0	21.3	21.0	19.0	13.8
Photosynthetic area (Cm²/plant)	864.5	2175.7	1963.1	2596.2	1689.9	936.2
Fresh wt. of shoot (g)/plant	81.3	141.8	103.8	97.9	88.6	78.9
Fresh wt. of root tuber (g)/plant	156.1	311.3	193.2	189.6	148.3	99.0
Total chlorophyll mg/g fresh weight	92.7	98.3	103.4	98.6	93.4	88.4
Total sugar content mg/g dry weight						
i) Leaves	120.7	168.4	151.2	139.2	124.6	97.3
ii) Root	376.8	490.6	497.7	350.13	314.7	236.8
Proline (mg/g fresh wt.)						
i) Leaves	15.6	26.5	32.1	40.0	44.2	47.3
ii) Root	7.1	12.9	15.8	25.7	32.2	33.2
Na+ (mg/g dry wt.)						
i) Leaves	30.6	50.7	80.7	94.3	104.5	128.4
ii) Root	11.4	22.5	36.5	50.9	58.1	66.4
K+ (mg/g dry wt.)						
i) Leaves	44.4	54.6	67.4	85.4	69.8	59.6
ii) Root	25.6	31.7	38.7	44.7	53.2	52.2

and changes in biochemical composition with special reference to chlorophyll, sugar and proline concentrations, have made this crop so salt tolerant that there was no reduction in fresh weights of the root tubers at 12000 ppm, and even at 16000 ppm salt concentrations of irrigation water. Uptake and transport of potassium being independent of high concentrations of ambiant sodium, proved to be beneficial for growth under saline environment (14,15, 20). It is argued that Na^+ is being utilized more efficiently, possibly substituting K^+ as well in metabolic functions (1).

Vegetative growth and certain biochemical parameters of upland and desi cotton as effected by different concentrations of salts in irrigation medium are given in Tables II and III. Heights of the plants show an inverse ratio with the increase in salinity, with the exception of desi cotton irrigated by 4000 ppm salinity (Fig. 1). Increasing concentrations of salt in general have progressively decreased the vegetative growth of desi as well as upland cotton. Though reduction in the height of the latter is less than the former. Chlorophyll content has increased in both kinds of cotton species at lower salinities but there appears a reduction while approaching higher salt concentration. Increase in salt concentration of irrigation water up to 12,000 ppm in upland cotton and 16,000 ppm in desi cotton has increased chlorophyll content of the leaves which has enhanced the efficiency of the photosynthetic apparatus, resulting in production of more photosynthetates (sugars) and ultimately more primary products (organic acids) for the synthesis of amino acids. Increase in the chlorophyll content at low salinities may be due to a greater supply of the compounds necessary for synthesis (26) or suppression of iron containing enzymes which otherwise inactivate the biosynthesis of chlorophyll (22). Increase in the sugar, proline, sodium and potassium content of the leaves in both cotton species is directly proportional to the increase in salt concentration of the irrigation medium.

Concentrations of sugar and proline increasingly progressed with the increase of salts in irrigation water. Participation of these two components in osmoregulation has also been suggested (23, 25). Difference in the pattern of sugar accumulation in the leaves of beet root with that of cotton at higher salinities may be due to difference in the storage mechanism of sugars in these plants. The presence of more chlorophyll sugar and proline in desi cotton, in comparison with that of upland cotton, shows that the latter is a comparatively more tolerant species.

The number of bolls produced per plant in both kinds of cotton has been adversely affected by the increase in salts of the irrigation water (Table V), though salinity has helped in reducing boll sheding. It is interesting to note an increase in the amount of

TABLE II. Growth Response of Desi Cotton Under Sea Water Irrigation.

Parameter	Control (-sea water)	Concentrations of Salts in Sea Water (ppm)				
		4,000	8,000	12,000	16,000	20,000
Height of plant (cm)	125.0	126.5	117.8	116.5	113.5	85.0
Total chlorophyll (mg/g fresh wt.)	72.1	79.5	85.6	88.1	75.6	70.4
Total sugars (leaves) (mg/g dry wt.)	39.6	55.8	70.3	83.8	94.6	107.6
Proline (leaves) (mg/g fresh wt.)	21.2	29.8	43.6	48.3	55.7	66.8
Na^+ (mg/g dry wt.) i) Leaves	10.5	17.2	25.3	33.2	48.7	55.3
ii) Root	4.4	5.6	12.5	20.7	38.5	47.6
K^+ (mg/g dry wt.) i) Leaves	36.8	42.6	49.6	56.2	58.2	58.7
ii) Root	18.4	20.2	27.4	33.8	38.8	41.5

TABLE III. Growth Response of Upland Cotton Under Saline Irrigation.

Parameter	Control (-sea water)	Concentration of Salts in Sea Water (ppm)				
		4,000	8,000	12,000	16,000	20,000
Height of Plant (cm)	103.0	102.5	79.0	78.8	72.5	56.3
Total Chlorophyll (mg/g fresh wt.)	65.4	71.3	77.5	69.8	64.2	59.4
Total Sugars (mg/g dry wt.)	35.1	57.3	61.6	76.9	80.4	83.3
Proline (Leaves) (mg/g fresh wt.)	18.4	25.3	28.3	34.6	36.8	46.0
Na^+ (mg/g dry wt.) i) Leaves	7.3	13.2	21.3	29.5	34.6	47.3
ii) Root	2.1	3.6	7.5	15.6	26.1	34.7
K^+ (mg/g dry wt.) i) Leaves	29.7	36.2	41.0	47.4	48.1	50.0
ii) Root	13.6	18.2	22.2	28.4	30.4	31.6

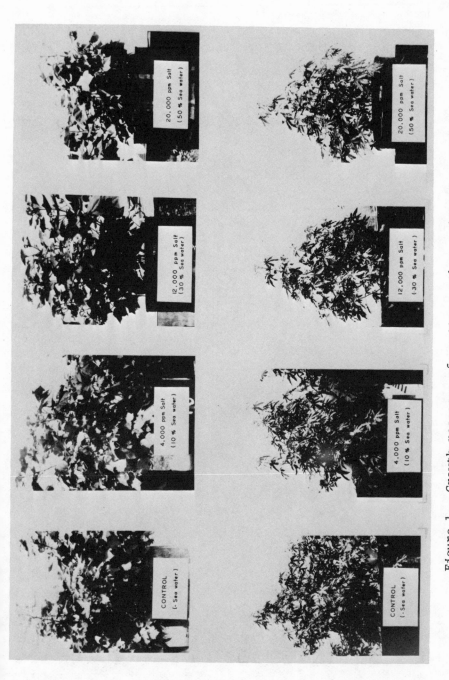

Figure 1.   Growth response of cotton under saline irrigation
(upper plants, G. hirsutum Var. M-100, lower plants
G. arborium Var. D-9).

TABLE IV. Yield of Desi and Upland Cotton Under Sea Water Irrigation.

Parameters	Control (-sea water)	Salt Concentration of Sea Water (ppm)				
		4,000	8,000	12,000	16,000	20,000
Gossypium aborium (Desi Cotton)						
No. of bolls retained/plant	68	60	53	45	29	27
No. of bolls shed/plant	15	5	5	10	12	15
No. of Total bolls/plant	83	65	58	55	41	42
No. of seed/boll	36	37	36	35	35	33
Weight of 100 seeds (g)	6.1	6.7	6.2	6.2	6.2	5.9
Weight of lint/plant (g)	87.5	69.7	69.8	59.5	34.2	34.1
Weight of seeds/plant (g)	150.0	148.5	118.6	97.5	62.5	52.4
Gossypium hirsutum (Upland Cotton)						
No. of bolls retained/plant	58	48	37	35	30	19
No. of bolls shed/plant	21	4	13	16	17	20
No. of total bolls/plant	79	52	50	51	47	39
No. of seed/boll	41	54	37	33	31	39
Weight of 100 seeds (g)	8.6	9.5	8.6	8.5	8.4	8.4
Weight of lint/plant (g)	43.4	87.2	67.9	54.1	20.7	7.2
Weight of seeds/plant (g)	205.2	246.9	118.0	97.8	78.3	46.1

lint per plant at lower levels of salinity in upland cotton. The number of seeds per boll has also increased at the 4000 ppm salinity level in the above mentioned species. In desi cotton, there has been a reduction in the weight of seeds and lint per plant, proportionate to the increase in salt of the irrigation water. However, the magnitude of reduction in yield at higher salinities is less in desi cotton than upland cotton.

Reduction in the number of bolls per plant in both kinds of cotton as a result of the increse in salt concentration of irrigation water could be attributed to the adverse effect of salinity on the viability and germination of pollen grain. Such an effect has already been noticed by us in wheat (1). Decrease in boll sheding in cotton at lower salinites as observed here has also been pointed out by Pasternak, et al (19). The role of Ca^{++} in stopping the formation of the absicission layer and thus reducing the amount of fruit sheding is well known in literature. The possibility of Na^+ playing a similar role in reducing the sheding of cotton bolls could not be ignored. Pasternak, et al (19) noticed an increase in the yield of cotton as a result of irrigation with brackish water containing 2300 ppm salts, but this was mostly due to the increase in the number of cotton bolls per plant. In spite of the fact that the number of bolls in upland cotton was reduced, due to saline irrigation, in the present investigation, the yield of lint per plant has shown an increase of up to 12,000 ppm salinity. It appears that with the reduction in the number of bolls per plant, the area of the leaves responsible for feeding photosynthate, per boll, has increased, thus translocating more photosynthate in these sinks. In desi cotton however, there is a reduction in the weight of lint and seed per boll at all the levels of salinities. Reduction in lint and seed weight per plant, at highest salt concentrations in desi cotton, is less, as compared to upland cotton. This shows that the former can produce better at higher levels of salinity than the latter (Table V). Reduction in cotton bolls per plant under saline conditions has been also reported by Longenecker (11). Increase in lint per plant at lower salinities in upland cotton is an interesting phenonenon and needs some more work for confirmation. Absence of this trend in desi cotton could be due to genetic or physiological variability of this species. We have also noticed a considerable increase in the activity of cellulose synthetase in upland cotton in comparison to that of desi cotton (unpublished data) under saline environment. From these findings it is inferred that the presence of higher amounts of Na^+ in the plant system might promote the activity of this enzyme, leading to synthesis of more lint in upland cotton.

Quartile and mean lengths of fibers (Table V) show slight increase in desi cotton irrigated by various dilutions of sea water, whereas these parameters in upland cotton show some reduc-

TABLE V. Fiber and Oil Analysis of Cotton Plants as Affected by Sea Water Irrigation*

Treatments	Upper Quartile Length (mm)	Mean Length (mm)	Short Fiber (%)	Fineness (Micronaire value)	Maturity Coefficient	Oil %
Gossypium arborium (Desi Cotton)						
Control (-sea water)	17.27	13.72	35.4	7.5	0.94	21.63
4,000 ppm	18.29	14.73	29.6	7.4	0.96	22.75
8,000 ppm	18.29	14.22	31.0	7.5	0.93	23.24
12,000 ppm	17.53	14.84	29.7	7.1	0.94	23.85
16,000 ppm	17.53	13.97	34.9	7.6	0.93	20.16
20,000 ppm	18.29	14.73	29.9	7.1	0.87	19.45
Gossypium hirsutum (Upland Cotton)						
Control (-sea water)	27.94	22.10	14.3	3.8	0.74	24.04
4,000 ppm	25.15	20.32	14.7	4.5	0.74	26.58
8,000 ppm	25.91	20.07	16.0	4.1	0.76	26.99
12,000 ppm	25.65	20.32	18.0	3.7	0.75	25.00
16,000 ppm	28.70	22.61	17.2	4.0	0.74	24.66
20,000 ppm	25.65	20.83	17.2	4.1	0.75	22.47

* The cooperation of Pakistan Central Cotton Committee, Ministry of Food and Agriculture is thankfully acknowledged for performing analysis presented in the table.

tion (with the exception at 16,000 ppm). Saline irrigation has helped in reducing the occurrence of short fibers in desi cotton although it has increased their occurrence in upland cotton. The data does not show any drastic effect of saline irrigation on fineness and maturity coefficient in any of the above mentioned cotton species. Similar results were obtained by Iyengar, et al (9) while irrigating cotton with 15,000 ppm salt irrigation water. There is some increase in the oil content of seeds at lower salinities (Table V) in our experiment, which starts decreasing at higher salinities and drops down even below the values of the control seeds.

On the basis of our data it can be concluded that beet root could be grown on sandy soils, by irrigation with chemically amended brackish water containing salts up to 12,000 ppm, without loss in the yield of root tuber. This limit of salinity could be further extended in the case of fodder beets, which are comparatively more salt tolerant than other varieties. As regards cotton, one will have to anticipate the permissible concentration of salts in irrigation water which still can give an economically feasible yield of fiber and oil under the above mentioned conditions, in different species. Increase in the yield of upland cotton at lower salinities is also of great interest. Salts reaching a concentration of 8000 ppm in irrigation water may give an economically feasible yield on sandy strata.

Breeding for salt tolerance is another approach for solving this problem (3, 8). If the plants obtained through breeding devices are cultivated using appropriate physiological techniques, it would be a great breakthrough in saline agriculture.

Acknowledgement

Receipt of financial grant from Pakistan Science Foundation for undertaking this project is thankfully acknowledged. Aquarist, Karachi Municipal Corporation deserves our special thanks for the supply of sea water to our research project.

References

1. Z. Abdullah, R. Ahmad and J. Ahmad, Plant & Cell Physiol. 19(1):99-106 (1978).
2. R. Ahmad and Z. Abdullah, in: "Advances in Desert and Arid Land Technology and Development", Vol I, (A. Bishay and W. G. McGinnies, eds) pp. 593-618, Harwood Academic Publishers, New York (1979).

3. M. Akbar and T. Yabuno, Japan J. Breed. 27:237-240 (1977).
4. L. S. Bates, Plant & Soil 39:205-207 (1973).
5. P. F. Brownell, Pl. Physiol 40:460-468 (1965).
6. P. F. Brownell and C. J. Crossland, Ibid., 49:7940797 (1972).
7. H. Boyko, "Salinity and Aridity", Dr. W. Jung Publishers,
    Netherlands (1972).
8. E. Epstein, in: "Plant Adaptation to Mineral Stress in Problem
    Soils" (M. J. Wright and S. A. Ferrari, eds.), Office of
    Agriculture, Technical Assistance Bureau, Washington, D. C.
    (1977).
9. E. R. R. Iyengar, J. B. Panya and J. S. Patolia, J. Pl. Physiol
    XXI(2):113-117 (1978).
10. V. A. Kovda, in: "Irrigation, Drainage and Salinity", an
    International Source Book, FAO/UNESCO (1973).
11. D. E. Longenecker, Soil Science 118(6):387-396 (1974)
12. S. Maclachlam and S. Zalik, Can. J. Bot. 41:1053-1062 (1963).
13. H. Marschner and J. V. Possingham, Z. fur Pflanzenphysiol.
    75(1):6-10 (1975).
14. E. V. Mass and G. J. Hoffman, J. Irrig. and Drainage, Div. ASCE
    103(IR2):115-134 (1977).
15. E. V. Mass and R. H. Neimen, in: "Crop Tolerance to Suboptimal
    Land Conditions" (G. A. Jung, ed.) Chapter 13, ASA Spec. Pub.
    227-299 (1978).
16. A. Meiri and J. Shalhavet, in: "Arid Zone Irrigation",
    (B. Yaron, E. Danfors and Y. Vaadia, eds.), Berlin,
    Heidelberg, New York: Springer-Verlag (1973).
17. N. Nelson, J. Biol Chem. 153:375-380 (1944).
18. A. Poljakoff-Mayber and A. Meiri, "The Response of Plants of
    Changing Salinity", Final Technical Report, The Volcani
    Institute of Agricultural Research, Bet-Dagan, Israel, (1969).
19. D. Pasternak, M. Twersky and Y. De Malach, in: "Stress
    Physiology in Crop Plants", (H. Mussell and R. C. Staples,
    eds.), John Wiley & Sons, New York (1979).
20. M. G. Pitman and W. J. Cram, in: "Integration of Activity in
    the Higher Plant", (D. H. Jennings, ed.), Cambridge Univ.
    Press, London (1977).
21. R. J. Poole, Pl. Physiol. 47:735-739 (1971).
22. B. A. Rubin, and E. V. Artsikhovskaya, Uspekhi Sovremennoi
    Biologii 57(2):317-334 (1964).
23. M. C. Shannon, Agronomy Journal 70:719-722 (1978).
24. G. F. Somers, in: "Stress Physiology in Crop Plants" (H.
    Mussell and R. G. Staples, eds.), John Wiley & Sons, New York
    (1979).
25. G. R. Stewart and J. A. Lee, Planta 120:279-280 (1974).
26. B. P. Strogonov, Nauka, Struktura i funkstsii kletok rastenii
    pri Zasolenii. Igdatel stvo "Nauka", Moscow (1970).

27. United States Laboratory Staff, "Diagnosis and Improvement of Saline and Alkali Soils", Agri. Hand Book, p. 60 (1954).
28. Y. Waisel and R. Bernstein, Bull Res. Counc. of Israel., Vol. 7D(2):90-92 (1959).
29. G. F. White, Environmental effects of arid land irrigation in developing countries. Man and Biosphere, Technical Notes (8) UNESCO, p. 31 (1978).

RESEARCH ON SEAWATER IRRICULTURE IN INDIA

E. R. R. Iyengar

Seawater Irriculture Discipline
Central Salt and Marine Chemicals Research Institute
Bhavnagar   364002   INDIA

SUMMARY

This paper deals with the problems of using sea water, as a supplemental source for irrigation on coastal dune sand, by taking advantage of regional rainfall for growing crops of food, fiber, vegetables and others. The responses of the crops to seawater or its dilutions exhibit wide varietal differences and forced selection (acclimatization) of varieties is possible by direct application of seawater (with no ill effect on the product, especially of grains.

Yield reductions, by salinity, to the extent of 30 - 50% or more are imminent; whereas, raising the yield potential by hybridization enhances the scope of using seawater for crop growth.

Studies on the Use of Seawater on Crops

About 70% of the earth's surface is covered by seawater. In 1909, Osterhoat concluded that seawater is fundamentally as important to plants as to animals. The total salinity is 35,000 ppm with an SAR value of 64.7. The ionic composition reflects the preponderance of sodium (504 meq/1) followed by magnesium (102/1) as cations and chloride (560 meq/1) and sulphate (56 meq/1) as anions. The carbonate and bicarbonate are of the order of 0.4 and 2.6 meq/1 respectively. On normal soil application, even diluted seawater retards plant growth and yield. However, seawater has been used, as such or after dilution, to grow crops and desert plants. Investigations carried out in Spain, Israel, Germany, India and USA have clearly shown that seawater can be used under certain predetermined conditions (1).

165

A comparative study made of silty clay soil and coastal dune sand indicated the suitability of sand over silty clay soils, wherein salt content of the soils equilibrated with seawater had 800 to 14,500 ppm salts, respectively. In sand, the permeability is ten times higher than silty clay soils, denoting the rapidity with which the accumulated salts get washed to deeper layers (2).

Effect of Salinity on Germination

As a prerequisite for the evaluation of seawater tolerance in crop plants, an attempt was made to ascertain the response of many crop varieties to various grades of seawater salinity at germination and early seedling growth stages. The crops tested were cereals (wheat, barley, rice, maize), millets (bajra, jowar, ragi), pulses (redgram, greengram, gram, lentil), oil seeds (sesamum, peanut, safflower, mustard, linseed), fodder (alfalfa, wheat grass, etc.), narcotics (tobacco) and fibre crop (cotton). The rate of imbibition is inversely proportional to the salts present in the salinity grades. Many of the crop varieties tolerated up to 10,000 ppm of seawater without significant reduction in germination percentage; this finding is specific to crops as well as to the varieties within the crop. Tolerance to seawater salinity is of the order of cereals > pulses, millets > oil seeds and grasses > legumes. There is significant reduction in seedling development, from 10,000 ppm upwards (3, 4).

To test inhibition of growth under salinity attributed to specific ionic effects over osmotic and amelioration effects of added deficient nutritive salts in seawater, studies were made on the rate of respiration and protein synthesis; also on phosphorus metabolism in seedlings of wheat, barley and gram. The seedlings were grown under definite seawater salinity with base nutrients. Addition of nutrients increased endogenous respiration in excised roots of wheat. Protein synthesis was markedly affected by salinity with increase in soluble protein nitrogen (Iyengar, unpub.) Tracer studies with ^{36}Cl and ^{32}P conducted under controlled conditions indicated interacting effects produced by added nutrients on the uptake of ^{36}Cl and in various phosphorus fractions. The P_i utilization to produce organic labile phosphorus compounds was limited by salinity (5).

Salinity Tolerance at Further Growth Stages

Although the seeds germinate in diluted seawater the seedlings do not establish well under saline water irrigations even with the addition of balanced nutrients. This led to the studies on crop

varieties grown with further dilutions after avoiding imposition of
salinity during germination and establishment of seedlings.

Pot experiments were conducted using earthen pots of 35 cm top
diameter and 60 cm height with provision for easy drainage - and
the inner side coated with asphalt. Coastal dune sand with the
following mechanical analysis was used: coarse sand 68%, fine sand
26%, silt and clay 3% and carbonates 3%. Seeds were sown in pots
at appropriate times and irrigated with tapwater for germination
and establishment of the seedlings to 35-45 days growth with
application of nutrient solution whenever required. Seawater of
desired concentration (10,000 and 15,000 ppm) for bajra-babapuri,
wheat-karchia and safflower-NP30 was used till the maturity of the
crops. Depending upon environmental conditions, the interval of
irrigation was varied (from 3-6 days). For each irrigation five
litres of medium per pot was applied. The growth characteristics
and yield were recorded. At the end of the experiments, soil
samples were collected at 0-10, 10-30 and 30-60 cm depths and
accumulation of salts in the substrate was determined (6-8).

From the experiments conducted it has been observed that
characteristic salinity symptoms were evidenced by plants treated
with seawater dilutions. Growth and yield were affected considerably by reduction in height, number of leaves, tillers, branches
and ear heads, ultimately contributing to reduction in yield.
Grains obtained from treated plants exhibited very little change in
chemical composition, viz., total carbohydrates, crude protein and
ash, etc., while heavy accumulation of minerals was seen in leaves
and stem with predominance of sodium. Upon amendment of seawater
with nutritive salts, accumulation of sodium in the vegetative
portions was lessened.

Build-Up of Crop Resistance to Higher Salinity

The total dissolved salts of seawater varies during different
seasons depending upon the dilution taking place in the sea. The
process of acclimatization of species of crops from lower to higher
salinity was attempted with a view to taking advantage of the
fluctuation in salinity of seawater. This process consisted of
treating the plants, year after year, from a lower salinity to
higher salinity, to induce crops to accept even direct irrigation
with seawater. This method of growing plants is similar to pot
experiments described earlier; thus, it was possible to induce
wheat-karchia to accept a salinity of 20,000 ppm - and direct
irrigation of bajra with seawater (24,000 - 35,000 ppm). In all
experiments, from early stages of growth till days 35-45, tapwater
was applied. The reduction in growth and yield was apparent with

very little change in the chemical composition of seeds (Table 1;
9).

Nutrient Needs

The study was initiated with base nutrient solution, Hoaglands
solution, and extended to seawater of 10,000 ppm. It was necessary
to evaluate the efficacy of common fertilizers like ammonium
sulphate, ammonium phosphate, calcium ammonium nitrate (CAN),
potassium sulphate, etc. Good ameliorating effects were produced
by using CAN ammonium phosphate and potassium sulphate. However,
the dose of fertilizer required was one and one-half to two times
higher than the normal application for the crops and had to be
distributed throughout the growing period from seedling stage to
grain setting.

Sand Bed Cultivation

The pot experiments may not generally satisfy the definite
requirements of the crops and many problems are confronted when the
crops are grown under field conditions. To overcome some of the
difficulties that may be encountered in the field, experiments were
conducted on artificially prepared sand beds. During the 'kharif'
(June-September) season (1968 to 1970), bajara-babapuri was sown at
the onset of monsoon and fertilizers like CAN, ammonium phosphate
and potassium sulphate were applied at double the normal NPK level
and distributed in 8-10 split doses. On cessation of rains, dilute
seawater (15,000 - 24,000 ppm) was used to irrigate the crop. The
growth and yield data per plant and the yield for the net area were
collected. Similarly, wheat-karchia and safflower NP30 were grown
during the 'rabi' (October-March) season (1968-70). Varieties of
tapioca, cotton, sorghum, etc. were tried on sand bed with diluted
seawater of 10,000 -20,000 ppm. Cotton was successful with 10,000
ppm having no effect on the quality of lint (10).

To evaluate the mode of application of fertilizers, an
experiment was conducted during the 'kharif' season (1969) with
bajra and in rabi season (1969-70) with wheat-karchia. The modes
of application were: 1) foliar spray (urea as N, KH_2PO_4 as P_2O_5);
2) soil application (CAN, ammonium phosphate and potassium
sulphate); and 3) half soil and the other half as foliar spray. The
dose of fertilizers for all treatments were constant. The last
treatment was greatly beneficial and followed by soil application.
Different sources of nitrogen to wheat-karchia were studied and it
was found that all sources of nitrogen (including CAN) were equally
beneficial under seawater irrigation (15,000 - 20,000 ppm).

TABLE 1. Nutritive Value of Bajra and Wheat Grown With Seawater*

Percentage of constituents present per 100 gram of seeds.

Constituents of the Seeds	Bajara		Wheat		Remarks
	Sea-Water	Normal Sample	Sea-Water	Normal Sample	
Moisture	7.7	12.4	8.6	12.8	The samples were tested
Protein	9.5	11.6	17.3	11.8	by the National Institute
Fat	5.0	5.0	1.5	1.5	of Nutrition (ICMR),
Minerals	2.1	1.3	1.5	1.5	Hyderabad-7.
Fiber	1.2	1.2	2.1	1.2	
Carbohydrates	74.5	67.5	69.0	71.2	
Calorific value	381.0	361.0	359.0	346.0	
Calcium (mg)	51.0	42.0	51.0	41.0	
Phosphorus (mg)	384.0	296.0	248.0	306.0	
Iron (mg)	20.0	5.0	10.3	4.9	
Thiamine (mg)	0.46	0.33	0.52	0.45	
Riboflavin (mg)	0.17	0.25	0.18	0.17	
Nicotiic Acid (mg)	1.8	2.3	4.9	5.5	

* E. R. R. Iyengar and Thomas Kurian. 45th Session, National Academy of Sciences, INDIA

Saurashtra University, Rajkot (1976).

Pot experiments with sesamum, rice, gram and alfalfa did not prove successful, even under very low dilution of seawater (5,000 ppm). Similarly, groundnut and fenugreck tried in sand bed were not successful. An attempt was made to increase salinity tolerance by following a method of presoaking/hardening-treatment of seeds with growth regulators like GA, ascorbic acid, kinetin, etc. Application of the above growth substances at early stages to increase salinity tolerance was also tried. Such treatments did not promote salinity resistance in crops or any improvement of yield.

Field Experiments

Tranformation of laboratory results for practical utilization necessitated conducting field experiments for many years in proper localities of the coastal regions.

An area of two hectares at Hathab (Bhavnagar), obtained on lease from the Forest Department, Government of Gujarat, was cleared, leveled and fenced.

During the rabi season (1970-71), field experiments with an acclimatized variety of wheat-karachia was carried out to assess the fertilzer requirement with seawater irrigation. Seeds were sown in beds of 6 x 2m at a distance of 25 cm from row to row and were irrigated with seawater having 8,000 to 12,000 ppm salts. Two levels each of nitrogen, phosphorus and potash (90 and 120 Kg N, 40 and 60 Kg P_2O_5 and 40 Kg K_2O per hectare) were applied as CAN, super phosphate and potassium sulphate, in eight split doses. There were eight treatment combinations and five replications. The layout of the experiment was a randomised block design. Seawater was applied at intervals of two to four days. Wheat plants could tolerate the diluted seawater irrigation from germination to maturity. From the growth and yield data, it was observed that applications of 90 Kg nitrogen, 60 Kg P_2O_5 and 40 Kg K_2O per hectare gives the maximum yield of grains of 6.25 quintals/hectare (quintal = 100 kg).

Field experiments with bajra-babapuri were carried out during the 'kharif' season (1971) to assess fertilizer requirements with seawater as a supplemental source of irrigation. Sand beds of 6.25 x 4 m were prepared in the field and bajra seeds (Penniselium lyphoides) seeds were used for sowing, at the onset of monsoon, and allowed to germinate and establish as rainfed crop. Two levels each of nitrogen, phosphorus and potash (90 and 120 Kg N, 40 and 60 Kg P_2O_5 and 0 and 40 Kg K_2O per hectare) in the form of ammonium sulphate, ammonium phosphate and potassium sulphate were applied in nine split doses. Since the rainfall was not well distributed,

occasional irrigation with fresh water was also applied. The
plants could grow up to 35 days as rainfed crop; subsequently,
seawater (15,000 - 24,000 ppm) irrigations were given till maturity
of the crop. Depending upon environmental conditions, the plants
received six to eight irrigations with seawater. The bajra plants
exhibited less injury symptoms even after seawater irrigation.

The total amount of grain and fodder obtained was 88.47 Kg and
747 Kg, respectively, for an area of 780 m^2. Though the application of fertilizers did not result in significant difference in the
growth and yield of bajra, maximum yield of 11.72 Kg/100 m^2 was
obtained by applying 120 Kg nitrogen and 60 Kg phosphorus (P_2O_5)
per hectare.

During the 1971 'rabi' season, sugarbeet, safflower and barley
were tried under field conditions to assess the varietal responses
to given salinity levels. The twenty sugarbeet varieties responded
well to seawater up to 20,000 ppm and the varieties Kewi Poly and
USA-7 gave better yields. This was a collaborative project with
the Indian Institute of Sugarcane Research, Lucknow (11).

Diluted seawater increased the yield of spinach by 27% over
control. Greenhouse experiments were performed using 5,000 ppm of
seawater by trickle irrigation after initial establishement of
seedlings with nonsaline water. When the fully grown plants were
explanted to determine the rate of synthesis at various temperatures, from 10 - 40°C, distinct peaks for seawater and control were
obtained (Fig. 1). For seawater the peak was between 18.5 and
20°C, while for the control it was between 22 to 24°C and a gradual
fall noted thereafter to 40°C. It is important to note that
temperature has a prominent role to play in photosynthesis and
respiration of plants under saline conditions (Fig. 2; Iyengar,
unpub.)

Studies made on different crop varieties indicate the possible
initial screening to the level of first approximation and selection, leaving much scope for improvement of yield of tolerant
varieties by breeding experiments.

The seawater tolerant variety of bajra (Pennisetum SWB) has
been crossed with male sterile lines, viz. 23A and 5071A. The F_1
crosses were tested in pots with seawater of 15,000 ppm and the
yield increase over SWB was 11.8% (SWB x 23A) and 10% (SWB x
5071A). Further crosses have been made to obtain F_2 and F_3
generations to assess their yield potential. Another set of 53
varieties was also tested and the yield of SWB bajra is best of the
varieties tested so far. Field Experiments have been extended to
the sub-humid climate of the eastern region in Orissa and 10
quintals/hectare yield have been obtained.

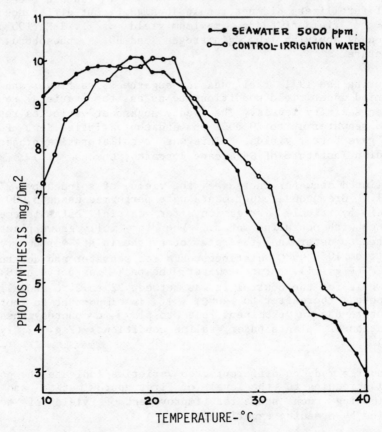

Figure 1.   Effect of Salinity on Photosynthesis at Different
            Temperatures.

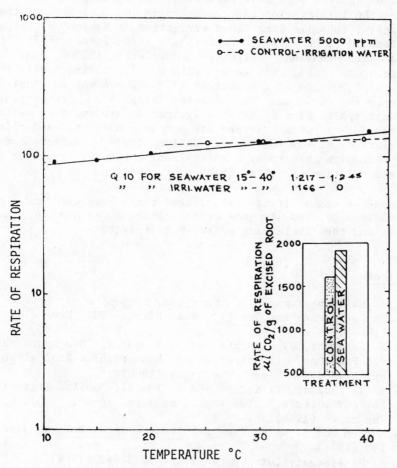

Figure 2.   Dark Respiration of Spinach at Different Temperatures.

The system of irrigation is a constraint in growing crops in sand with seawater. In the Gulf regions silt content is high in seawater (about 4%). However, clear seawater was obtained with a device which consisted of embeding a filtration unit in the shore in the intertidal region so that the waves pass over it and the seawater is filtered and carried to a sump well through the conveyance assembly (12). The clear seawater applied on sand is percolated quickly and heavy flooding is impractical on a large scale as the infiltration rate in sand is 10 times higher than in clay soils, thus limiting flood irrigation. Sprinkler irrigation was attempted to grow wheat Karchia in 1976-77 and 1977-78 (Rabi season) with constant salinity of seawater 15,000 ppm, while Bajra-Babapuri (SWB) was grown with a fluctuating salinity of 18,000 to 24,000 ppm of seawater in 1978. There was no significant difference in yield due to sprinkler irrigation. Experiments in design and application of trickle irrigation are underway with the hope that such systems may provide adequate moisture and maintain salt tolerance at the root zone. Soil analysis of different depths showed that with the proper quantity/application of seawater there is no accumulation of salts at the root zone.

From the above it is established that seawater and coastal dune sand can be used to grow crops, taking advantage of regional rainfall and the fluctuating salinity of seawater.

References

1. E. R. R. Iyengar and M. R. Narayana, Symposium on "Science and Nation during the 3rd Five Year Plan", ASWI, New Delhi (1964).
2. M. R. Narayana, V. C. Mehta and D. S. Datar, "New Approaches to Old Problems" (H. Boyko, ed.), Monographiae Biologicae XVI, pp. 314-321, W. Junk, The Hague (1966).
3. E. R. R. Iyengar, T. Kurian and A. Tewari, "Saline Irriculture for Agriculture," (H. Boyko, ed.) pp. 24-40, W. Junk, The Hague (1968).
4. T. Kurian and E. R. R. Iyengar, Ind. J. Agri. Res. 5(3):145-150 (1971).
5. E. R. R. Iyengar, Curr. Agri. 2(1 & 2):39-42 (1978).
6. E. R. R. Iyengar and T. Kurian, Ind. J. Agri. Res. 5(4):249-255 (1971).
7. J. S. Patolia, T. Kurian, J. B. Pandya and E. R. R. Iyengar, Curr. Agri. 2(1 & 2):35-38 (1978).
8. E. R. R. Iyengar and J. S. Patolia, Sand Dune Res. (Japan) 27(1):34-37 (1980).
9. E. R. R. Iyengar and T. Kurian, Agri. Section, 56th Science Congress, Bombay (1969).

10. E. R. R. Iyengar, J. B. Pandya and J. S. Patolia, Ind. J. Plant Physiol. 21(2):1-6 (1978).
11. E. R. R. Iyengar and J. B. Pandya, Sand Dune Res. (Japan) XIV(1):45-54 (1977).
12. G. L. Natu, D. M. Patel, E. R. R. Iyengar and R. V. Vyas, Ocean Engng. 7:49-53 (1980).

POTENTIAL USES OF MICROALGAI

CHAIRMAN'S REMARKS

S. Aaronson

Department of Biology
Queens College
of the City University of New York
Flushing, NY   11367   USA

The increasing world shortage of fossil fuel and raw materials for industry requires that we examine all potential sources for industrial raw materials and minimize the use of petroleum for industrial raw materials. Microalgae, like metaphytes, may be grown in commercial quantities for use in a variety of industries such as the food, chemical, paint, etc. Unlike metaphytes, all of the microalgal biomass is useful. In the Orient microalgae have been grown on a limited scale in large containers with yields up to 20,000 tons/yr. In the papers which follow, I will describe the potential of microalgae as biomass or as a biomass product of a process necessary to maintain environmental quality. Dr. Dubinsky reviews the use of microalgae - to produce food for aquatic organisms, sewage treatment, biomass for biogas, fertilizers and soil conditioners, or chemicals for industry, i.e., glycerol, carotenoids, vitamins, lipids, amino acids, carbohydrates and volatile substances. Dr. Ben-Amotz describes the production of carotenoids and animal feed protein meal from Dunaliella species while Dr. Mitsui describes the use of selected microalgae, especially blue-green bacteria, as nitrogen fixers and food for marine invertebrates and fish. Dr. White critiques the potential of microalgae to serve as a source for products for industry.

REVIEW OF THE POTENTIAL USES OF MICROALGAE

Z. Dubinsky

Dept. of Life Sciences
Bar-Ilan University
Ramat-Gan, Israel

S. Aaronson

Biology Dept., Queens College
City University of New York
Flushing, New York 11367

SUMMARY

The effectiveness of microalgae in storing solar energy as
reduced carbon compounds is well known. Phytoplankton are also
known to be the main primary producers in marine and many other
aquatic ecosystems. However, so far, microalgae have rarely been
used directly by man.

At present the use of microalgae in aquaculture is increasing,
mostly as food for aquatic organisms such as oysters, shrimp and
fish for direct human consumption -or as food for zooplankton in an
artificial food chain. Such systems, particularly in SE Asia, are
often combined with sewage treatment. Microalgae may indeed be
utilized in sewage treatment in high irradiance regions. In such
photosynthetic sewage ponds, the oxygen evolved by the algae is
utilized by aerobic bacteria, which in turn, mineralize the sewage,
making necessary nutrients such as phosphorus, nitrogen and carbon
available to the algae. The algal biomass can be extracted as a
source of chemicals for industry; glycerol, carotene, vitamins and
lipids, amino acids, carbohydrates, volatile substances, and the
high protein residue fed to poultry. Microalgae are being considered for energy production, either via photosynthetic hydrogen
production or by fermentative production of biogass. Either the
algae directly, or the digested sludge can be used as fertilizers
and soil conditioners. Microalgae may also be used as traps for
toxic or polluting molecules.

The combination of some of the above mentioned applications is
studied for life support systems in space missions. The main

obstacles in the way to widespread use of microalgae seem to be the
costliness of harvest and the lack of sufficient knowledge concerning the economics of the various proposed systems.

INTRODUCTION

We suggest that microalgae may serve as a supplemental source
for useful agricultural products (1-4) at the same time that their
growth does not make demands on land or mineral resources or
require large amounts of scarce or depleted energy supplies that
are needed for conventional agriculture. Until recently, man has
found that microorganisms may supply him with needed products such
as pharmaceuticals, industrial raw materials, etc. not available by
other cheaper methods (5) or humans have produced microbial biomass
in times of war to provide products such as lipids, not available
by other means (6). In the last twelve years, as the world population has increased and food supplies have not kept pace with this
population growth, microorganisms in the form of single cell
protein (SCP) have been examined for their potential in supplying
human food and/or animal feed (7-9). Microalgae were also considered, but never seriously (9), although they can be grown in
bulk as effectively as other microorganisms (4, 10). It is our
purpose to describe some of the ways microalgal biomass may serve
as sources of molecules for human, animal, agricultural, industrial
and other purposes.

Human food or food supplement

Microalgae have served as human food in times of famine (11,
12) and also in times of plenty (Table 1; 13,14). The blue-green
bacterium, Spirulina, may still be eaten in the Lake Chad area of
Africa when other foods are scarce (11, 12) and the freshwater red
algae Lemanea mamillosa is presently eaten, after frying, in India
(13). Microalgae were used as food by the Aztecs in Mexico (15,
16) in the past and macroalgae, and probably microalgae as well,
have served as food in Asia for millenia. In recent years microalgae, along with other microorganisms, have been suggested as SCP
to supplement human and animal foods (7,9). They have proven
unpleasant or toxic for humans but no attempt has been made to
render algal SCP harmless or more palatable because it is presently
not competitive with conventional human foods or food supplements.
Microalgae for human use has suffered from adverse economical
analyses due to the costs of separation and drying (1,2), problems
of digestibility (18-21) and palatability (22 - 25) and public
acceptance (26,27).

TABLE 1. Microalgae, lichens and blue-green bacteria eaten by man.

Organism	Common Name	Country	Reference
Blue-green bacteria			
Nostoc collinum	"Star jelly"	India	Tanaka (162)
Nostoc commune	"Star jelly"	Japan	Tanaka (162)
		Mexico	Ortega (16)
Nostoc flagelliforme	"Earth-hairs"	China	Elenkin (163)
Nostoc verucosum	"Reed falling star"	China	Tanaka (162)
	"Rock flower, rock egg"	Siam	Smith (164)
Phormidium tenue		Japan	Namikawa (165)
Phylloderma sacrum	Suizenji-nori	Japan	Namikawa (165)
Spirulina plantensis	Tecuitlatl, Cocol, Cocolin, Amomoxtle	Aztecs Mexico	Farrar (15) Ortega (16)
Spirulina platensis	Dike, Die, Douhe	Tchad	Brandily (11) Leonard & Compere (12)
Green Algae			
Oedogonium sp.		India	Tiffany (166)
Spirogyra sp.		India	Tiffany (166)
		Burma	Biswas (167)
Stigeoclonium amoenum		Hawaii	Tanaka (162)
Lichens			
Alectoria spp.			Tanaka (162)
Cetralia spp.			" "
Cladonia spp.	Reindeer Moss		" "
Evernia spp.			" "
Gyrophora spp.	Rocktripe	Arctic	Hedrick & Lyon (168)
Parmelia			Tanaka (162)
Umbilicaria			" "

Food in aquaculture

The importance of phytoplankton as the main primary producers in most aquatic foodwebs is well known and they are estimated to contribute about one third of the 10^{18} Kcal annual global gross primary production (28-30). This vast storage of light energy transduced by photosynthesis into potential chemical energy in reduced carbon compounds is accomplished by a mere one-thousandth of the total standing crop of green plants (31). This remarkable difference in performance was attributable to the lack of non-photosynthesizing structures such as trunks, roots, etc., which in terrestrial plants accounted for most of their standing crop (32). This primary productivity provided, either directly in herbivores, or indirectly in carnivores, the food for the 70×10^6 metric tons of aquatic organisms harvested by man, both by fishing from the 'wild' and farmed (Table 2). It seems that only marginal, if any, increases in world fish catch from the wild may be expected in the future. Many previously highly productive areas of the oceans are either fully exploited, or, as in the case of the Peru current and the N.E. Atlantic, have collapsed due to overfishing. This prediction seems reasonable in the light of the stabilization of catches during the last decade despite the addition of vessels and the improvements in equipment (34).

Significant increases in yields in the future will thus depend on artificial stimulation of algal growth, a concept underlying most aquacultural systems, freshwater and marine. As algal growth rates are mostly limited by the supply of nutrients such as nitrogen, phosporus and carbon (35-37), these have to be provided very much as they are in conventional agriculture (38). The nutrients required for complete or semi-synthetic culture media for algae

TABLE 2. Global production of aquatic organisms[*] from the wild and from aquaculture in marine and freshwater environments.[**]

	10^6 tons yr^{-1}	
	Total Catch	Aquaculture
Saltwater (marine and brackish)	60	1 (1.6%)
Freshwater	10	4 (40%)

[*]Exclusive of the over 10^6 tons of seaweed, grown mostly in S.E. Asia or harvested from the wild elsewhere.
[**]Ackefors and Rosen (29) and FAO (33).

(39) are, at least as energy (= money), costly to produce as the commonly used agricultural fertilizers (40-42). Synthetic media are therefore used only for the growth of algae for rearing larvae in the most intensive aquacultural installations where product quality and quantity might justify the expense (43, 44).

An alternative line of research aimed at creating a constant algal bloom to increase productivity of valuable, higher trophic level organisms is that of artificial upwelling. In tropical seas, in spite of favorable temperatures and plentiful sunlight, phyto-plankton productivity is extremely low because the water body is permanently stratified and nutrients are confined to the cool deep waters. At these depths, photosynthesis is limited by a lack of light. At present, two such experimental systems are being tested, one in California (45) aimed at seaweed production for biogas generation and the other in St. Croix (U.S. Virgin Islands). The latter has been in operation since 1972 and has resulted in the expected increase in phytoplankton production of the diatom (Chaetoceros curvisetus) on which various shellfish were success-fully reared (46). In the proposed full scale systems, energy for pumping the water will be obtained by tapping the temperature difference between the surface and deep waters (46, 47).

Ryther and his coworkers at Woods Hole have been studying systems where sewage treatment and mariculture are coupled (48, 49). The plentiful nutrients in the sewage stimulated algal growth in otherwise nonproductive seawater. The cultures, up to outdoor ponds of $150m^2$ (50), were usually diatom dominated, although species composition could not be controlled (51). The algal cul-tures were directly fed to shellfish eliminating the costly step of algal harvesting and this resulted in good growth and high yields.

In S.E. Asia similar concepts, involving various degrees of integration between aquaculture and sewage treatment, have been practiced with great diligence for the last 4,000 years. At present, these systems are being carefully scrutinized (52) and they are being viewed as a major solution to both insanitation (53) and protein shortages in many developing countries (54-57). A demonstration scale, integrated unit consisting of a farmhouse, sewage-oxidation and fish raising ponds and fields is now under construction by Edwards' group at the Asian Institute of Technology in Bangkok. Similarly, animal waste treatment has also been com-bined with algal growth and aquaculture in both small, simple units in Asia (57-59) and larger and more elaborate units elsewhere.

Animal feed supplement

As interest in using microalgae as human food or food supple-ment declined, interest in using microalgae for animal feed increased. Microalgae have been successfully fed to fish, poultry, swine, rabbits, mollusks (See 2 for references), also silkworm larvae and bees (60). A major component of the cost of producing microalgae are the chemicals required for the culture medium. These nutrients (mainly nitrogen, phosphorus and carbon) can be obtained by the algae from sewage instead of the artificual culture medium; the use of sewage for algal growth medium has been demonstrated by Oswald's group in California (61-63), Ryther's group in Massachusetts (48-51) and Shelef's group in Israel (64-67). In the process, algae are produced, sewage is treated, and valuable water is reclaimed. Such a combination greatly reduced the cost of the algae (1-3) but made their use for human nutrition unacceptable because of microbiological safety and aesthetic considerations. These considerations may, however, be lessened by moving the algae one or more links from humans along the food chain.

The protein values of microalgae differ in various test animals (68-70), protein values also depend on the algal species and the processing but were mostly better than that of other plant proteins but below eggs or fish meal. Feeding experiments with broilers gave lower growth rates than those of controls whenever algae replaced more than 25% of the protein sources in the feed mixture. The high carotenoid content of microalgae resulted in deeply pigmented egg yolks (712) which were preferred by most consumers. The orange pigmentation of skin in broilers was less desirable (68,72) and might affect acceptability. Toxicological studies with rats, broilers and carp were favorable with algae grown on mineral media or on domestic sewage. It seemed that in some cases, the digestive system of fish and poultry acted as a very efficient barrier against contamination of the edible muscle tissue by pollutants such as heavy metals found in industrial wastes (73).

Sewage and/or animal waste-grown algae can find their best use by addition to feed mixtures, replacing part of the expensive protein sources such as fish and soybean meal presently in use. Alternatively, the algae could be extracted first for chemicals as proposed in the case of Dunaliella or suggested by Dubinsky et al. (1,3) and the residual cake could then be fed to animals.

It is interesting to note that the only place where microalgae were grown on a large scale with the purpose of being used as food additive, mostly for cattle but also for poultry and in sericulture (60) was the Uzbek Soviet Socialist Republic.

Chemicals

The increasing cost and depletion of fossil fuels (74) and the increasing need of industry for raw materials (75) strongly suggest that new sources for chemicals for industry be examined, even those like microalgae. Microalgae like other microorganisms seemed so unpromising in the past because they appeared to be uneconomic when petroleum was cheap and plentiful. This and the world's supply of organic raw materials for industry seemed unlimited. These idyllic conditions are no longer true; our resources are limited and many of our sources for natural materials are overused and/or destroyed by pollution.

A.   Chemicals for the food industry: Microalgae are rich in proteins (including enzymes), carbohydrates, fatty acids, and vitamins (2) and if grown under sufficiently rigorous conditions may become a source of amino acids, vitamins, antioxidants or coloring, flavoring, thickening, and clarifying agents for the food industry. At present most of these materials are produced by bacteria or fungi grown on "cheap" plant carbohydrates or from spent yeast. It is conceivable that the economic situation may dictate that these carbohydrates be used for other human activities. Then microalgae could be grown in mass culture or immobilized on surfaces, fibers or spheres as other microbial cells (76) and these systems might then be harnessed at the expense of sunlight and minerals to produce the vast array of enzymes, carbohydrates, small organic molecules, etc. that can be released on a large and possibly continuous scale by microorganisms such as microalgae (77; see also next section).

B.   Pharmaceuticals: Microorganisms such as bacteria or fungi have been exploited for almost a century to provide useful drugs, antibiotics, and other pharmacologically active compounds (78,79). Among these are antibiotics against bacteria, fungi, viruses, nutrient supplements (vitamins, amino acids), steroid transformations, removal of allergenic compounds (80); anticlotting, debridement, cardiotonic, hypotensive, antiinflammatory, neuromuscular blocking, smooth-muscle relaxing, antifertility, diabetogenic, hypocholesteremic, and thickening agents (81); animal hormones (82); salivation-inducing agents, enzyme inhibitors, enzymes (79). Microalgae like macroalgae may produce a wide variety of pharmacologically-active compounds. Antibiotics active against bacteria, fungi and even viruses have been isolated from marine algae, especially seaweeds (macroalgae) (83-86). Antibacterial and antifungal agents were also found in microalgae (Table 3). Among the few antibiotic compounds indentified in microalgae were fatty acids, carbohydrates, nucleosides, peptides, terpenes, acrylic acids. The fatty acids in microalgae were often poly-unsaturated and easily oxidized and their antimicrobial activity was probably due primarily to the toxicity of these oxidized fatty acids (86). Little attention has been paid to algae as a source for new antimicrobial agents yet it is pertinent to note that algae in nature often prevent other microorganisms from adhering to their surface

TABLE 3. Microalgae producing antibiotics[*] (See 86 and 185).

	Compound
Prokaryota	
Blue-green bacteria	
Hydrocoleus sp.	Terpene, carbohydrate
Lyngbya majuscula	" "
Trichodesmium erythraeum	" "
Eukaryota	
Bacillariophyceae	
Asterionella notata	Unidentified
" japonica	Nucleosides, fatty acids
Bacillaria paradoxa	Unidentified
Bacteriastrum elegans	Fatty acids
Chaetoceros lauderi	Polysaccharide
" "	Fatty acids
" "	Acid polysaccharide
" peruvianus	Fatty acids
" pseudocurvisetus	Unidentified
" socialis	Fatty acids
Cyclotella nana	Unidentified
Fragillaria prinata	Peptides
Gyrosigma spenceri	Unidentified
Liomophora abbreviata	"
Lithodesmium undulatum	"
Navicula incerta	
Nitzschia longissima	Unidentified
" ascicularis	"
" seriata	"
Rhizosolenia alata	
Skeletonema costatum	Fatty acids
Thalassiosira decipiens	Fatty acids
" nana	"
Thalassiothrix frauenfeldi	Unidentified
Chlorophyceae	
Dunaliella sp.	"
Spirogyra sp.	"
Chrysophyceae	
Stichochrysis immobilis	"
Cryptophyceae	
Hemiselmis sp.	"
Rhodomonas sp.	"
Dinophyceae	
Gonyaulax tamarensis	Terpene, carbohydrates
Prorocentrum micans	" "
Goniodoma sp.	Unidentified
Prymnesiophyceae	
Coccolithus sp.	Terpene, carbohydrates
Isochrysis sp.	" "
Monochrysis (=Pavlova) sp.	" "
Phaeocystis pouchetii	Acrylic acid
Prymnesium parvum	Terpene, carbohydrate

by producing antimicrobial molecules. See reviews by Sieburth (87), Tassigny and Lefevre (88), and Hellebust (89). When microalgae produced blooms in nature, the number of bacteria, especially gram-positive bacteria in the bloom water declined significantly (90) and the bloom-forming algae were found to produce substances that inhibited bacterial growth and the degradation of organic matter in the euphotic zone of a lake (91).

Microalgae, may like macroalgae, produce phycocolloids (agar, carrageenan, alginates, etc.). These compounds were found to have hypocholesteremic properties (92). Folk medicine described several microalgal prescriptions that may be useful. Acetabularia major was used to alleviate the symptoms of gall and other stones (93, 94). Nostoc was used to relieve the symptoms of gout, cancer and fistula (95). Spirogyra and Oedogonium were used to alleviate piles in India (96) and Scenedesmus obliquus extract was more recently used in the postoperative care of the coagulation surface (97) and a S. obliquus paste was useful in the treatment of vaginal and cervical inflammation (98, 99).

Acetylcholine-like molecules were found in the dinoflagellates, Gymnodinium veneficum (100), and acetylcholine, choline o-sulfate and an unidentified choline ester were reported in Amphidinium carteri (101). A norepinephrine-like molecule was found in Noctiluca (102). The green alga, Scenedesmus acutus, contained 10 major and 20 minor volatile amines; among the amines were methylamine, dimethylamine, ethylamine, ethanolamine, putrescine, cadaverine, spermidine, N-(3-aminopropyl)-1,3-diaminopropane, N-(4-amino-butyl)-1,4-diaminobutane, 2-phenylethylamine, tyramine, piperazine and γ-butyrolactam (103). Ethanolamine, putrescine and spermidine were found in a Chlorella sp. (104). Kneifel (82) reviewed the occurrence of amines in macroalgae and microalgae; di- and polyamines were common in most algae and reached their highest concentration in green microalgae. Alkaloids were not common in algae except for saxitoxin and gonyautoxin in dinoflagellates (82). A quaternary amine was found in the colorless dinoflagellate, Crypthecodinium cohnii (105).

C. Carbohydrates: Many microalgae accumulate large quantities of intracellular saccharides and polysaccharides as reserve materials or to compensate for higher external osmotic pressures. A green microalga, Dunaliella, is currently being exploited for the production of glycerol (106). Seaweed (macroalgae) currently supply phycocolloids (polysaccharides such as agar, carrageenan, etc.) for food additives. If the world's supply of seaweed diminishes as the result of over-exploitation and/or pollution, it may become necessary to look to the mass culture of microalgae such as the red alga, Porphyridium, which produces a sulfated galactan (107). Polysaccharides are also used in the petroleum industry; microalgae may provide the long chain polymers with flocculating properties that are needed for oil drilling.

D.   Proteins: Microalgae contain from 46 to 60% of their dry weight as protein and their proteins contain all of the essential amino acids (2). Microalgae are also highly efficient producers of protein in large quantity (4). If these qualities become economically desirable for the production of protein for the chemical (amino acids, proteins) or food industry (amino acids, flavoring, etc.) it is likely that they will be examined seriously by industry. Microalgal protein is not, however, likely to become useful for humans in the near future for the reasons described in an earlier section.

E.   Lipids: Microalgae contain large quantities of fats and oils (Table 4; 108). These are used in a variety of industrial purposes such as the manufacture of surfactants, fatty nitrogen compounds, rubber, surface coatings, grease, textiles, plasticizers, food additives, cosmetics and pharmaceuticals. Algal lipids for industrial use could spare the use of petroleum products which might then be used for energy purposes and permit plant and animal fats to be used mainly for human consumption or cooking use. See Dubinsky et al. (1), Aaronson (2) and Berner et al. (108) for details on algal lipids, their value, and the economics of their production.

F.   Volatile substances: Macroalgae contain a variety of volatile molecules that are mainly responsible for the characteristic odor of seaweeds; among these molecules are thiol compounds, organic acids, alcohols and aldehydes, terpenes, phenols, hydrocarbons and amines (109). Similar molecules are to be found in microalgae (Table 5).

Blue-green bacteria produce geosmin (110,111) and methyl isoborneol was found in Lyngbya sp., (112) both have an earthy odor. β-Cyclocitral, with a tobacco-like odor, was attributed to a bloom of Microcystic wesenbergii (113). Aldehydes were identified in Cryptomonas ovata (114) and Synura petersenii (115); in S. petersenii the aldehydes were identified as acetaldehyde, furfuraldehyde, n-heptanol, acetone, and valeraldehyde. Formaldehyde, acetaldehyde and methylethylketone were found in Chlamydomonas globosa (116). Many microalgae produced a fishy odor with amines (117). Cyanidium caldarium contained methylheptone, geranylacetone, β-ionone, dihydrotrimethyl-naphthalene and butenyllidene-trimethylcyclohexene (118). Diterpenes were found in the marine green alga, Chlorodesmis fastigigiata (119). Chlorella spp. yielded limonene, myrecene and eucalyptol (120). Velev and Sakarijan (121) were able to isolate essential oils from the green

TABLE 4. Total lipids of microalgae in vitro, in nature and in sewage oxidation ponds[*]

Algae	No. of species	Range of total lipids (% dry wt)	Mean total lipid (% dry wt)
Blue-green bacteria	19	1-29	8
Bacillariophyceae	29	2-44	17
Chlorophyceae[**]	46	0-79	17
Chrysophyceae	2	29-35	33
Cryptophyceae	5	1-44	19
Dinophyceae	10	3-36	17
Euglenophyceae	2	10-37	21
Phaeophyceae[***]	17	0-9	3
Prasinophyceae	3	3-21	9
Prymnesiophyceae	10	2-48	21
Rhodophyceae[***]	14	0-14	3
Xanthophyceae	5	2-34	15

[*]Based on various lipid extraction procedures

[**]A few macroalgae included

[***]Mostly macroalgae

TABLE 5. Essential oils or volatiles in microalgae

Organism	Compound	Odor	Ref
Prokaryotes			
Blue-green bacteria	Geosmin	Earthy	110-113
" " "	β-Cyctocitral	Tobacco-like	"
" " "	Alkylsulfides	Putrefactive	"
Eukaryotes			
Many Algae	Amines	Fishy	82
Ankistrodesmus spp.	Proazulenes	Herbaceous	123
Chlorella spp.	"		120
Chlorella vulgaris	Fresh cell concrete	Balsam with	122
Scenedesmus spp.	Mixture	Grass-tobacco	121
Scenedesmus acutus	Fresh cell concrete	Balsam with animal note	122
" "	Dry " "	Woody with decay note	122
" obliquus	Fresh " "	Balsam with medicinal note	122
" "	Dry " "	Woody with oak-moss note	122

alga, Scenedesmus acutus, equivlent to 3% of the dry wt. and containing terpenoid alcohols, carbonyl esters, ethers and phenols;
the mixture had the odor of grass-tobacco with an oak-moss note and
could be used in the Chypre and Fougere compositions. Further
analysis of the S. acutus oil demonstrated n-alkanes from n-
dodecane to n-octocosane, saturated fatty alcohols, diacetone
alcohol, nitrogen compounds (indol, aliphatic amines); free fatty
acids (122). Liersch found that essential oils made up to 0.8% of
Chlorella spp. (120), up to 1.2% of Scenedesmus spp. and up to 4.2%
of Ankistrodesmus spp. (123).

   G. Pigments: All microalgae contain significant amounts of
carotenes and xanthophylls which could satisfy the needs for these
pigments for coloring poultry, eggs, human food, animal feed, carp
and goldfish, and as a dietary supplement for vitamin A (2). These
pigments are presently obtained form plant sources or a fungus,
Blakeslea trispora (6), which produces large quantities of β-
carotene. Microalgae, more specifically Dunaliella spp., may prove
economically useful for example, D. bardawil produces glycerol, β-
carotene and algal meal (124).

   H. Miscellaneous: Microalgae contain plant growth factors
(77) and they have also been used in small quantity to prepare
radioactive biochemicals for research and medicine from isotopically labeled CO_2, water, etc.

Microalgae as a source of useful molecules on a continuous or discontinuous basis

   Microorganisms have proven useful for the production by
secretion or excretion of a variety of large and small organic
molecules for the food and pharmaceutical industries. See Demain
(78) and Woodruff (79) for reviews. No published work is available
that might indicate the usefulness of microalgae in this area. We
suggest, however, that microalgae may be harnessed to produce useful molecules. Microalgae may excrete large quantities of organic
molecules. (See Table XIV in Aaronson et al. (2) for details.)
Among these molecules are small molecules: sugars, nucleic acid
derivatives, cAMP, amino acids, amines, fatty acids, volatiles and
macromolecules: polysaccharides, nucleic acids, peptides, proteins
(including enzymes). See Hellebust (89) and Aaronson (77) for
reviews. Microalgae may be induced to produce large quantities of
extracellular molecules in the same way as other microorganisms but
without the expenditure of expensive natural raw materials and
energy. This production might become continuous with the efflux of
useful molecules and biomass at the expense of inorganic salts and
solar energy. Furthermore, some of this production might be
coupled to domestic wastewater treatment or smokestack efflux of

CO_2 where the saleable end products might be useful biomass, products for industry, and reuseable water for agriculture in arid lands and/or cooling water for industry. Microalgae as other microorganisms may be induced to excrete desired moleucles under a variety of environmental or life cycle manipulations such as stage in life cycle, senescence, nutrient deprivation, chemical or physical stress (77). The production of useful molecules, as in other microorganisms, may be enhanced by the selection of deregulated mutants.

Soil conditioner and fertilizer

Microalgae, more specifically blue-green bacteria, grown in high rate sewage oxidation ponds, have been suggested as a source of fertilizer following their digestion for methane because of their ability to fix nitrogen (125). Microalgae have been found to increase the water-stable aggregation of soil particles and the water-holding capacity of soils (126). Microalgae contain large quantities of water absorbing macromolecules such as polysaccharides and proteins as well as quantities of organic nitrogen and phosphorus easily converted to fertilizers used by plants in the process of decay. It is possible that microalgal biomass derived from high rate sewage oxidation ponds (a product of necessary wastewater treatment and reclamation) may find a market as a soil conditioner and fertilizer as the current conditioners and fertilizers increase in cost, as the need for soil conditioners increases, and as proof is provided that this product does not transmit disease or toxic waste materials.

Sewage treatment

The method of wastewater purification by oxidation lagoons is well known in areas where temperature and light are not limiting and has been successfully used in the past (127, 128). In these lagoons, the incoming organic matter was decomposed by aerobic bacteria which then released the mineralized compounds in forms suitable for uptake by algae. When exposed to light in the shallow lagoons, the algae incorporated carbon, nitrogen and phosphorus-containing molecules, as well as many other nutrients, while increasing their biomass. In the photosynthetic process oxygen was evolved which was utilized by the bacteria, accelerating their catabolic activity (62, 129). Such lagoons required long retention times of up to 90 days to achieve water quality acceptable for disposal into the environment. Long retention times are unacceptable in areas where land is scarce and valuable. High rate oxidation ponds (HROP) as proposed and built by Oswald in California (61,63,130,131) and by Shelef in Israel (64-67) were elaborations

of the same principle. Induced circulation of the sewage in the
ponds prevented the sinking of the algae out of the illuminated
layers of the pond and allowed better nutrient exchange between the
algae and the sewage. In Israel, with pond depths of about 0.4m in
a folded channel design, residence times of 2.5 days in the summer
and 10 days in the winter were achieved. The present operating
systems cover areas of up to 27,000 m^2 (Richmond, CA) (10,65). The
performance of such systems in terms of sewage treatment and of
sanitary engineering criteria of the effluent was satisfactory
(61,63-67,130,131), but we believe that the econimics might be
dramatically improved if the algae produced, rather than treated as
a nuisance in the effluent, could be utilized (1,3).

As traps for toxic or polluting molecules

     Microalgae, as other microorganisms, may prove useful in the
uptake of heavy metals in industrial waste outfalls by accumulating
the toxic metals in their cell bodies in a waste trap and then
being harvested to remove the toxic compound(s) from the fresh or
salt water. Among the metal ions that accumulate as much as
several thousand fold in microalgae are zinc, mercury, cadmium,
coper, uranium, and lead. Microalgae also accumulate pesticides
and other polluting hydrocarbons. This concentrating capability
of microalgae may be useful in "scrubbing" waste waters of industry
or possibly smokestack effluent to remove and concetnrate toxic
materials. The algal product, however, may have no further
economic use unless it concentrates useful amounts of toxic
molecule or can be used as biomass for biogas production. Microalgae may also remove excess nitrate and/or phosphate or sulfite
from domestic industrial or feed lot or paper mill waste water.
This type of microalgal "scrubbing" or organic pollutants may be
coupled with bacterial oxidation in the high rate sewage oxidation
pond which has proven useful for the sewage treatment of domestic
or feed lot wastes (132). The resulting algal biomass may be used
for any of the products mentioned in earlier sections of this
review.

     Microalgae may prove useful in the uptake of heavy metals thus
tending to accumulate these toxic metals in their cell bodies and
removing them from the surrounding fresh or salt waters. Chlorella
fusca absorbed Zn^{2+} (133), and Cd^{2+} (134). Synedra ulna absorbed
Hg^{2+} (135). Anabaena and Ankistrodesmus braunii accumulated Cd^{2+}
and Cu^{2+} (136); as did Scenedesmus obliques (137). Chlorella
pyrenoidosa took up Hg^{2+}. Cd^{2+} and Ni^{2+} (138); uranium was accumulated by several Scenedesmus spp., Chlamydomonas spp., Chlorella
spp., Dunaliella tertiolecta, Synechococcus elongata, Platymonas
sp., and Porphyridium cruentum (139); Pb^{2+} was taken up by
Phaeodactylum tricornutum and Platymonas subcordiformis (140).
Microalgae may also accumulate pesticides and other polluting

hydrocarbons: PCB's (141), toxiphene (142) and methoxychlor (143)
by Chlorella pyrenoidosa;   aldrin and dieldrin   by Anabaena
cylindrica, Anacystis nidulans, Nostoc muscorum (144); and chlori-
nated naphthalenes by Chlorococcum sp. (145).

The exact mechanism of pollutant uptake is not clear and may
vary from agent to agent and species to species.   Heavy metal
uptake (Cu^{2+}) may be faster and more ion accumulates in dead or
metabolically poisoned cells than in live cells (134,146).

This uptake capability may be useful in scrubbing waters free
of pollutants but the accumulated pollutants and the algal biomass
in which they have been concentrated have limited use, if any, for
the pollutants may interfere with biogas production from algal
biomass and if they do not, the pollutants may disseminate into the
environment again.

Space applications

During the peak of space research activity in the USA (147),
as well as in the USSR, microalgae were studied as the central
component of bioregenerative life support systems for prolonged
space missions.   The algal cultures would absorb respiratory CO_2
and evolve O_2 with sunlight acting as the energy source for the
process.   At the same time, nitrogen and phosphorus and other
essential elements present in the human wastes would be taken up by
the multiplying algae. Excess algae would be harvested and used as
a food supplement (147-150).   This idea fell out of favor even
before the prolonged manned space mission concepts were shelved and
was replaced by abiotic life support systems such as those used in
the Apollo flights.   Important reasons for abandoning algae for
space systems were nutritional and fears of the   long term
reliability of biological systems.

Energy

The worldwide energy crisis has reminded us that all kinds of
fossil fuels are ultimately the results of photosynthesis (151).
This realization brought into focus the possiblity of using biomass
as an immediate source of energy (152-157) and new publications
such as Biomass Digest and Biomass Refining Newslettter. This biomass goal might be achieved by "energy farms" of fast growing
trees, utilization of agricultural crop residues and petroleum
plantations of high hydrocarbon plants (158) or by the fermentation
of any biomass to generate alcohol, methane or any other combusti-
ble biogas. Among biomasses studied in that context were seaweeds
and microalgae.   It seemed, however, that these energy solutions
were bound to compete for soil, water and fertilizer with conven-

tional food producing agriculture, at least if they were to have
any real impact on the vast energy needs of a country like the U.S.
For example, to provide 5% of the energy needs of th U.S. in 1990,
an area equal to that of Texas of algal cultures would be required
(37).

The economics of any algal biomass production for energy
generation alone looked unfavorable (1,3,36), a conclusion also
reached in the Dynatech study (159) commissioned by the U.S.
Department of Energy. If algae are produced in a combined facility
for wastewater treatment of the type proposed by Oswald, Shelef or
Ryther, the economics may be quite different, especially in the
light of the current upward trend in oil prices. In the combined
facility, the limiting factor would be the available amount of
human and animal wastes that can be diverted for that purpose (36).

Growth in extreme environments

Microalgae, like other microorganisms, and unlike conventional
plants and animals, may be grown under extreme environmetal conditions (Table 6) including unusually high and low temperatures, high
and low pH, and high salinities. These unusual capabilities of
some microalgae make it possible to grow them in bulk, as for
example Dunaliella spp., in high salinity, with the production of
useful molecules such as glycerol, carotenes and microalgal meal
(106,124). This flexibility is particularly useful in arid areas
of the earth.

Problems

The growth of microalgae on a large scale is not without
problems. Among these are: (a) determining the conditions which
will sustain dominance of the desired microalgal growth over that
of "wild" species under nonsterile field conditions (36); (b)
determining the conditions that will maintain maximal production of
desired moleucle(s) under (a) above; (c) determination of the
biological-engineering conditions that optimize (a) and (c) above
on large field scale projects, i.e. optimal algal densities and
flow rates to prevent sinking, avoid anaerobic conditions, permit
good nutrient supply, and optimize biomass and/or product yield;
(d) finding the most economical harvesting method in energy, time,
and cost. Among the methods examined to date: alum and poly-
electrolyte flocculation, electroflotation, microstraining, filtration, centrifugation, no one is without its economic, biomass, time
or methodological drawbacks (36,160); (e) determining the final
processing which will yield the most useful product or molecules
after harvesting, i.e. sun spray, or steam drying, pelletizing,
extraction procedures.

TABLE 6. Microalgae that grow in extreme environments[*]

Very low temperatures (below 10°C)

Chloromonas pichinchae	Nostoc sp.
Cylindrocystis brebissonii	Prasiola crispa
Fragillaria sublinearis	Raphidonema nivale
Koliella tartrae	

Very high temperatures (over 40°C)

Acanthes exigua	Cyanidium caldarium
Blue-green bacteria	Mougeotia
Chlorella spp.	

Low pH (below pH 3)

Chlamydomonas acidophila	Euglena mutabilis
Chlorella spp.	Zygogonium sp.
Cyanidium caldarium	

High pH (above pH 10)

Blue-green bacteria	Euglena gracilis
Chlorella spp.	Staurastrum pingue

High salt concentration

Chlorella spp.	Dunaliella spp.

[*]See Soeder and Stengel (169) and Kushner (170) for details.

Economics of microalgal biomass and products

The products of microalgal biomass must compete in quality and price with conventional material. Because of our lack of experience with algal products, they must offer significant economic and/or quality benefits to induce the consumer to use them. At present, we think that there is not enough economic return from the production of microalgal biomass for a single product, i.e. protein, lipid, etc., to warrant exploitation at current prices for the product unless that product commands unusually high prices eg. Chlorella as a health food in Japan (161). Coupled, however, to domestic or feedlot wastewater treatment, the production of microalgal biomass produces products such as lipids, defatted algal meal and reutilizable water which together offer economically advantageous products in addition to a necessary service —wastewater treatment. Based on 1978 prices, Dubinsky and coworkers (3) calculated that microalgal biomass would yield a profit only if algal oil and algal meal were sold separately and the value of sewage treatment and reutilizable water was factored into the calculation. In 1980 the price of soybean oil and meal (suggested as price references) increased 50% and 53% respectively, while costs have probably increased about 30% in two years making the net yield from microalgal production more profitable. Thus it apears economically feasible at present to couple the use of microalgae in domestic and feedlot wastewater treatment and the production of useful compounds for industry and animal feeds. This should not, however, be construed to indicate that these are the only uses for microalgae. If the cost of production and the value of the product warrant it, other microalgal products may become competitive on the world market.

ACKNOWLEDGMENTS

This writing was supported by grants from the National Science Foundation No. PFR-7919669 to S.A. and from the National Institutes of Health 5-S05-RR-07064 to Queens College.

REFERENCES

1. A. Dubinsky, T. Berner and S. Aaronson, Biotech. Bioeng. Symp. 8:51-68 (1979).
2. S. Aaronson, T. Berner and Z. Dubinsky, in: "The Production and Use of Microalgal Biomass," G. Shelef, C.J. Soeder and M. Balaban, eds., Elsevier/North Holland Biomedical Press, Amsterdam (1980).
3. Z. Dubinsky, S. Aaronson and T. Berner, Ibid. (1980).

4. S. Aaronson, This volume.
5. M. V. Pape, in: "Microbial Energy Conversion," H. G. Schlegel and J. Barnea, eds., Pergamon Press, Oxford, pp. 515-530 (1977).
6. C. Ratledge, in:"Economic Microbiology," Vol. 2, A. H. Rose, ed., Academic Press, London, pp. 263-302 (1978).
7. R. J. Matales and S. R. Tannenbaum, eds., "Single-Cell Protein I," M. I. T. Press, Cambridge, MA (1968).
8. S. R. Tannenbaum and D. I. C. Wang, eds., "Single-Cell Protein II," M. I. T. Press, Cambridge, MA (1975).
9. J. K. Bhattacharjee, Appl. Microbiol. 13:139-161 (1970).
10. J. C. Goldman, Water Res. 13:1-19 (1979).
11. M-Y. Brandily, Sciences et Avenir 152-516-519 (1959).
12. J. Leonard and P. Compere, Bull. Jard. Bot. Nat. Belg. 37 (Suppl):1-23 (1967).
13. M. Khan, Hydrobiologiya 43:171-175 (1973).
14. H. W. Johnston, Tuatara 22:1-114 (1976).
15. W. V. Farrer, Nature (Lond.) 211:341-342 (1966).
16. M. M. Ortega, Revta Lat.-Am. Microbiol. 14:85-97 (1972).
17. V. R. Young and N. S. Scrimshaw, see ref. 8, pp. 564-568.
18. E. Tamura, H. Baba, A. Tamura, N. Matsuno, Y. Kobatake and K. Morimoto, Ann. Rep. Nat. Inst. Nutr., pp. 31-33 (1958).
19. H. Hayami and K. Shino, Ann. Rept. Nat. Inst. Nutr., Japan (1948).
20. R. O. Matthern, "Survey of Algae studies, under unconventional food research," Publ. U.S. Army, Natick Laboratories, MA (1962).
21. C. I. Waslien, CRC Crit. Rev. Fd. Sci. Nutr. 6:77-151 (1975).
22. H. Tamiya, in: "Proceedings of Symposium on Algology," UNESCO, New Delhi, India, pp. 379-389 (1959).
23. R. C. Powell, G. M. Nevels, and M. E. McDowell, J. Nutr. 75:7-12 (1961).
24. H. D. Payer, "Algae project, Institute of food research and product development," Kasetsart University, and the Ministry of Economic Cooperation, Fed. Rep. Germany, 66 pp. (1971).
25. G. Clement, see ref. 8, pp. 467-475 (1975).
26. W. A. Vincent, Process Biochem. 4:45-47 (1969).
27. E. W. Becker and L. V. Venkatasaman, "Algae Project," Central Food. Tech. Res. Inst., Mysore, India 35 pp. (1978).
28. E. P. Odum, "Fundamentals of Biology," W. B. Saunders, Philadelphia, PA (1971).
29. H. Ackefors and C. G. Rosen, Ambio 7:132-143 (1979).
30. R. H. Whittaker and G. E. Likens, in: "Primary Productivity of the Biosphere," H. Lieth and R. H. Whittaker, eds., Springer, New York, pp. 305-328 (1975).
31. H. W. Johnston, Tuatara 13:90-104 (1965).
32. J. M. Ryther, Organic production by plankton algae and its environmental control, in: "Pymatuning Symposia in Ecology," No. 2, Univ. of Pittsburgh, PA (1960).

33. Culture of algae and seaweeds, Food and Agriculture Organization, Rome, Fish. Rep. 188:34-35 (1976).
34. M. A. Robinson, World fisheries to 2000, Marine Policy (January) 19-32 (1980).
35. W. J. Oswald, in: Proc. IBP/PP Tech. Meeting, Trebon, Centre for Agr. Publ. and Docu., Wageningen, Netherlands (1970).
36. W. J. Oswald and J. R. Benemann, in:"Biological Solar Energy Conversion," A. Mitsui, S. Miyachi, A. San Pietro and S. Tamura, eds., Academic Press, New York, pp. 379-398 (1977).
37. J. C. Goldman and J. H. Ryther, Ibid., (1977).
38. B. Hepher, Limnol. Oceanogr. 7:131-136 (1962).
39. R. Ukeles, in:"Marine Ecology," Vol. III, Part 1, pp. 367-466, O. Kinne, ed., J. Wiley & Sons, London (1976).
40. M. J. Perelman, Environment 14:8-13 (1972).
41. D. Pimentel, L. E. Hurd, A. C. Belliotti, M. J. Forster, I. N. Oka, O. D. Sholes and R. J. Whitman, Science 182:443-449 (1973).
42. D. Pimentel, W. Dritschilo, J. Krummel and J. Kutzman, Science 190:754-761 (1975).
43. J. E. Bardach, J. H. Tyther and W. O. McLarney, "Aquaculture," Wiley Interscience, New York (1972).
44. H. H. Webber, Helgolander wiss Meeresunters, 20:455-463 (1970).
45. W. J. North, in: "Biological Solar Energy Conversion," A. Mitsui, S. Miyachi, A. San Pietro and S. Tamura, eds., pp. 347-362, Academic Press, NY (1977).
46. O. A. Roels, S. Laurence and L. Van Hemelryck, Ocean Management 5:199-210 (1979).
47. R. D. Fuller, Ocean Management 4:241-258 (1978).
48. J. H. Ryther, Science 130:602-608 (1959).
49. J. H. Ryther, W. M. Dunstan, K. R. Tenore and J. E. Huguenin, Bio-Science 22:144-152 (1972).
50. J. C. Goldman and J. H. Ryther, in:"Biological Control of Water Pollution," J. Tourbier and R.W. Pierson, eds., pp. 197-214, Univ. of Pennsylvania Press, Philadelphia, PA (1976).
51. J. C. Goldman and J. H. Ryther, Biotech. Bioeng. 18:1125-1144 (1976).
52. M. G. McGarry and C. Tongkasame, J. Water Pollut. Control Fed. 43:824-835 (1971).
53. J. T. Dale, Ibid., 51:662-665 (1979).
54. F. W. Bell and E.R. Canterbery, "Aquaculture for the Developing Countries," Ballinger Publ. Co., Cambridge, MA (1976).
55. P. Edwards, Joint SCSP/SEAFDEC Workshop on Aquaculture, 2:307-319 (1977).
56. P. Edwards, in: "The Production and Use of Microalgae Biomass," F. Shelef, C. J. Soeder and M. Balaban, eds., Elsevier/North Holland Biomedical Press, Amsterdam (1980).

57. P. Edwards, Aquaculture (1980).
58. N. De Pauw, H. Verlet and L. de Leenheer, in: "The Production and Use of Microalgae Biomass," G. Shelef, C. J. Soeder and M. Balaban, eds., Elsevier/North Holland Biomedical Press, Amsterdam (1980).
59. L. V. Venkataraman, "Rural oriented fresh water cultivation and Production of algae in India," (In Press).
60. A. M. Muzafarov, M. I. Mavlain and T. T. Taubaev, Mikrobiologiya 47:179-184 (1978).
61. H. F. Ludwig, W. J. Oswald, H. B. Gotaas and V. Lynch, Water Pollution Control Federation J. 23:1337-1355 (1951).
62. W. J. Oswald and H. B. Gotaas, Trans. Am. Soc. Civil Eng. 122:73-105 (1957).
63. W. J. Oswald and C. G. Golueke, in: "Algae, Man and Environment," D. F. Jackson, ed., pp. 371-390, Syracuse Univ. Press, Syracuse, N. Y. (1968).
64. G. Shelef, R. Moraine, A. Meydan and E. Sandbank, in: "Microbial Energy Conversion," H. G. Schlegel and J. Barnea, eds., pp. 427-442, Gottingen (1976).
65. G. Shelef, M. Ronen and M. Kremer, Prog. Water Technol. 9:645 (1977).
66. G. Shelef, R. Moraine, T. Berner, A. Levi and G. Oron, in: "Photosynthesis 77: Proceedings of the Fourth International Congress on Photosynthesis," D. O. Ham, J. Coombs, and T. W. Goodman, eds., pp. 657-675, The Biochemical Society, London (1978).
67. R. Moraine, G. Shelef, A. Meydan and A. Levin, Biotechnol. Bioeng. (In Press) (1980).
68. Von O. P. Walz, F. Koch and H. Brune, Tierernahrg. u. Futtermittelkde 35:55-75 (1975).
69. W. Pabst, M. D. Payer, I. Rolle and C. J. Soeder, Fd. Cosmet. Toxicol. 16:249-254 (1978).
70. S. Mokady, S. Yannai, P. Einav and Z. Berk, Nutr. Rep. Intl. 19:383-390 (1979).
71. W. L. Marusich and J. C. Bauernfeind, Poultry Sci. 49:1555-1566 (1970).
72. W. L. Marusich and J. C. Bauernfeind, Ibid., 49:1566-1579 (1970).
73. S. Yannai, S. Mokady, K. Sacks, S. Kantarow and Z. Berk, Nutr. Rep. Intl. 19:391-400 (1979).
74. M. K. Hubbert, in: "The Environment and Ecological Forum, 1970-1971," U.S. Atomic Energy Commission, Oak Ridge, TN (1972).
75. I. S. Shapiro, Science 202:287-289 (1978).
76. H. H .Weetall, Food Prod. Dev. 7:40-52 (1973).
77. S. Aaronson, "Chemical communication at the Microbial Level," CRC Press, Boca Raton, FL (In Press) (1980).
78. A. L. Demain, Biotechnol. Lett. 2:113-118 (1980).

79. H. B. Woodruff, Science 268:1225-1229 (1980).
80. B. J. Abbott, Adv. Appl. Microbiol 20:203-257 (1976).
81. H. W. Matthews and B. F. Wade, Ibid., 21:269-288 (1977).
82. H. Kneifel, in "Marine Algae in Pharmaceutical Science,"
    H. A. Hoppe, T. Levring and Y. Tanaka, eds., pp. 365-401,
    Walter de Gruyter, Berlin (1979).
83. P. R. Burkholder, L. M. Burkholder and L. R. Almodovar, Bot.
    Mar. 2:149-156 (1960).
84. I. S. Hornsey and D. Hide, Br. Phycol. J. 9:353-361 (1974).
85. M. Aubert, J. Aubert, M. Gauthier, see ref. 82, pp. 267-291
    (1979).
86. K. W. Glombitza, see ref. 82, pp. 303-342, (1979).
87. J. McN. Sieburth, in: "Advances in Microbiology of the Sea,"
    M. R. Droop and E. J. F. Wood, eds., pp. 63-94, Academic
    Press, London (1968).
88. M. Tassigny and M. Lefevre, Mitt. Int. Ver. Limnol. 19:26
    (1971).
89. J. A. Hellebust, In: "Algal Physiology and Biochemistry,"
    W. D. P. Stewart, ed., pp. 838-863, University of
    California Press, Berkeley (1974).
90. R. J. Chrost, Acta Microbiol. Polon. 7:125-133 (1973).
91. R. J. Chrost, Ibid., 7:167-176 (1975).
92. G. Michanek, see ref. 82, pp. 203-235 (1979).
93. J. Brachet and S. Bonotto, eds, "Biology of Acetabularia,"
    Academic press, N. Y. (1970).
94. S. Bonotto, Adv. Mar. Biol. 14:123-250 (1976).
95. H. A. Hoppe, see ref. 82, pp. 24-119 (1979).
96. A. Misra and R. Sinha, Ibid, pp. 237-242 (1979).
97. A. Gabik, Cesk. Gynek. 35:4-6 (1970).
98. K. Balak and O. Rydlo, Gynekologie 38:95-96 (1973).
99. K. Balak, O. Rydlo and M. Vojta, Ibid, 39:378-379 (1974).
100. B. C. Abbott and D. Ballantine, J. Mar. Biol. Assoc. U.K.
    36:169-189 (1957).
101. R. F. Taylor, M. Ikawa, J. J. Sasner, Jr., F. P. Thurberg and
    K. K. Anderson, J. Phycol. 10:279-283 (1974).
102. E. Ostlund, Acta Physiol. Scand. 31(Suppl.) 112:1-67 (1954).
103. I. Rolle, R. Payer and C. J. Soeder, Der Spermidingehalt von
    Scenedesmus acutus (276-3a) und Chlorella fusca (211-8b)
    Arch. Mikrobiol. 77:185-195 (1971).
104. M. Steiner and T. Hartmann, Planta 79:113-121 (1968).
105. L. Provasoli and K. Gold, Arch. Mikrobiol. 42:196-203.
106. A. Ben-Amotz and M. Avron, in: "The Production and Use of
    Microalgae Biomass," G. Shelef, C. J. Soeder and M. Balaban,
    eds., Elsevier/North Holland Biomedical Press, Amsterdam
    (1980).
107. J. Ramus, in: "Biogenesis of Plant Cell Wall Polysaccharides,"
    F. Loewus, ed., pp. 333-359, Academic Press, New York (1973).
108. T. Berner, Z. Dubinsky and S. Aaronson, This volume.

109. T. Katayama, in: "Physiology and Biochemistry of Algae," R. A.
     Lewin, ed., pp. 467–473, Academic Press, N. Y. (1962).
110. L. L. Medsker, D. Jenkins and J. F. Thomas, Environ. Sci.
     Technol. 2:461–464 (1968).
111. T. Kikuchi, T. Mimura, K. Harimaya, H. Yano, T. Arimoto, Y.
     Masada and T. Inoue, Chem. Pharm. Bull. a21:2342–2343 (1973).
112. J. A. L. Tabachek, and M. Yurkowski, J. Fish. Res. Bd. Can.
     33:25–35 (1976).
113. F. Juttner, Z. Naturforsch. 31c:491–495 (1976).
114. R. P. Collins and K. Kalnins, J. Protozool. 13:435–437 (1966).
115. R. P. Collins and K. Kalnins, Lloydia. 28:48–52 (1965).
116. R. P. Collins and G. H. Bean, Phycologia 3:55–59 (1963).
117. V. Hermann and F. Juttner, Anal. Biochem. 78:365–373 (1977)
118. F. Juttner, Z. Naturforsch. 34c:186–191 (1979).
119. R. J. Wells and K. D. Barrow, Experientia 35:1544 (1979).
120. R. Liersch, Arch. Microbiol. 107:353–356 (1976).
121. T. K. Velev and A. Sakarijan, Sonderdr. Parfum. Kosmet.
     57:317–320 (1976).
122. T. K. Velev and.G. D. Zolotovich, in: "VI Intl. Congr.
     Essential Oils, Abstr. 23 (1974).
123. R. Liersch, Arch. Microbiol. 108:313–315 (1976).
124. A. Ben-Amotz, This volume.
125. J. R. Benemann, B. Koopman, J. Weissman and W. J. Oswald, in:
     "Microbiol Energy Conversion," H. G. Schlegel and J. Barnea,
     eds., pp. 399–412, Pergamon Press, Oxford (1977).
126. D. Bailey, A. P. Mazurak and J. R. Rosowski, J. Phycol. 9:99–101
     (1973).
127. A. F. Bartsch, J. Water & Pollution Cont. Fed. 33:239–249 (1960).
128. J. E. Svore, in: "Waste Stabilization Lagoons," U.S. Pub. Health
     Service Publ. No. 872 (1961).
129. D. W. hendricks and W. F. Potte, J. Water Pollut. Control Fed.
     46:333–339 (1974).
130. G. Shelef, W. J. Oswald and C. G. Golueke, "Kinetics of algal
     systems in waste treatment," 185 pp., San. Eng. Res. Lab.,
     College of Engineering, U. of California, Berkeley, Report
     No. 68-4 (1968).
131. W. J. Oswald, Solar Energy 15:107–117 (1973).
132. A. S. Watson, ed., "Aquaculture and Algae Culture," Noyes Data
     Corp., Park Ridge, N. J. (1979).
133. H. Wihlidal and E. Broada, Z. Allgem. Mikrobiol. 18:447–451
     (1978).
134. S. Mang and H. W. Tromballa, Z. Pflanzenphysiol. 90:293–302
     (1978).
135. M. Fujita, K. Iwasaki and E. Takabate, Environm. Res. 14:1–13
     (1977).
136. V. Laube, S. Ramamoorthy and D. J. Kushner, Bull. Environm.
     Contam. Toxicol. 21:763–770 (1979).
137. J. R. Cain, D. C. Paschal and C. M. Hayden, Arch. Environm.
     Contam. Toxicol. 9:9–16 (1980).

138. U. Gerhards H. Weller, Z. Pflanzenphysiol. 82:292-300 (1977).
139. T. Sakaguchi, T. Horikoshi and A. Nakajima, J. Ferment. Technol. 56:561-565 (1978).
140. M. Schulz-Baldes and R. A. Lewin, Biol. Bull. 150:118-127 (1976).
141. J. C. Urey, J. C. Kricher and J. M. Boylan, Bull. Environm. Contam. Toxicol. 16:81-85 (1976).
142. D. F. Paris, D. L. Lewis and J. T. Barnett, Ibid., 17:564-572 (1977).
143. D. F. Paris and D. L. Lewis, Ibid., 15:24-32 (1976).
144. C. W. Schauberger and R. B. Wildman, Bull. Environm. Contam. Toxicol. 17:534-541 (1977).
145. G. E. Walsh, K. A. Ainsworth and l. Faas, Ibid., 18:297-302 (1977).
146. T. Horikoshi, A. Nakajima and T. Sakaguchi, Nippon Nogei Kagaku Kaishi 51:583-589 (1977).
147. R. L. Miller and C. H. Ward, "Algal bioregenerative systems," SAM-TR-66-11 U.S. Air Force School of Aerospace Medicine, Brooks Air Force Base, TX (1966).
148. C. G. Golueke, W. J. Oswald and P. H. McGauhey, Sewage Ind. Wastes 31:1125-1142 (1959).
149. W. J. Oswald, C. G. Golueke and D. O. Horning, J. Sanit. Engng., Div. Am. Soc. Civ. Engnrs. 91:SA4, 23-45 (1965).
150. G. Shelef, W. J. Oswald and P. H. McGauhey, Ibid., 96:SA1, 91-110 (1970).
151. T. H. Odum, "Environment, Power and Society," Wiley-Interscience, N. Y. (1970).
152. R. F. Ward, in: "Proceedings of Fuels from Biomass at University of Illinois," J. T. Pfeffer, ed., U. of Illinois Press, Champaigne-Urbana, IL (1977).
153. Anon., "Fuels from Biomass Program," (DOE/ET-0022/1) U.S. Dept. Energy, Washington, D.C. (1978).
154. Anon., "Update: January 1978, Fuels from Biomass Program," (DOE/ET-0022/1) U.S. Dept. of Energy, Washington, D.C. (1978).
155. C. C. Burwell, Science 199:1041-1047 (1978).
156. E. S. Lipinsky, Ibid., 199:644-651 (1978).
157. E. T. Hayes, Science 203:233-239 (1979).
158. M. Calvin, Bioscience 29:533-537 (1979).
159. Anon., "The economics and engineering of large-scale algae biomass systems," Dynatech R/D, M.I.T., Cambridge, MA (1977).
160. J. R. Benemann, J. C. Weisman, B. L. Koopman and W. J. Oswald, Nature (Lond.) 268:19-23 (1977).
161. O. Tsukada, T. Kawahara and S. Miyachi, in: "Biological Solar Energy Conversion," A. Mitsui, S. Miyachi, A. San Pietro and S. Tamura, eds., pp. 363-366, Academic Press, N.Y. (1977).
162. S. Tanaka, "Cyclopedia of Edible Plants of the World," Keigaku Publ. Co., Tokyo (1976).
163. A. A. Elenkin, On some edible freshwater algae, Priroda 20:964-991 (1931); Cited in Johnston (1970).

164. H. M. Smith, J. Siam. Soc. Nat. Hist. Suppl. 9:143 (1933).
     Cited in Johnston (1970).
165. S. Namikawa, Bull. Coll. Agric. Tokyo Univ. 7:123-124 (1906).
     Cited in Johnston (1970) (31).
166. L. H. Tiffany, "Algae, the Grass of Many Waters," C. C. Thomas,
     Springfield, IL (1958).
167. K. Biswas, Sci. and Cult. 19:246-249 (1953) Cited in Johnston
     (31).
168. U. P. Hedrick and J. B. Lyon, eds., "Sturtevant's Notes on
     Edible Plants," Albany, N. Y. (1953).
169. C. Soeder and E. Stengel, in: "Algal Physiology and
     Biochemistry," W. D. P. Stewart, ed., pp. 714-740, U. of
     California Press, Berkeley (1974).
170. D. J. Kushner, ed., "Microbial Life in Extreme Environments,"
     Academic Press, London (1978).

THE POTENTIAL USE OF DUNALIELLA FOR THE PRODUCTION OF GLYCEROL, β-CAROTENE AND HIGH-PROTEIN FEED

Ami Ben Amotz   and   Mordhay Avron

Israel Oceanographic   Biochemistry Department
& Limnological Research   Weizmann Institute of
Tel-Shikmona, P.O.B. 8030,   Science
Haifa, Israel   Rehovot, Israel

SUMMARY

The unicellular wall-less genus Dunaliella contains several species of photoautotrophic green algae which constitute the most extreme halotolerant eukaryotic organisms known. They can adapt to and grow in media containing salt in concentrations ranging from dilute to saturated. The mechanism responsible for this unique halotolerance involves the production and maintenance of intracellular glycerol concentrations which are proportional and close to iso-osmolar with the salt concentration of the growth medium. Cells grown in a medium containing 4 M sodium chloride are composed of about 50% glycerol on a dry weight basis. A species of Dunaliella, recently isolated and named Dunaliella bardawil, accumulates in addition to glycerol up to 8% β-carotene on a dry weight basis. This alga, therefore, provides us with a method whereby solar irradiation, via the process of photosynthesis, is used to drive a reaction in which a useful chemical substance, glycerol, is the major product of the biological photoreduction of carbon dioxide. The remaining dry algal material contains two other important chemical products, β-carotene and protein. The optimization of growth in open ponds is under current investigation to determine whether commercial production of the algae would be economically justified.

INTRODUCTION

Utilization of the photosynthetic machinery for the production of energy, chemicals and food has a particular appeal because it is the most abundant energy-storing and life-supporting process on

earth.   Starting with the basic photosynthetic reaction of converting carbon dioxide and water into organic carbon and oxygen, with
solar irradiation as the energy source, all photosynthetic plants
and algae have a basic biochemical pathway producing a variety of
organic chemicals.   Aquatic plants, including algae, are among the
most efficient converters of radiant energy; photosynthetic efficiencies for microalgae under laboratory conditions with low light
intensity have been reported to be around 20% (1).   Goldman (2, 3)
has recently summarized the theoretical upper limit in the light
conversion efficiency of large-scale algal cultures grown under a
situation in which light is made the growth rate limiting factor.
Maximal yields of 30-40 g dry wt m^{-2} day^{-1} under optimal conditions
are calculated.

The unicellular halophilic alga Dunaliella seems to be an
ideal plant for biological collection and conversion of solar
energy into chemical energy.   Dunaliella has a few specific advantages and lacks some of the drawbacks common to terrestrial plants
in utilizing solar energy for biomass production.   The halophilic
alga does not require fresh water and grows well in a wide range
of saline waters.   Lacking a typical polysaccharide cell wall,
Dunaliella invests a much smaller fraction of the photosynthetic
products in difficult to digest structural constituents than do
most other algae and plants.

Dunaliella can grow in salt water on arid land where there is
maximum availability of solar energy and where the land is not
utilizable for any other kind of potential crop.   It can be grown
in a population density resulting in the presentation of an optimum
absorbing surface area to unit land area ratio throughout the year.
Most significantly, the major photosynthetic product in Dunaliella
is glycerol, the concentration of which varies in direct proportion
to the extracellular salt concentration, reaching a maximum of
around 80% of the algal dry weight (4-6).   The use of Dunaliella
for direct conversion of solar energy into the useful chemical
products glycerol and β-carotene (7) is, therefore, of particular
interest.   The purpose of this manuscript is to examine the basic
features of Dunaliella, and to describe the potential for glycerol
and β-carotene production in outdoor cultures of Dunaliella
bardawil.

The genus Dunaliella contains species whose normal habitats
range from seawater of around 0.4 M NaCl to salt lakes containing
NaCl at concentrations up to saturation (5 M).   Figure 1 illustrates the effect of the salt concentration of the medium on the
growth of two species of Dunaliella; D. salina, a small flagellate
of 50 μm^3, and D. bardawil, a large flagellate of 500 μm^3.   A
remarkable adaptation to a wide range of salt concentrations is
evident in both species.   The levels of maximum numbers of cells

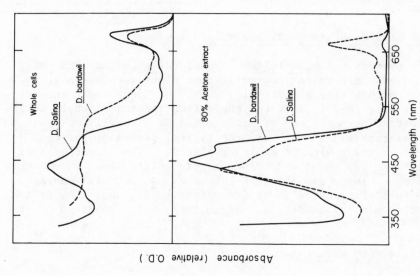

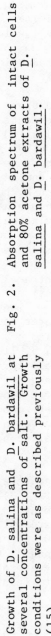

Fig. 2. Absorption spectrum of intact cells and 80% acetone extracts of D. salina and D. bardawil.

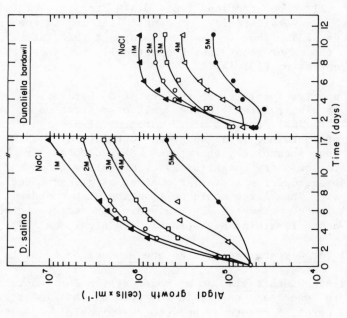

Fig. 1. Growth of D. salina and D. bardawil at several concentrations of salt. Growth conditions were as described previously (15).

per ml of 10^7 for D. salina and 10^6 for D. bardawil represent a
similar biomass production in accordance with the differences in
cell volume.

Production of β-carotene by D. bardawil

A few strains of Dunaliella accumulate large amounts of β-
carotene under certain environmental conditions such as high NaCl
concentration (8), nitrogen deficiency (9) and high light intensities (10). Figure 2 illustrates absorption spectra of whole cells
and 80% acetone extracts of D. salina and D. bardawil. D. salina
contains chlorophylls a and b and the typical prominent carotenoids
found in other green algae and higher plants. The recently isolated D. bardawil (7) shows under appropriate cultivation an unusual
accumulation of large amounts of β-carotene. D. bardawil can accumulate up to 8% β-carotene on a dry weight basis, while D. salina
grown under similar conditions accumulates only about 0.4% β-
carotene

Intracellular Composition

The unique ability of Dunaliella to survive in highly saline
water bodies depends on the photosynthetic production and accumulation of high intracellular concentrations of glycerol (Fig. 3;
4, 5, 11). A linear relation between the concentrations of intracellular glycerol and extracellular salt is maintained over a broad
range of salt concentrations from 0.5 M to 5 M. This has been
observed in several species of Dunaliella and in one species of
Asteromonas (12). Thus it has been proposed that internal glycerol
serves as the major solute which osmotically balances the external
salt concentration (13, 14).

The species-specific capability of D. bardawil to produce β-
carotene is also related to the salt concentration of the medium
(Fig. 3). Maximal amounts of β-carotene of about 8 g per g chlorophyll are obtained in algae grown outdoors at 4.5 M NaCl. D.
bardawil can accumulate, therefore, 8% β-carotene in addition to
50% glycerol on a dry weight basis. D. salina grown under similar
conditions accumulates a similar content of glycerol but only 0.4%
β-carotene. Thus large-scale cultivation of D. bardawil may yield
at least three commercially valuable products; glycerol, β-
carotene and a remaining high protein dry algal meal (7).

Potential for Glycerol and β-Carotene Production by D. bardawil

Utilizing Dunaliella as a model system for converting solar
energy into chemical energy in the form of the industrial chemicals, glycerol, β-carotene and other valuable products, raises
basic questions concerning the evaluation of photosynthetic yield

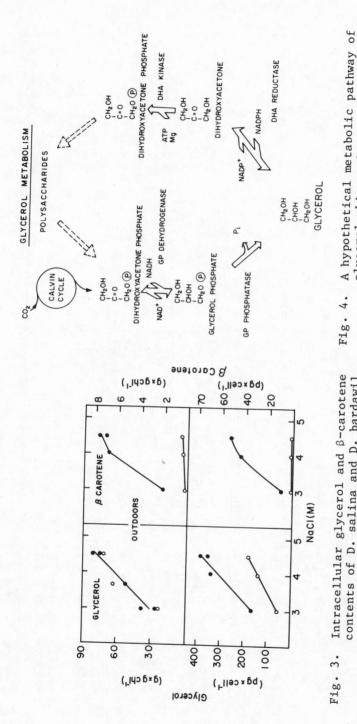

Fig. 4. A hypothetical metabolic pathway of glycerol biosynthesis and degradation in *Dunaliella*.

Fig. 3. Intracellular glycerol and β-carotene contents of *D. salina* and *D. bardawil* maintained at different salt concentrations, outdoors.

o–o *D. salina*; ●–● *D. bardawil*.

TABLE 1.   Potential for Glycerol and β-Carotene Production by Dunaliella bardawil

	Productivity g x m⁻² x day⁻¹	
	Glycerol	β-carotene

1. Theory

a. From Cell Growth Potential

$$\frac{(2 \times 10^9 \text{ cells x liter}^{-1} \text{ x day}^{-1})\ (200 \times 10^{-12}\text{g glycerol x cell}^{-1})}{(200 \text{ liter x m}^{-2})} =$$

	80	—

$$\frac{(2 \times 10^9 \text{ cells x liter}^{-1} \text{ x day}^{-1})\ (33 \times 10^{-12}\text{g carotene x cell}^{-1})}{(200 \text{ liter x m}^{-2})} =$$

	—	13

b. From Available Solar Irradiation

$$\frac{(2 \times 10^3 \text{ K cal x m}^{-2} \text{ x day}^{-1})^a\ (0.08 \text{ stored energy x avail. energy}^{-1})^b}{(0.25 \text{ g cell matter x K cal}^{-1})^c\ (0.4 \text{ g glycerol x g cell matter}^{-1})^d} =$$

	16	—

$$\frac{(2 \times 10^3 \text{ K cal x m}^{-2} \text{ x day}^{-1})\ (0.08 \text{ stored energy x avail. energy}^{-1})}{(0.25 \text{ g cell matter x K cal}^{-1})\ (0.08 \text{ g carotene x g cell matter}^{-1})} =$$

	—	3.2

2. Observed Thus Far

$$\frac{(0.2 \text{ g glycerol x liter}^{-1})\ (100 \text{ liter x m}^{-2})}{(2/5 \text{ harvested x day}^{-1})} =$$

	8	—

$$\frac{(0.015 \text{ g carotene x liter}^{-1})\ (100 \text{ liter x m}^{-2})}{(2/5 \text{ harvested x day}^{-1})} =$$

	—	0.6

[a] Average visible solar energy.
[b] Photosynthetic conversion efficiency.
[c] Heat of combustion of Dunaliella ≅ 4.3 K cal g⁻¹ of volatile material.
[d] Assuming 40% glycerol on a dry weight basis.

limitations in general. Table 1 illustrates calculations and observations of the potential for glycerol and β-carotene production by D. bardawil. The theoretical cell growth potential indicates extrapolation from typical growth rate measurements under saturating light.

Solar energy strikes the earth at a very low flux of about 2000 Kcal m^{-2} day^{-1}. In addition, photosynthetic conversion efficiencies are rather low; thus under ideal conditions the most efficient plant can convert at best about 10% of visible solar irradiation into stored energy or organic matter (2, 3). In reality, photosynthetic conversion efficiencies of natural terrestrial systems are considerable lower and seldom exceed 1-2%, primarily because other factors such as light availability, nutrients, water, etc. are limiting. Clearly, production of glycerol and β-carotene will be limited by the available solar irradiation and by the light conversion efficiency rather than by the cell growth potential. Maximal potential yields of about 16 g glycerol m^{-2} day^{-1} and 3 g β-carotene m^{-2} day^{-1} under sunlight conditions and with a conversion efficiency of 8% can be calculated. In practice, observed productivities of glycerol and β-carotene have been lower and did not exceed 8 g glycerol m^{-2} day^{-1} and 0.6 g β-carotene m^{-2} day^{-1} (7).

Glycerol Metabolism

The biochemical pathway of glycerol synthesis and degradation in Dunaliella is of obvious interest. When the alga grows in high salt concentrations, the endogenous concentration of glycerol in the alga attains several molar. Thus the enzyme activities involved in the pathway modulating glycerol concentration may be expected to be unusually high. Three novel enzymes have been described in Dunaliella; these enzymes have not been described elsewhere and are most likely involved in glycerol metabolism (Figure 4). The first is $NADP^+$-dependent dihydroxyacetone reductase which catalyzes the interconversion of dihydroxyacetone and glycerol (6, 15). The classical NAD^+-dependent glycerol phosphate dehydrogenase has also been described in Dunaliella (16). The second unique enzyme is dihydroxyacetone kinase which is highly specific toward dihydroxyacetone (17), and the third is glycerol-1-phosphatase (18) which specifically dephosphorylates α-glycerolphosphate.

Taking into consideration the presence of these enzymes in Dunaliella, a reasonable hypothetical scheme for the osmoregulatory metabolism of Dunaliella is shown in Figure 4. Glycerol synthesis is suggested to depend on the supply of its carbon skeleton from photosynthetic metabolites of the Calvin cycle or from stored non-osmotically active substances such as starch. Glycerol may accumulate by the production of triose phosphate followed by the reduction to α-glycerolphosphate and dephosphorylation. Conversion of glycerol back to polysaccharides may proceed via oxidation to dihydroxyactone and phosphorylation to dihydroxyacetone phosphate.

References

1. G. Shelef, W. J. Oswald and P. H. Mc Gauhey, J. Sanitary Eng. Div. Am. Soc. Civ. Engrs. 96:SAI, 91-110, (1970).
2. J. C. Goldman, Wat. Res. 13:1-19, (1979).
3. J. C. Goldman, Wat. Res. 13:119-136, (1979).
4. A. Ben-Amotz, and M. Avron, Plant Physiol., 51:875-878, (1973).
5. A. Ben-Amotz and M. Avron, in: "Energetics and Structure of Halophilic Microorganisms, S. R. Caplan and M. Ginzberg, (eds.), Elsevier, Amsterdam, pp. 529-541, (1978).
6. L. J. Borowitzka and A. D. Brown, Arch. Microbiol., 96:37-56, (1974).
7. A. Ben-Amotz, and M. Avron in: Algal Biomass: Production and Use. G. Shelef and C. J. Soeder (eds.), Elsevier, Amsterdam, In Press (1980).
8. N. P. Maysuk, and M. I. Radchenko, Hydrobiol J., 6:40-46 (1970).
9. A. J. Aasen, K. E. Eimhjellen and S. Liaan-Jensen, Acta Chem Scan. 23:2544-2545 (1969).
10. L. A. Loeblich, Dissertation, University of California, San Diego.
11. L. J. Borowitzka, D. S. Kessly and A.D. Brown, Arch. Microbiol. 113:131-138 (1977).
12. A. Ben-Amotz, and M. Avron, in: "Genetic Engineering of Osmoregulation." (D. W. Rains and R. C. Valentine and A. Hollaender, eds), Plenum Press New York, pp. 91-99 (1980).
13. A. Ben-Amotz, in: "Biochemical and Photosynthetic Aspects of Energy Production," (A. San Pietro, ed.), Academic Press, London, New York, pp. 191-207 (1980).
14. A. D. Brown, and L. J. Borowitzka (1979). in: "Biochemistry and Physiology of Protozoa," (M. Levandowsky and S. H. Hunter, eds.), Vol. 1, 2nd Edition, pp. 139-150, Academic Press, New York, (1979).
15. A. Ben-Amotz, A. and M. Avron, Plant Physiol., 53:628-631 (1974).
16. K. Wegmann, Biochim. Biophys. Acta 234:317-323 (1971).
17. H. R. Lerner, I. Sussman and M. Avron, Biochim. Biophys. Acta, 615:1-9 (1980).
18. A. Ben-Amotz, I. Sussman and M. Avron, Experientia, (In Press) (1981).

UTILIZATION OF MARINE BLUE-GREEN ALGAE AND MACROALGAE IN WARM WATER

MARICULTURE

A. Mitsui, R. Murray, B. Entenmann,
K. Miyazawa and E. Polk

School of Marine and Atmospheric Science
University of Miami
Miami, Florida 33149 USA

SUMMARY

A number of marine blue-green algae have been tested as food for brine shrimp and rotifers. Preliminary studies indicate that several strains are an equal or superior food source when compared with algae currently used for cultivation of these animals.

Several marine blue-green algae and marine macroalgae are a good food source in the warm water mariculture of a Tilapia hybrid without using supplemental animal protein. Preliminary study of the nutritional value of marine blue-green algae has shown a wide variation in amino acid composition and total protein.

INTRODUCTION

At the 1st International Workshop on Biosaline Research in 1977, the concept and future direction of a saltwater based bio-solar energy system for food and feed production was discussed (1-3). The present paper briefly describes our recent research with special emphasis on the development of warm seawater cultivation of fish and shrimp using marine photosynthetic organisms as feed. This system is being developed as one part of an overall program for the multiple utilization of marine photosynthetic organisms (Fig. 1).

Isolation and Growth of Marine Blue-Green Algae and Photosynthetic Bacteria

During the past 8 years we have isolated a large variety of blue-green algae (cyanobacteria) from tropical and subtropical marine environments. Included among these are unicellular, colonial, heterocystous and non-heterocystous filaments. Further, the range of sizes in the cultures is also wide, from 1μ single cells to 2 to 5 mm colonies to sheets 2 to 3 cm in length. In addition, we have also isolated many marine photosynthetic bacteria (4). Some of the isolated forms have very high growth rates and many strains show a nitrogen fixing capability. The growth doubling times of Chromatium sp. Miami PBS 1071, using ammonia or molecular nitrogen as the nitrogen source, are 1.75 and 2.00 hours, respectively (5; Ohta and Mitsui, unpub.). This strain also grew at a wide range of salinites (Fig. 2). We also isolated many other rapidly growing strains of photosynthetic bacteria and blue-green algae, with doubling times ranging from 2-15 hours under nitrogen fixing conditins. Included among these are strains which grow only at high salinities, others which grow at normal seawater salinities and many with a wide range of salinity tolerance (6). These high growth rates are of key importance to the successful development of feed production systems.

Concept of Mariculture of Fish, Shrimp and Shell Fish using Photosynthetic Marine Organisms

A program for using these marine photosynthetic organisms in mariculture of fish, shrimp and shellfish has been formulated as shown in Figure 3. The tropical and subtropical algae and bacteria are cultured in seawater and either fed directly to fish and shellfish or fed indirectly to fish, shrimp and lobster through the cultivation of other organisms such as rotifers, brine shrimp and copepods. It should be emphasized that all of these experiments are conducted without supplementing animal protein. Marine algae are used as the sole food source for these animals. Overall there are 4 major advantages to such a system: (1) the principal energy source for the system would be solar radiation; (2) the system would be based on salt water, which is available worldwide; (3) the production would be in temperate to tropical areas which would accelerate growth of the animal under cultivation; and (4) the use of N_2 fixing algae which would eliminate the need for nitrogen enrichment in ponds.

Preliminary Results of Mariculture using Marine Blue-Green Algae

The green algae and some diatoms have been fairly well studied for their uses in aquaculture. Some work involving blue-green

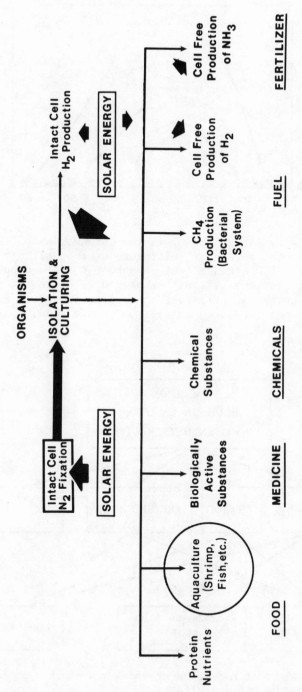

Figure 1. Multiple utilization of marine photosynthetic organisms.

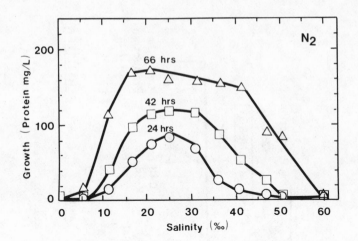

Figure 2.   Salinity for growth of marine Chromatium sp. Miami PBS
            1071 in molecular nitrogen as the nitrogen nutrient
            source (Ohta and Mitsui, unpub.) Culture conditions: pH
            7.5; various salinities as indicated in figure; light
            intensity 150 $\mu E/m^2/sec$.; temperature $28^{\circ}C$. Experiments
            were not conducted in optimum growth conditions.

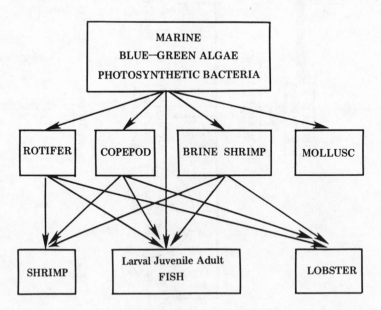

Figure 3.   Use of marine blue-green algae and photosynthetic
            bacteria in aquaculture.

algae has been reported, with favorable results when the algae were used to rear fish (8-11), brine shrimp (12) and rotifers (13). These studies, however, largely involved fresh water algae, and little has been reported on marine blue-green algae.

We are investigating the multiple use of various strains of marine blue-green algae from our collection for the cultivation of a number of animals. Currently, feeding experiments involving brine shrimp, rotifers and Tilapia, a euryhaline fish, are underway.

Brine Shrimp: The brine shrimp, Artemia salina, is used as a food source in numerous ways in aquaculture. The brine shrimp are hatched out of cysts, counted and maintained in small beakers of seawater at 28°C at density of one animal per ml. Unialgal cultures of the algae undergoing investigation are fed twice daily to brine shrimp. After 4 days, surviving animals are counted and measurements made of total body length. We have also tested other algae widely used in the rearing of salt water shrimp larvae, such as Dunaliella sp., Thalassiosira fluviatilis and Skeletonema costatum in control experiments.

Of the 48 marine blue-green algae tested, 4 promote excellent growth (Table 1) when compared to other results using similar conditions (12, 14, 15). The brine shrimp reached a total length of approximately 2mm or more in 4 days, starting from an initial length of 0.4 to 0.5 mm.

Rotifers: Rotifers, Branchionus plicatilis, are used in the rearing of a number of marine fish larvae and have recently been used as a food for various larval stage of salt water shrimp (16, 17).

We have also used rotifers in our algal tests. In this case 10 egg bearing females are placed into 10 ml of water kept at 28°C. They are fed the various algal strains for a 2 to 3 day period and the total number of animals is then counted (Table 2).

A species of marine Chlorella is used as the control in all trials. From this information the instantaneous growth rate and population doubling time are determined. Among 21 different blue-green algal strains tested, two show great promise (BG II6S and BG 231). Several marine blue-green algae are significantly better than the marine Chlorella sp. Fourteen of the blue-green algae supported approximately equal growth, while 4 showed reduced growth of the rotifer population.

Fish: Perhaps the most successful results were obtained by using the algae to feed a hybrid produced by crossing Tilapia

TABLE 1: Growth of <u>Artemia salina</u> on selected marine blue-green
          and other algae (Murray, et al, unpubl.). Marine blue-
          green algal strains which are better and/or equal to
          <u>Skeletenoma costatum</u> are listed. Means are results of 3
          replicates. Survival after 4 days. Initial mean body
          length, 0.4 to 0.5 mm.

Algal Strains	Final mean length (mm)	Mean survival (%)
Miami BG 389	2.36	90
Miami BG LF1	2.07	78
Miami BG 243	1.99	87
<u>Dunaliella</u> sp.	1.97	85
Miami BG 326	1.94	96
Miami BG 543	1.74	63
Miami BG 364	1.64	61
<u>Anacystis</u> sp.	1.62	87
<u>Thalassiosira fluviatilis</u>	1.51	97
Miami BG 339	1.51	63
Miami BG 231	1.49	72
<u>Skeletonema costatum</u>	1.43	87

TABLE 2: Growth of rotifers on blue-green algae and
          <u>Chlorella</u> sp. (Murray, et al, unpub.).
          Means are results of 3 replicates.

Algal Strain	Initial No. of Rotifers	Final No. of Rotifers	Instantaneous growth rate	Doubling Time (days)
Miami BG II6S	10	187	1.46	0.47
Miami BG 231	10	179	1.44	0.48
Miami BG 232	10	127	1.27	0.55
Miami BG 326	10	117	1.23	0.56
Miami BG 315	10	114	1.22	0.57
Miami BG 347	10	110	1.20	0.58
Miami BG 7	10	109	1.19	0.58
<u>Chlorella</u> sp.	10	88	1.09	0.64
Miami BG 339	10	85	1.07	0.65
Miami BG 115S	10	83	1.06	0.65
Miami BG 379	10	81	1.05	0.66
Miami BG II7S	10	77	1.02	0.68
Miami BG 157	10	75	1.01	0.69

mossambica males with T. aurea females. The F_1 progeny are 95%
males, have a wide salinity range and exhibit accelerated growth.
Their growth at $32^o/oo$ salinity equals or exceeds that in fresh
water in fish under 2 grams. Larger fish have survived at salinities ranging up to $70^o/oo$, although their growth rate at higher
salinites was not examined.

Initially, the blue-green algae were presented to fish maintained in aquaria; however, growth was best when the fish were
introduced directly into the carboys used for algal culture. By
using smaller fish, gains in body weight in excess of 300% in 7
days were recorded. This allows a fairly rapid screening of the
algae. The results of feeding experiments in which the algae were
cultured in enriched seawater with either reduced supplemental
nitrogen or no supplemental nitrogen are shown in Figures 4 and 5.
The control feed was a formulated ration consisting largely of fish
meal, soy meal, corn, wheat and vitamin and mineral supplements.
As can be seen, the growth response of the fish fed only certain
strains of blue-green algae was equivalent to those fed a mixture
of animal and vegetable protein. If these algae can support growth
of larger fish and be mass produced at a low cost, they should
serve to lower feeding expense in fish pond operations.

Analysis of Food Value of Marine Blue-Green Algae

Some preliminary work has already been done in an attempt to
identify unusual amounts of proteinaceaous compounds or amino acids
in these blue-green algae. Gas chromatography and amino acid analysis was used to survey whole cell hydrolysates. The amino acid
profiles obtained showed a wide variation among individual algal
strains, for example, of the 11 amino acids identified as essential
for salt water shrimp by other workers (18), 9 are present in these
marine blue-green algae. The other 2 amino acids could not be precisely determined with our methods because of low sensitivity and
poor separation. Analysis of total protein revealed a range of 25
to 60% on a dry weight basis (19).

We propose to make a thorough nutritional study of the best
algal strains after feeding experiments, including total protein,
amino acids, lipids, carbohydrates, vitamins and other substances.

Preliminary Results of Fish Culture with Macroalgae

We have begun this past summer a program of direct feeding of
macroalgae to Tilapia hybrids. A hybrid Tilapia was allowed to
graze on the Rhodophytes, Solieria tenera, Gracilaria sp. and
Agardhiella sp., and Chlorophyte, Ulva lactuca (32% protein) (Table
3). Purina Experimental Marine Chow (25% mixture of plant and

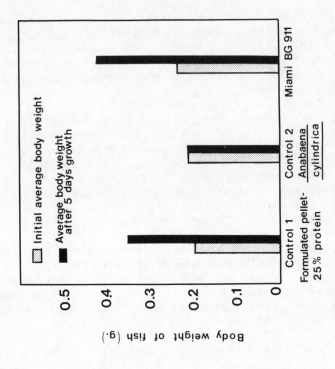

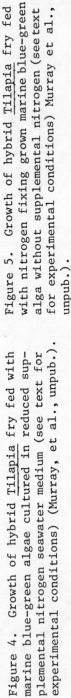

Figure 5. Growth of hybrid Tilapia fry fed with nitrogen fixing grown marine blue-green alga without supplemental nitrogen (see text for experimental conditions) Murray et al., unpub.).

Figure 4. Growth of hybrid Tilapia fry fed with marine blue-green algae cultured in reduced supplemental nitrogen seawater medium (see text for experimental conditions) (Murray, et al., unpub.).

TABLE 3. Growth of fish on macroalgae (Entmann and Mitsui, unpub.).

Algae used	Average weight of fish (g) (Initial)	Average weight of fish (g) (Final)	Cultivation period (days)	No. of fish per tank	Average weight increase per day per fish (g)
Gracilaria and Agardhiella	8.7	21.5	30	10	0.43
Solieria tenera	13.3	22.5	21	19	0.44
Gracilaria and Agardhiella	14.8	36.5	36	10	0.60
Ulva lactura	12.7	17.3	10	30	0.46
Ulva lactuca	47.2	55.0	13	14	0.60

animal protein) was used as a control food. Tests were run with the fish and algae growing together, allowing them maximum exposure to that food source. The Purina ration was fed to the Tilapia daily, averaging about 6% of their body weight per feeding. After several weeks, epiphytes often grew on the sides of the tanks (mainly Enteromorpha). The Tilapia would also browse on these fouling filamentous greens; therefore, experiments were ended before these algae became a significant part of their diets. No fish mortalities ever occurred. It appears that algae-fed Tilapia grew (at worst) about half as well as those fed Purina ration. Fish fed on both the red and green algae exhibited similar growth rates. A feed conversion ratio of 1:1 was achieved with Ulva; that is, 1 g. of Ulva (dry weight) produced a 1 g. increase in the weight of the Tilapia. This is very promising for aquaculture work on a large scale, especially considering the ease and rapidity of growth exhibited by Ulva.

Conclusion

In preliminary studies, marine blue-green algae could be used for mariculture without supplementing animal protein and nuitrients. These experiments are still underway on a laboratory scale. Following a more extensive study on culture conditions, an outdoor study will be made.

The direct feeding experiments using marine macroalgae for Tilapia, without the use of animal protein, were already conducted outdoors. It is hoped that the growth rates of fish fed the algae can be improved through proper environmental control in further studies made on an even larger scale.

Acknowledgements

This work is partly supported by the U.S. National Science Foundation (Grant Nos. AER 75-11171 and PFR 76-17159). Any opinions, findings and conclusions expressed in this publication are those of the authors and do not necessarily reflect the views of the National Science Foundation.

References

1. A. Mitsui, in: "Summary Report, 1st International Workshop on Biosaline Research", (A. San Pietro, ed.) pp. 34-41, Indiana Univ. (1977).

2. A. Mitsui, in: "The Biosaline Concept", (A. Hollaender, J. C.
    Allen, E. Epstein, A. San Pietro and O. Zaborsky, eds.) pp.
    177-215, Plenum Press, New York (1979).

3. A. Mitsui, in: "Microbial Conversion Systems for Food and
    Fodder Production and Waste Management." (T.G. Overmire, ed.)
    pp. 3-31, KISR, UNESCO, UNEP, ICRO and IFIAS (1978).

4. A. Mitsui, Y. Ohta, J. Frank, S. Kumazawa, C. Hill, D. Rosner,
    S. Barciela, J. Greenbaum, L. Haynes, L. Oliva, P. Dalton,
    J. Radway and P. Griffard, in: "Proceedings of the 2nd
    International Conference on Alternative Energy Sources,"
    (T. N. Veziroglu, ed.), Hemisphere Publ. Co., Washington, D.C.
    (In Press).

5. Y. Ohta and A. Mitsui, in: "Proceedings of the VIth
    Internatioal Fermentation Symposium and Vth International
    Symposium on Yeasts," (J. E. Zajic, ed.), Pergamon Press,
    Toronto, (In Press).

6. A. Mitsui, in: "Proceedings of the 5th International Ocean
    Development Conference," IODC Organizing Committee, Tokyo Bi
    29-52 (1978).

7. A. Mitsui, in: "Proceedings of Bio-Energy '80," Bio-Energy
    Council, Washington, D. C. pp. 486-491 (1980)

8. C. F. Hickling, "Fish Culture," Faber and Faber, London (1962).

9. M. R. Ahmad, Hydrobiol. 29:388-392 (1966).

10. J. G. Stanley and J. B. Jones, Aquac. 7:219-223 (1976).

11. H. Durand-Chastel, The Spirulina algae, in: "European Seminar
    on Biological Solar Energy Conversion Systems. (D. O. Hall and
    P. M. Vignais, eds.) Grenoble-Autrans, France (1977).

12. J. Person-Leyruyet, Aquac. 8:157-167 (1976).

13. K. Hirayama, K. Takagi and H. Kimura, Bull. Jap. Soc. Sci.
    Fish., 45(1):11-16 (1979).

14. P. Sorgeloos, Aquac. 1:385-391 (1973).

15. L. V. Sick, Mar. Biol. 35:69-78 (1976).

16. G. H. Theilacker and M. F. McMaster, Mar. Biol. 10:183-188
    (1971).

17. A. P. Scott and S. M. Baynes, Aquac. 14:247-260 (1978).

18. K. L. Shewbart, W. L. Mies and P. D. Ludwig, Mar. Biol.,
    16:64-67 (1972).

19. A. Mitsui, in: "Proceedings of the 3rd International Ocean
    Development Conference," 3:11-30, Seino Printing Co., Tokyo
    (1975).

PROCESSING OF BY-PRODUCTS NECESSARY FOR AN ECONOMIC SYSTEM

Don H. White

Department of Chemical Engineering
University of Arizona
Tucson, AZ   85721   USA

As   a chemical engineer   and as   a discussant for   the session
"Potential Uses of Microalgae" I would like to direct my remarks to
what are the   essential ingredients   of an economic   and successful
biosalinity   project.   This   workshop   on   biosaline   research has
covered topics, such as food and economic plants, potential uses of
microalgae,   and present   and future   applications.   This   implies
eventual industrial,   environmental and other   useful applications,
so it is necessary to consider the engineering and economic aspects
of any given project.   Such considerations   are not only necessary
for any eventual application but also it can accelerate research by
pointing toward promsing projects.

What are the goals of biosaline research?   From the topics and
presentations   at   this workshop,   some goals   are   directed toward
seeking   basic   scientific   knowledge;   however,   most   topics   are
utility-oriented.   It appears   that biosaline research is directed
toward regions that are short of potable and irrigatable waters but
have access to brackish to sea-water levels of salinity.   For such
regions it would appear that biosaline research can be   directed in
one of three   approaches, depending   upon the   specific   region, as
follows:

1.   If water is extremely scarce and capital   for purchase of
outside   commodities not   available, then biosaline   research could
attempt   to provide some   of the   basic food   needs.   However, the
regions   of the World that are   in such difficulties now or   in the
foreseeable future are limited.

2.   Seek   plants   and   microorganisms   that   under   biosalinity

stresses produce products of extremely high value.   Some examples
were presented at this workshop, ranging from carotenes to oysters.

3.   For the majority of situations, where the economics are
marginal, plan from the initial research upon an integrated process
whereby at least two valuable products are recovered and the
remainder of the biomass be used as fuel.   It is difficult for
stressed plants to compete with our "normal" plants, but given
coproducts plus unique arid or biosaline regional needs, many
economic systems should evolve.

One reasonably good engineering system that can be applied to
many biosalinity projects is an integrated system for food/animal
feed/fuel coproducts.   Such products are needed in essentially all
parts of the World. The essential ingredients of such a system are
as follows:

1.   Harvest or process a food product for human consumption.

2.   Recover a valuable animal feed coproduct, or convert some
of the cellulosic components to sugars, proteins and/or ethanol.

3.   Pelletize (densify) the remaiing biomass residue for use
as a solid fuel.

In the short remaining time as a discussant, brief comments
are given to illustrate the above approach.   With respect to the
use of algae, the economics of most recovery systems are marginal.
Oswald's use of lakes as a reactor makes the most sense.   In
addition, since such recovery systems are marginal, two promising
approaches mentioned in this workshoip are as follows:

1.   Utilize the algae to grow fish, thus avoiding the costly
mechanical recovery step.   In some cases, cleaning the algae fish
ponds by circulating the dirty water (fish excreta) to oyster beds
makes sense.

2.   Develop strains of algae ala Dubinsky for use to clean
municipal wastes.   Again, in some instances, the stresses of
municipal wastes may generate valuable pharmaceutical products via
algae.

The same principles should apply to the processing of kelp,
rather than (as some research in progress) convert it to synthetic
natural gas.   First, even the processing of coal to synthetic
natural gas, which is two-fold more economic than the slow
biomethanation processes, is only marginal in cost.   Second, it
makes no sense to degrade purposely the valuable alginic acids of
kelp into a lower value fuel product. Only in remote regions where

no liquid or solid fuels can be obtained, should this approach be taken. For several years now the integrated process approach for food/feed/energy coproducts from biomass has been studied theoretically, technically and experimentally by chemical engineering at the University of Arizona. An integrated packaged process, suitable for small communities and agricultural groups (such as co-ops is under development. Also, a unique approach to an integrated process for coproducts of a cultural protein feedstock, an animal feed and a home-stove burning fuel is under development.

THE POTENTIAL FOR MICROALGAL BIOMASS

S. Aaronson

Department of Biology
Queens College
City University of New York
Flushing, NY 11367   USA

SUMMARY

The potential for obtaining economically significant amounts of microalgal biomass is examined in the light of the following: photosynthetic efficiency of microalgae; experience with large scale production of microbial and microalgal biomass; microalgae as human and animal food; and the economic aspects of microalgal biomass production. Microalgal biomass may soon prove economically useful as a product of a necessary process to maintain environmental quality wastewater treatment, while also providing biomass for animal feed and products for industry.

INTRODUCTION

The limits of the earth's arable lands (1), the continuing need for for agricultural products and/or raw materials for industry (2), the need for agricultural products for animal feed and human food (3), the growth of world population (4), and the increasing cost and depletion of fossil fuels (5), all point to the need for new sources of agricultural products that will not tax the earth's declining agricultural and energy resources. I suggest that microalgae may be (6-8) a supplemental source for useful agricultural products at the same time that their growth does not make demands on land or mineral resources or require large amounts of scarce or depleted energy supplies (6) needed for conventional agriculture. Furthermore, the growth of microalgae in high rate domestic waste sewage oxidation ponds (9), can provide microalgal biomass for industrial materials (6), or biogas generation (9-11) from sunlight and CO_2 evolved during primary sewage oxidation or

231

from industrial activity (12). This type of microalgal growth will
also reduce the eutrophication potential of wastewater and provide
re-utilizable water for agriculture and/or industrial cooling (8).
Microalgae may also be grown on arid lands in the tropics or
subtropics in saline or alkaline waters and at relatively high
temperature to $45^\circ C$ (6); conditions that are not useful for
conventional agriculture. This type of microalgal biomass production may necessitate the additional of minerals (nitrates, ammonia
and phosphate) and CO_2 unless this production is coupled to a
source of these nutrients such as sewage.

Photosynthetic efficiency of microalgae

In terms of photosynthetic efficiency, microalgal yields are
greater than macroalgae and similar to higher plants (Table 1).
Pirt (13) has recently estimated that up to 18% of the solar energy
can be stored in algal cells in contrast to the 6% of higher plants
in conventional agriculture (14).   Photosynthetic efficiencies of
36-46% (reflecting species differences) of the white light used
were claimed for microalgae on continuous culture in the laboratory
(15).   Not only do microalgae appear to be competitive with higher
plants in terms of photosynthetic efficiency but they also appear
to be competitive in terms of biomass produced per unit area as
well (Table 2).

Microalgal biomass

The biomass of microorganisms such as yeast and bacteria has
been produced in large quantity (10-200 x 10^3 tons/yr) in the last
14 years at the expense of organic substrates (Table 3) and at
least one microalga, Chlorella, has been produced in relatively
large quantity (20 x 10^3 tons/yr) on light and CO_2 in Japan.

Unlike higher plants, the microalgal biomass has a uniform
cell content and chemistry as there are no leaves, stems or roots
with their different chemical composition like higher plants.
Microalgal and metaphytan biomass usually have little ash content
(less than 10% dry wt) in contrast to the larger amount of ash (up
to 50%) of macroalgae. Microalgae may be selected for the richness
of their protein, lipid or carbohydrate content (Table 4) and the
use to which their biomass may be put.   The content of major cell
biochemicals may be modified by a variety of environmental (Berner
et al, unpub. data). Microalgae can be grown on a large scale in a
variety of outdoor ponds on different parts of the earth under
varying light and temperature conditions (Table 5).   Golman (17)
recently reviewed the outdoor mass culture of microalgae and
suggested that yields of 15-25 g of dry wt m^{-2} day^{-1} could be
attained for reasonably long periods of time.

TABLE 1. Yields and photosynthetic efficiencies for several
          crops (6).

	Yield (t ha^{-1} y $^{-1}$)	Total photosynthetic efficiency (%)
Theoretical maximum U.S. average (annual)	224	6.6
Microalgae	17-92	0.8 -2.3
Macroalgae	0.8-65	0.04-2.2
Higher Plants	13-112	0.8 -3.2
"Energy Farm" (lumber)	25	

TABLE 2. Yield of Microalgal biomass compared with various crops*

Crops	Total Biomass (tons dry wt. acre^{-1} yr^{-1})
Microalgae**	31
Corn	≈5.
Sugar Beets	4-10
Wheat	≈2
Rice	≈6
Trees	≈5
Alfalfa	≈6
Slash Pine	≈8
Sugar cane	20
Napier grass	19

*Adapted from Weisz and Marshall (16).
**Calculated using yield data for outdoor microalgalculture
   of Table 5: mean = 19g dry wt. m^{-2} day $^{-1}$ as suggested
   by Goldman (17).

TABLE 3. Biomass (Single cell protein) production (1965–1979).

Country	Capacity (Ton/yr)	Micro-organism	Substrate	Ref.
Finland	10,000	Fungi	Sulfite liquor	19
France	20,000	Yeast	Gas Oil	19
Italy	100,000	Yeast	n-Paraffins	19
Japan	60,000	Yeast	Ethanol	19
	20,000	Chlorella	CO_2	18
Rumania	60,000	Yeast	n-Parafins	19
Sweden	10,000	Yeast	Potato starch	18
United Kingdom	4,000	Yeast	n-Parafins	19
	60,000	Bacteria	Methanol	19
United States	10,000	Yeast	Ethanol	19
	10,000	Yeast	Sulfite liquor	18
	10,000	Yeast	Whey	18
USSR	200,000	Yeast	n-Parafins	19
	10,000	Yeast	Sulfite liquor	19

TABLE 4. Range of major biomolecules in microorganisms and conventional foods (see 7 for details)

Organism or Food	Range (% Cell dry wt.)			
	Protein	Carbohydrates	Lipids	Total Nucleotides
Bacteria	47–86	2–36	1–39	1–36
Blue-green bacteria	36–65	8–20	2–13	3–8
Microalgae	46–60	2–7	1–76	3–6
Fungi	13–61	25–69	1–30	5–13
Egg	49	3	45	
Meat muscle	57	2	37	1
Fish	55		38	
Milk	27	38	30	
Corn	10	85	4	
Wheat	14	84	2	
Soy Flour	47	41	7	

Microalgae as human and animal food

Microalgae have served as human food in times of famine (21, 22) and also in times of plenty (23, 24). The blue-green bacterium, Spirulina, may still be eaten in the Lake Chad area of Africa when other foods are scarce (22) and the freshwater red alga, Lemanea mamillosa, is presently eaten, after frying, in India (23). Microalgae were used as food by the Aztecs in Mexico (25, 26) in the past and macroalgae, and probably microalgae as well, have served as food in Asia for millenia.

In recent years microalgae, along with other microorganisms, have been suggested as single-cell protein (SCP) to supplement human and animal foods. They have proven unpleasant or toxic for humans (29) but no attempt has been made to render algal SCP harmless or more palatable because it is presently not competitive with conventional human foods or food supplements. Microalgal SCP has, however, proven useful as an animal feed supplement and microalgae are used extensively as part of the food chain of invertebrate larvae and adults in aquaculture (Dubinsky and Aaronson, this volume). As the price of fish and soybean meal, currently used as a protein supplement in domestic animal feed, continues to rise and world demand for high quality protein mneals continues to grow (Table 6), microalgal SCP may become economically useful especially as it is a cost-saving by-product of a necessary wastewater process (8, 30). Microalgal SCP is sufficient for proper nutrition for it contains adequate to rich amounts of the essential and non-essential amino acids as well as most fat-and water-soluble vitamins needed by animals (7). Microalgae can be grown in large quantity in outdoor ponds or tanks in many climates and environments (Table 5) and the annual protein yield is better than most other protein sources (Table 7). It may be argued that microalgae accumulate toxic materials such as pesticides and heavy metal ions which may render them toxic; these same toxic materials however, accumulate in widely accepted food crops and animal feed materials if they are exposed to air or water containing them (31, 32).

Economic aspects of microalgal biomass production

The value of microalgal biomass as a protein meal may be indicated by the current price of soybean meal which is hovering in the range of $240 to $272/ton (New York Times, October 18, 1980) and the data in Table 8 that the need and value of the world imports of vegetable oil residues has increased significantly (55%) in 4 years (1974-1978) for developed and developing countries.

In the production of microalgal biomass, as in other industries, experience indicates that economics of scale can be

TABLE 5. Yields of microalgae in mass culture*

Alga	Place	Yield Range $(g\ dry\ wt\ m^{-2}\ da^{-1})$
Chlorella	Cambridge, MA	2-11
"	Essen, Germany	4
"	Tokyo, Japan	4
"	" "	16-28
"	" "	14
"	Jerusalem, Israel	12-16
"	" "	27-60
"	Japan	21
"	Taiwan	22
"	"	18-35
"	Rumania	22-36**
Diatoms	Woods Hole, MA	13
"	Fort Pierce, FL	25
"	Woods Hole, MA	10
Microactinium	Richmond, CA	13
"	" "	32
"	" "	12
Phaeodactylum	Plymouth, England	10
Scenedesmus	Tokyo, Japan	14
"	Dortmund, Germany	28
"	Trebon, Czechoslovakia	12-25
"	Tulitz, Poland	12-16
"	Rupite, Rumania	23-30
"	Firebaugh, CA	10-35
"	Bankok, Thailand	15-35
Spirulina	Bankok, Thailand	15-18
"	Mexico City, Mexico	10-20
Selenastrum bibrajanum	Rumania	20-40**

*Adapted from data cited in Goldman (17) except as shown.
**Polesco-Ionaseco (20).

TABLE 6. World* demand and supply for high quality protein meals**

	10^6 tons soy bean meal equivalents		
	1955	1970	1990‡
World demand	21.2	52.6	120
World supply:	20.3	51	120
as Fishmeal	1.9	8.0	15
By Product Meals	11.1	17.4	30
Soy Bean Meal	7.3	25.6	50

*Excluding China.
**Adapted from Taylor and Senior (18).
‡Projected, assuming a rate of population increase of 2.6%/year.

expected. For most chemical industries the cost of production decreases by 25 to 40% as the productive capacity increases; in some industries such as brewing there have been production cost decreases up to 60% (33). Maclennan (34) has argued that a capacity of 10^5 ton/yr in the animal feed grade SCP would be economically practical but this practicality will also depend on the costs and nutritional advantages of specific animal feed SCP for a particular animal industry. Note here that the cost of producing poultry is tightly coupled to the cost of SCP as well as other components of the computer-programmed nutrient mix (35).

The economics of microalgal biomass will also depend on the other uses that are made of this biomass and the state of the world economy. Of the 100 most important organic chemicals (by production volume) there are six which may be synthesized by microbial or chemical synthesis (36). These six are: ethanol, acetic acid, isopropanol, acetone, n-butanol and glycerol. Until recently, the microbial synthesis of these six was less economical than chemical synthesis; citric acid, glutamic acid, vitamin C and vitamin B_{12} are presently made by microbial synthesis (36). As the price of petrochemicals increases and the need for new sources for industrial chemicals also increases (2), it becomes likely that new sources of raw materials will be needed, perhaps involving microbial sources such as microalgae. In a separate paper in this volume, we detail the potential uses of microalgae and some of the problems inherent in obtaining microalgal biomass.

TABLE 7. Protein productivity of microalgae compared with other protein sources.[1]

Protein Source		Protein Yield (Kg dry wt ha^{-1} y^{-1})	Ref.
Microalgae			
Chlorella	(54% protein)	37,449	1
Diatoms	(33% protein)	22,886	1
Scenedesmus	(43% protein)	29,821	1
Spirulina	(57% protein)	39,530	1
Clover leaf		1,680	38
Grass		670	38
Peanuts		470	38
Peas		395	38
Wheat		300	38
Milk from cattle on grassland		100	38
Meat from cattle on grasland		60	38

1. Data compiled from mean yield (19 g dry wt m^{-2} da^{-1}) of microalgae in Table 2 this paper and mean protein content for these algae in Table 3 in Aaronson et al (7).

TABLE 8. Value of vegetable oil residues imports in the world (37)

	Value ($10^9)		
	World	Developed Countries	Developing Countries
1974	2.19	2.04	0.15
1976	2.64	2.42	0.22
1978	3.39	3.02	0.38*

*Estimated

Acknowledgement

The writing of this paper was supported by a grant, PFR-7919669 from the National Science Foundation.

REFERENCES

1. R. H. Whittaker and G. E. Likens, in: "Primary Productivity of the Biosphere, (H. Lieth and R. H. Whittaker, eds.) pp. 305-328, Springer, N. Y. (1975).
2. I. S. Shapiro, Science 202:287-289 (1978).
3. Third World Food Survey, "Freedom from Hunger Campaign Basic Study" (FAO, Rome), Vol. 11 (1963)
4. National Academy of Sciences "Rapid Populations Growth" Vols. 1 and 2, Johns Hopkins Press, Baltimore (1971).
5. M. K. Hubbert, in: "The Environment and Ecological Forum 1970-1971," U.S. Atomic Energy Commission, Oak Ridge, TN (1972).
6. Z. Dubinsky, T. Berner and S. Aaronson, Biotech. Bioeng. Symp. 8:51-68 (1979).
7. S. Aaronson, T. Berner and Z. Dubinsky, in: "The Production and Use of Microalgae Biomass in Algae Biomass," (G. Shelef, C. J. Soeder and M. Balaban, eds.) Elsevier/North-Holland Biomedical Press, Amsterdam (1980).
8. Z. Dubinsky, S. Aaronson and T. Berner, Ibid. (1980).
9. Anon. "Fuels from biomass program." DOE/ET-0022/1, U.S. Dept. of Energy, Washington, D.C. (1978).
10. C. C. Burwell, Science 199:1041-1047 (1978).
11. P. H. Abelson, Ibid., 208:1325 (1980).
12. E. Stengel, Ber. Bot. Ges. 83:589-606 (1970).
13. S. J. Pirt, Biochem. Soc. Trans. 8:479-481 (1980).
14. J. A. Bassham, Science 197:630-638 (1977).
15. S. J. Pirt, Y. K. Lee, A. Richmond and M. Watts Pirt, J. Chem Technol. Biotechnol., 30:25-34 (1980)
16. P. B. Weisz and J. F. Marshall, Science 206:24-29 (1979).
17. J. C. Goldman, Water Res. 13:1-19 (1979).
18. I. J. Taylor and P. J. Senior, Endeavor 2:31-34 (1978).
19. M. Moo-Young, Process Biochem. 11:32-34 (1976).
20. L. Polesco-Ionasesco, Acta Bot. Horti. Bucur. pp. 183-190 (1974).
21. M-Y. Bandily, Sciences et Avenir 152:516-519 (1959).
22. J. Leonard and P. Compere, Bull Jard. Bot. Nat. Belg. (Suppl.) 37:1-23 (1967).
23. M. Khan, Hydrobiologiya 43:171-175 (1973).
24. H. W. Johnston, Tuatara 22:1-114 (1976)
25. W. V. Farrer, Nature (London) 211:341-342 (1966).
26. M. M. Ortega, Revta Lat.-Am. Microbiol 14:85-97 (1972).

27. R. J. Matales and S. R. Tannenbaum, "Single-Cell Protein,"
    M.I.T. Press, Cambridge, MA (1968).
28. J. K. Bhattacharjee, Appl. Microbiol. 13:139-161 (1970).
29. V. R. Young and N. S. Scrimshaw, "Single-Cell Protein II,"
    (S. R. Tannenbaum and D. I. C. Wang, eds.) M.I.T. Press,
    Camabridge, MA (1975).
30. R. Moraine, G. Shelef, A. Meydan and A. Levin, Biotech.
    Bioeng. (In Press).
31. Anon. "Environmental Quality and Safety," pp. 17-34, Vol. 3,
    (F. Coulston and F. Korte, eds.) G. Thieme Publishers,
    Stuttgart (1974).
32. M. M. Chaudry, A. I. Nelson and E. G. Perkins, J. Am. Oil
    Chem. Soc. 55:851-853 (1978)
33. A. T. Bull, D. C. Ellwood and C. Ratledge, Symp. Soc. Gen
    Microbiol. 29:1-28 (1979).
34. D. G. MacLennon, in: "Continuous Culture 6, Applications and
    New Fields, (A. R. Dean, D. C. Ellwood, C.G.T. Evans and J.
    Melling, eds.) p. 69, Ellis Horwood Ltd., Chichester,
    (1976).
35. S. R. L. Smith, Phil. Trans. R. Soc. 290:341-354 (1980).
36. M. V. Pape, in: "Microbial Energy Conversion, (H. G. Schlegel
    and J. Barnea, eds.) pp. 515-530, Pergamon Press, Oxford
    (1976).
37. 1798 Yearbook of International Trade Statististics,
    United Nations, N. Y. (1979).
38. W. A. Vincent, Process Biochem. 4:45-47 (1969).

STRESS BIOLOGY

CHAIRMAN'S REMARKS

Richard C. Staples

Boyce Thompson Institute for Plant Research
Tower Road
Ithaca, N. Y. 14853 USA

Halophilic bacteria make salt domes pink and the ocean waves glow.   Mangrove swamps dot the seashore, and beets grow better on salt.   Saline-tolerant organisms grow in a variety of ecological niches, and their strategies of adaptation reflect their place. Research on the various mechanisms of tolerance furnishes valuable clues for developing microorganisms and crop plants capable of producing food and chemicals where none could be grown before.

A great deal of information is available concerning microorganisms that are halophilic.   Among the extreme halophiles are the Halobacteriaceae which grow and survive only in brines containing 18-35% NaCl.   The halobacteria are obligately aerobic, gram--negative rods, and Dr. J. K. Lanyi discusses several ion pumps located in their membranes which provide the bacterium with a unique capability for survival in an unusually demanding environment.

Halophilic plants grow directly either in ocean water or in tidal marsh environments where the ocean salt is partially diluted. Studies on the adaptive mechanisms of halophytic plants have suggested ways that mesophytic plants might be developed for saline tolerance.   It has yet to be shown that these mechanisms, quite different from those by which mesophytes cope with salt, can operate to advantage in the genetic background of a crop plant.

The most conspicuous effect of salinity on mesophytic plants is growth retardation.   While crop plants are vulnerable to saline conditions in several ways, we do not understand how excess salts cause some of the injuries.   The complex nature of higher .plants with their interrelated parts too easily suffers a cascade of secondary effects, and we need to know how cells and tissues, as well as whole plants, respond to salt stress. Dr. A. Poljakoff-Mayber discusses some of the biochemical effects of salinity stress on higher plants while Dr. D. W. Rains evaluates the response of single cells and tissue cultures of higher plants under salinity stress.

If plants are to be bred for salinity stress, the characteristics which contribute to salinity tolerance must be identified. And the various environmental situations for which tolerance is sought must also be clearly defined. Thus crops might be irrigated with the moderately saline water so readily available in desert areas, or ocean water might be used along sandy seacoasts. Yet plants bred for these situations where NaCl is the predominant form of salt would be unacceptable for crops to be grown on saline or sodic soils where other ions predominate. Dr. M. C. Shannon discusses some of the many problems involved in breeding plants tolerant to salt.

Finally, the complex results derived from research must be assembled to provide novel solutions for crop production under saline conditions. Dr. J. Gale offers a provocative design for a protected environment and desalinated irrigation system usable in the Negev Desert. Clearly adapting plants to the desert is an economic as well as technical challenge.

BIOCHEMICAL AND PHYSIOLOGICAL RESPONSES OF HIGHER PLANTS TO SALINITY STRESS

Alexandra Poljakoff-Mayber

Department of Botany
The Hebrew University of Jerusalem
Jerusalem,
ISRAEL

SUMMARY

The interest in the biochemical and physiological effects of salinity on plants is twofold: (1) practical - prevention of damage. Even low substrate salinity sauses stunting and inhibition of growth, resulting in poor yield of cultivated crops; and (2) basic scientific - a better understanding of the nature of salinity damage and of adaptations for salt tolerance may provide a better understanding of some basic life phenomena, e.g. the flexibility of, and the interrelationships between, various control mechanisms functioning in the plant, or the allocation of energy generated by respiration between energy consuming processes.

Salinity may inhibit growth either through disturbances in the water balance and reduction of turgor or through depletion of energy required for the metabolism involved in growth, or both. These disturbances may result either from difficulties in water uptake and transport within the plant or from toxic effects caused by excess of mineral ions in the tissues.

Salinity affects respiratory enzymes and in some plants salinity induces a shift in the respiratory pathways from the EMP to the PPP. Inhibition as well as stimulation of oxygen consumption has been reported. The amount of phophorylated compounds decreases due to salinity, and depression of oxidative phosphorylation has been reported. Phosphorylation proper may be affected or ATPase activity, or some other ATP cleaving mechanism, might be stimulated by salinty. The changes in ATPase activity are linked, aparently, with ion uptake and thus with osmotic adaptation.

Osmotic adaptation is linked also to nitrogen metabolism and amino acid balance. In many cases, but not always, salinity induces an increase in total amino acids and accumulation of proline and glycinebetaine.

It is not clear which of the changes observed represent salinity damage and which of them have a favorable, adaptive value. Comparison between the effects of salinity on halophytes and glycophytes does not provide an answer.

Hormones play, apparently, an important role in regulating the responses to salinity, and the level of at least two of them - ABA and cytokinins - changes drastically with increasing salinity. However, the information available does not point to clear cut relationships between these changing hormone ratios and changes in the physiology and behavior of the plant.

The existing methods for studying the intracellular distribution of various solutes cannot give clear answers, as homogenization and extraction result in the mixture of cellular contents. Development of methods which will permit one to differentiate between the contents of the different cellular compartments will enable a more rational interpretation of the data already available concerning the osmotic adaptation and the distribution of metabolic intermediates. Other relevant lines of investigation are proposed.

INTRODUCTION

Botanists and agriculturists, from age long observations on natural plant populations and on behavior of agricultural crops, have learned to estimate the range of change in the external environmental conditions that a certain plant, or a population of plants, can cope with. These conditions are broadly defined as "normal." It is expected that under these conditions a plant will be able to germinate, grow and reach a reasonable size, and be able to flower and to produce fruits and seeds. If, however, something happens contrary to expectations, e.g., germination is delayed, the stand is uneven, growth is stunted, or there is little or no yield, something is considered to be wrong in the environment and the plants are considered to be under "stress." Stress apparently interferes with the normal biochemical and physiological processes in the seed or in the plant, so that some abnormality develops which expresses itself externally, as noted above (1).

Development of conditions of stress in nature is a normal event which occurs periodically; heat waves, drought, mineral deficiency, disease epidemics and salinity in the soil have been

known to man for centuries. Yet following each of these calami-
ties, some plants survive the period of stress. Usually some
individual plants show some sort of immunity; these plants are
presumed to be able to develop defense mechanisms against the
harmful effects of stress.

Observations on the various biotopes in different areas of the
world show that there are plant populations characteristic for a
given biotype. Certain morphological, anatomical and physiological
features are typical for the population in each biotope. Plants
are therefore described as hydrophytes, or xerophytes, as halophytes or glycophytes, etc., definitions which in fact symbolize
the adaptations of the plants in order to cope with the special
conditions of the biotope.

Soil salinity has long been a big problem for mankind. It is
generally agreed that many socieites have salted themselves out of
their habitats by mismanaging agricultural practice. In modern
times the development of the medical sciences and social care have
resulted in a tremendous growth of population, accompanied by
problems of food supply. With the development of soil science and
the extended use of modern agricultural technology, man must meet
the challenge of increasing food production and of using for
agriculture, areas and sources of water formerly considered unsuitable for the purpose. Much has been achieved due to the development of various agrotechnical practices. Attempts are also being
made to develop salt tolerant varieties of various crops. To
succeed in this it is necessary to understand what is the actual
damage caused to plants by salinity and what are the mechanisms by
which individual plants, certain cultivars or plant communities
withstand the damage of salinity and develop normally. These
problems are nowadays under thorough investigation from both the
agricultural and physiological points of view.

The practical approach to the research of the effects of
salinity is accompanied by a genuine scientific curiosity to
understand the process of life in its endless variability of forms
and possibilities.

The problem of salinity and its effect on plants has been
reviewed from many standpoints (2-12). Reviews dealing with
drought tolerance or desiccation are also relevant to the problems
of salinity (13-19). This paper is not intended as a review but
rather as a lead to discussion; I will therefore permit myself to
cite literature selectively, in order to illustrate the points I
want to raise. I shall, also, permit myself to rely mainly on the
contents of recent reviews which already cite most of the relevant
literature.

Nature of the damage caused by salinity stress

The damage caused to the plant by salinity can be osmotic, toxic or nutritional (20). The osmotic effect resulting from soil salinity may cause distrubances in the water balance of the plant, reduction of turgor and through this inhibition of growth, either directly or through effect on other processes such as stomatal closure and reduction of photosynthesis. Counteracting this osmotic effect, by osmotic adaptation of the plant, involves excessive accumulation of ions, or synthesis of other osmotica, and thus is difficult to differentiate from the toxic effect of the accumulating ions. However, the toxic effect may be partly separated from the osmotic effect, experimentally, by use of external non-ionic osmotica (21; and literature cited therein).

The toxic effect, caused either by high total ionic strength or by specific ions, may affect various aspects of metabolism (4, 6), hormonal balance (22, 23) and structural development (24).

Solute accumulation in plant cells has a general nutritional aspect (25). However, inorganic solute accumulation may induce an internal imbalance between the various nutritionally obligatory ions as manifested by deficiency signs. Metabolically these effects may be considered toxic in nature.

The recognition of the importance of turgor for extention of growth (14, 15; and literature cited therein) naturally resulted in a thorough investigation of osmotic adaptation of plants under salinity or water stress (see literature cited in 9, 10, 12). However, the transport of the absorbed ions within the plant or the tissue is also of great importance. This is important for the osmotic adaptation of the shoot (12) but it is apparently equally important for regulation of the ionic content in the successive layers of a tissue such as the root cortex. Electron microscopic studies of Nir and Poljakoff-Mayber (24) showed distinct changes occurring in epidermal and subepidermal layers of roots of Tamarix exposed to high salinities. At the time no interpretation was offered. I would now like to suggest that the outer layer "protect" the inner ones by accumulating ions and delaying their movement inwards, until the death of the cell (salting out of the cytoplasm?). When the structure of the outer cells disintegrates the next layer, inwards, takes over the "protective" role. This is noticeable by the very distinct structural changes in the density of the cytoplasm and organization of the nucleus. These changes, death of cells and structural modifications of the nucleus, progress inwards with length of exposure to the high salinity, although the cuttings survive and their leaves show no loss of turgor.

In addition to osmoregulation and transport the Na/Ca ratio is regarded by Hyder and Greenway (26) and Greenway and Munns (12) as being of great importance. This ratio is apparently an important factor in regulating membrane permeability and thus ion uptake, translocation and compartmentation.

It seems, therefore, that the damage caused to higher plants by salinity is manifested in many ways and affects water and hormonal balance and metabolism, and also induces structural changes. A comparison of the different effects of salinity on glycophytes and halophytes gave us the impression that there is some similarity in the repsonses of the two groups of plants. But in halophytes most of the typical effects appear at much higher salinities than in glycophytes. Such high salinities are above the concentrations causing maximal growth rates and should, in fact, be regarded as stress conditions even for halophytes.

Osmotic adaptation and compartmentation

Under natural conditions, with progressive drying of the soil, the water potential of the root medium decreases. The plant usually responds by decreasing its water potential. This is usually considered as an osmotic adaptation which permits the maintenance of the potential gradient for water uptake and of a positive turgor. Maintenance of turgor is an obligatory condition for growth of the plant. There is apparently a critical level of turgor pressure below which extension growth is impossible (13). It follows, therefore, that growth stops before the turgor is reduced to zero. Any reduction in turgor due to salinity, even if temporary, is liable to induce growth inhibition (27). Osmotic adaptation allows, therefore, maintenance of turgor so that a decrease in the water potential of the tissue is not necessarily accompanied by a decrease in the relative water content (28-31). The water potential of leaves at zero turgor differs in plants differing in tolerance (32) as does the change of growth extension per unit in turgor(33).

Osmotic adaptation, under salinity, is usually achieved by ion uptake from the external solution or by internal synthesis of organic solutes. It is usually the halophytes which maintain the gradient for water uptake by ion accumulation (9, 34). The inability of some glycophytes to adapt osmotically under saline conditions, is sometimes ascribed to their inability to accumulate enough ions from the external surroundings (12). Sodium and potassium are the two main cations responsible for osmotic adaptation. A high concentration of these ions strongly inhibits most enzymatic reactions (4, 11). However, as enzymatic reactions do continue in the cytoplasm, it was hypothesized that compartmentation of ions inside the cell will prevent the exposure of the

enzymes to their toxic effect. Most of the sodium is apparently
located in the vacuole, which constitutes over 80% of the cell
volume, while potassium is more abundant in the cytoplasm (35, 36).
However, osmoregulation in the cytoplasm is believed to be achieved
mainly by increases in the concentration of organic solutes,
especially proline and glycinebetaine (11, 21, 37-41).

Most of the data on the distribution of ions and organic
solutes in the different compartments of the cell were derived from
indirect evidence based on compartment analysis, on the ratio
between highly vacuolated and slightly vacuolated tissue, etc. (35,
36). Lerner (42) has recently shown that treatment of thin slices
of plant tissue with polycations can induce pores, selectively, in
the plasmalemma and thus the cytoplasmic compartment is leaked of
solutes whose Stoke's radius is smaller than NAD, but larger than
triosephosphates. Measuring the sodium and potassium leaking out
from red beet slices he reported $[Na^+]$ of 62 mM and $[K^+]$ of 34 mM
($Na/K = 2$) in the cytosolic compartment, and $[Na^+] \simeq 124$ mM and $[K^+]$
of 4.8 mM ($Na/K \simeq 25$) in the vacuoles. The vacuole compartment was
emptied out by inducing total leakage of the tissue with toluene
treatment (43, 44). Further refinement of this method will enable
direct measurement and identification of small molecular weight
solutes present in the cytosolar and vacuolar compartments.

Preliminary histochemical localization of sodium in the cell
using the pyroantimonate method (45) showed large amounts of
precipitate in the cytoplasm, mainly attached to membranes (Fig. 1)
while some of it was located extracytoplasmically, in the cell wall
(Fig. 2). Localization of chloride by silver acetate precipitation
showed a similar distribution but the precipitate of chloride in
the cell wall was much more abundant than that of sodium. Neeman's
data for chloride localization are practically identical to those
of Stelzer and Lauchli (46). Neeman's findings for sodium and
chloride localization were very similar for maize roots and
coleoptiles and for roots of Tamarix and Atriplex. These results
and those of Lerner on the red beet storage tissue seem to indicate
that considerable amounts of sodium are found in the cytoplasm
although the Na/K ratio in cytoplasm is much lower than in the
vacuole. If the sodium is immobilized somehow on certain membranes
in the cytoplasm and not on others, this also may be regarded as a
form of compartmentation.

Osmotic adaptation of plants to decreasing external water
potentials or even to atmospheric drought is therefore achieved by
accumulation of osmotica and lowering of the internal water
potential. This apparently facilitiates water uptake by the plant
cells and maintenance of turgor. However, growth of these plants
is stunted, in spite of the osmotic adaptation, especially under
saline conditions. This growth inhibition is often explained as

being due to: (a) utilization of the photosynthates not for growth but for osmoregulation; (b) diversion of part of the energy derived by respiration to synthesis of the organic osmotica or to the maintenance of the ion uptake mechanisms, or to damage repair needs instead of to the usual cellular events; (c) damage to the enzymic protein exposed to the low water potentials and to the relatively high ionic strengths; and (d) partial closure of stomates and hence interference with CO_2 uptake. This occurs even with complete osmotic adaptation. Therefore, although osmotic adaptation may prevent the damage due to osmotic stress, the plant is still exposed to various toxic and nutritional damaging effects (31).

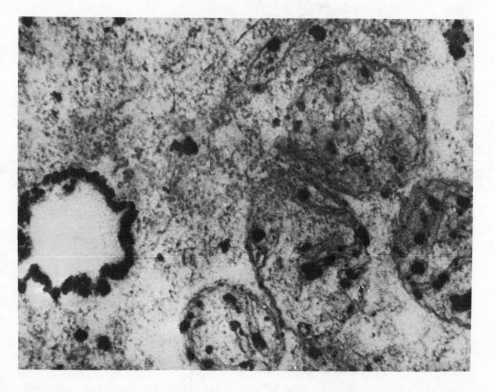

Figure 1.   An electron micrograph of a cortical cell of a maize root exposed to NaCl and treated with pyroantimonate. Dark grains in the mitochondria and on the vesicular membrane are pyroantimonate precipitates supposedly with Na; x 68,000 (From 45).

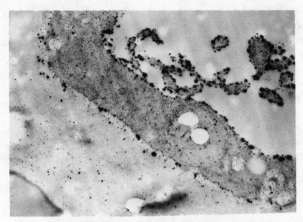

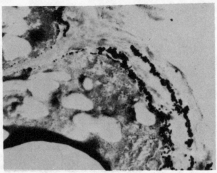

Figure 2.(a)   Electron micrograph of a part of a cell of a
               maize coleoptile, showing extra cytoplasmic
               Na-pyroantimonate precipitates, in the cell
               wall and on the membranes; x ≃ 40,000.

Figure 2.(b)   Electron micrograph of cells from maize col-
               eoptile showing AgCl precipitate in the cell
               wall and on the membranes; x ≃ 23,000.
               (From 45).

Effect of salinity on enzymes and on the respiratory metabolism of plant tissues

As was stated above, one of the explantions for the stunted
growth caused by salinity is the diversion of energy derived from
respiration to processes other than growth.   These processes could
be considered as maintenance and the relevant fraction of the
respiration devoted to it is referred to as "maintenance
respiration."

The rate of respiration can be considered as a rough criterion for the intensity of metabolism. The reports in the literature, concerning the effects of salinity on rate of respiration are rather contradictory. Nieman (47) measured oxygen consumption of leaf section of twelve crops and reported stimulatory effects of NaCl. Boyer (48) measured CO_2 evolution by infrared gas analysis and found that whenever cotton plants were exposed to NaCl, respiration always decreased. Porath and Poljakoff-Mayber (49, 50) found progressive inhibition of respiration (oxygen consumption) by increasing salt concentration, in pea root tips aged overnight and supplied with glucose during measurements. When, however, they fed $^{14}C-U$-glucose to the root tips and measured $^{14}CO_2$ evolution, salinity induced an increase in the percent ^{14}C (of the absorbed) evolved as CO_2. At the same time, the percent of the absorbed (calculated from the external) ^{14}C, decreased with increasing salinity (51). When $^{14}C-1$-glucose and $^{14}C-6$-glucose were fed to the root tips the C_6/C_1 ratio decreased with increasing salinity and the percentage of the glucose metabolized via the pentose phosphate pathway (PPP) increased (50). The decrease in C_6/C_1 was due to both a decrease in C_6 and an increase in C_1. At the same time, the activity of malate dehydrogenase (MDH) decreased and the activity of glucose-6-phosphate dehydrogenase (G6PDH) increased with increasing salinity. It must be pointed out that the roots were exposed to salinity only during growth. Respiration measurements, homogenization of the tissue and enzyme assays were carried out in absence of salt. It seems, therefore, that during the prolonged exposure of the roots to saline substrate some irreversible change occurred in the structure of the enzynmes, which persisted and affected their activity even when the salt was washed out or greatly diluted (52).

It appeared, therefore, as though salinity diverts the metabolism, at least of exogenously supplied glucose, from the usual glycolytic pathway to the PPP. This was considered as salinity damage.

As all the above reported experiments were carried out with peas of the cultivar Laxton Progress which is exceptionally salt sensitive, a comparison was made with other pea cultivars. Alaska and Dan cultivars are more salt tolerant and in them no increase in the PPP was observed on exposure to salinity, and C_6/C_1 ratios were practically unaffected. A more detailed investigation of the effect of salinity on the various glycolytic and PPP enzymes was conducted (Fig. 3 and 4), unpub.) The activity of the key enzyme PFK decreases with increasing salinity not only in the sensitive cultivar Laxton Progress but also in the relatively tolerant cultivar Alaska and in the wild pea, P. fulvum, but the activity of this enzyme increases in cultivar Dan and in P. elatius.

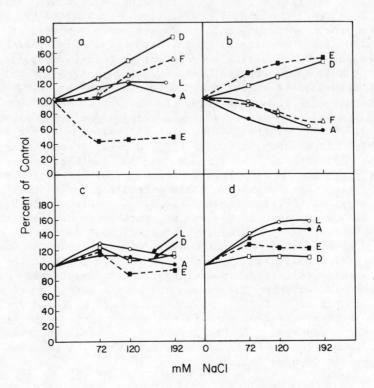

Figure 3.   Specific activity of enzymes of the glycolytic
             pathway in root tips of various pea cultivars
             and Pisum species grown in different concen-
             trations of NaCl. Results as percent of con-
             trol grown in Hoagland.

             A, Alaska; D, Dan; E, P. elatius;
             F, P. fulvum; L, Laxton Progress;

             a. glucose phosphate isomerase
             b. phosphofructokinase
             c. pyruvate kinase
             d. aldolase. (Poljakoff-Mayber, et al, unpub.)

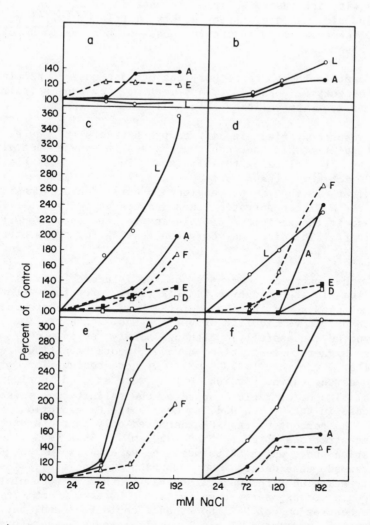

Figure 4. Specific activity of the PPP enzymes in root tips of var-
          ious   pea cultivars and   Pisum species grown at different
          NaCl concentrations. Results as percent of control grown
          in Hoagland. A, D, E, F, and L as in Figure 3.

a. ribose-5-P-isomerase   d. 6-phosphogluconic dehydrogenase
b. xylulose-5-P-epimerase   e. transketolase
c. glucose-6-P-dehydrogenase   f. transaldolase

(Poljakoff-Mayber, et al, unpub.)

The activity of glucose phosphate isomerase increases in all
cultivars with increasing salinity but not in P. elatius. Pyruvate
kinase and aldolase are affected very little by salinity and no
marked differences between the responses of the various cultivars
could be observed.

The pentose phosphate pathway enzymes are more sensitive than
glycolytic enzymes to salinity and most of them are stimulated by
exposure of roots to salinity (Fig. 4).

The response varied in different cultivars and there was no
clear cut relationship between the changes in enzyme activity and
sensitivity or tolerance to salinity. The only possible enzyme
response which may perhaps serve as a marker for salt sensitivity
of a cultivar is the 6-phosphogluconic acid dehydrogenase: the
lower is the NaCl concentration that induces on in vitro treatment
an increase in the activity of this enzyme, the more sensitive is
the cultivar. This idea may be worth testing for crops other than
peas.

For the purpose of comparison the glucose metabolism was
investigated in the halophyte Tamarix (53). Cuttings were taken
from a tree growing on the shore of the Dead Sea and were rooted in
the laboratory in vermiculite at various NaCl concentrations. The
cutting rooted successfully, although slowly, even in 400 MM NaCl
but not in higher concentration; however, rooted cuttings survived
even after transfer to 500 mM. Oxygen consumption of roots
decreased as the salinity to which they were exposed during growth
increased. The C_6/C_1 ratio dropped as the salinity was raised, due
to increase in the C_1 evolved as CO_2 while C_6 evolution did not
change. The percent of the PPP increased only in roots exposed to
NaCl concentrations higher than 250 mM. This was the NaCl
concentration above which $\Delta\Pi$ between the external solution and the
root decreased considerably and approached zero; i.e., the roots
apparently suffered from water stress. Ethanol accumulated in
Tamarix roots on exposure to salinity of up to 250 mM (54). In
higher concentrations of NaCl no ethanol could be found, but lactic
acid formation persisted (Table 1). It appears, therefore, that
increasing salinity induced changes in the ratio between the
various pathways of glucose metabolism in Tamarix roots and as the
growth of the cuttings was inhibited, these changes could be
considered as salinity damage.

Salinity also affects oxidative phosphorylation and the
ATP/ADP ratio in peas (55). Similar results were reported for
mitochondria isolated from green tissue of Suaeda (56). Exposure of
pear roots, or of mitochondria isolated from them, to NaCl induced
the development of some ATP cleaving agent although the mitochondria themselves were properly coupled and had a high
respiratory control (unpub.)

Flowers (57, 58) has found that in some halophytes ATPase activity was inhibited by salinity while in Salicornia it was stimulated. In all cases the effect of salinity was dependent also on other conditions, such as pH.

TABLE 1. Effect of salinity on uptake of external glucose, on respiration, on anaerobic metabolism of Tamarix roots, and on dry weight of leaves. Roots were incubated in 0.01% glucose. Oxygen uptake of control roots (100%) 157 µl O_2/mg N h^{-1}.

Computed from data of Ephron (54).

Salinity in growth Medium; mM NaCl	^{14}C Glucose Absorbed; % of Present Externally	O_2 uptake; % of Control	Ethanol Content; mg/100 g	Lactic acid Content; µg/100 g	Dry Weight of leaves as % of Control
0	30	100	n.d.	n.d.	100
24	17	65	41	9.3	97
72	16	48	n.d.	n.d.	52
120	4	60	48	9.3	--
168	3	56	63	9.2	56
240	3	50	56	11.2	--
288	3	46	0	11.9	26

Udovenko (8) tends to regard the increased respiration, caused by salinity as at least partly uncoupled, as the effect of DNP on the respiration of tissue exposed to salinity was only 50% of that obtained with the control tissue.

Several attempts have been made to evaluate "maintenance respiration" and to determine whether it is increased by exposure to salinity. Lambers (59) is of the opinion that the energy metabolism, at least in roots of higher plants, depends to a large extent on environmental conditions. Growth more often determines the efficiency of assimilate utilisation than the reverse. He found that in salt sensitive Plantago coronpus, exposure to salinity induced an increase in the fraction of the respiration used for maintenance as well as that used for growth. He concluded that the increase in respiration is the result of reduced growth and not its cause. Increased respiration was observed also in the salt tolerant species P. maritima whose growth was not inhibited by salinity. Luttge et al (60) showed that ion transport is linked to energy generating respiration. It appears that when a plant root has reduced energy supply, ion uptake also decreases. Schwartz and Gale (unpub.) measured gas exchange in the light and dark of plants exposed during growth to low levels of salinity. They divide respiration into two fractions: (1) proportional to assimilation – a basic fraction used for maintenance; and (2) a fraction that is affected by the prescence of salt (salt maintenance respiration). It was found that in some plants exposure to salinity increased salt maintenance respiration while in others (maize) no increase was observed. Their calculations for Xanthium showed that 30% of the reduction in dry weight increment caused by low level salinity could be ascribed to the increased respiration (C-efflux). Raising the ambient CO_2 concentration resulted in a higher rate of photosynthesis and reduced the growth inhibition caused by salinity. However, in maize, the increase in the ambient CO_2 concentration had very little effect on growth.

It appears, therefore, that salinity may have a dual effect on the respiratory mechanism of a plant: (a) it can interfere with normal metabolic pathways by altering properties of some enzymes and thus reduce the energy charge and the supply of energy to the usual life processes; and (b) it can cause a diversion of energy from the normal metabolic and growth processes to "maintenance" needs such as synthesis of osmotica, compartmentation, or even for repair processes, as will be suggested later.

It is generally accepted that the enzymes isolated from halophytes are as sensitive to the toxic effects of NaCl as those isolated from glycophytes (61, 62). This seems to me no longer valid. The data of Flowers et al (63) for the two fractions of malate dehydrogenase from Suaeda and the data of Kalir (64) for peroxidase and superoxide dismutase in Halimione (Figs. 5 and 6), open other possibilities. I would like to suggest that not all enzymes behave identically and enzymes located in certain places or on certain membranes in the cell may be salt tolerant while others will be salt sensitive.

Effect of salinity on the metabolism of nitrogen containing compounds

The main aspects that have been studied and will be refered to here are protein synthesis, free amino acid pool composition, and

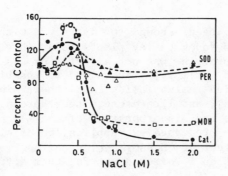

Figure 5. Specific activity of four enzymes isolated from leaves
          of Halimione portulacoides grown at different levels of
          of NaCl salinity. MDH, malate dehydrogenase; Cat.,
          catalase;   Per,   peroxidase;   and   SOD,   superoxide
          dismutase (From 64).

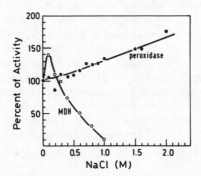

Figure 6. In vitro effect of NaCl (added to the reaction mixture)
          on specific activity of malate dehydrogenase   (O) and
          peroxidase (O) isolated from leaves of Halimione portu-
          lacoides grown in Hoagland's solution without NaCl (64).

accumulation of quarternary ammonium bases. Although the effect of
salinity on protein synthesis may be considered as damage, the
changes induced in the other two could be considered as a kind of
an adaptation to salinity which helps the plant to overcome the
effects of stress.

Salinity has been shown to interfere with the uptake of
inorganic ^{15}N into young barley plants (65), whereas the incorporation of ^{15}N into protein was not affected or was even stimulated.
On the other hand salinity was shown to inhibit the uptake of
externally supplied amino acids and their incorporation into
protein (11, 66-68), and literature cited therein). The inhibition
by NaCl of uptake and incorporation of ^{14}C-leucine into protein
could be partially reversed by application of kinetin. However,
the inhibitory effect of Na_2SO_4 could not be reversed by kinetin
(69). No explanation is offered yet for this difference.

The interference of NaCl with protein synthesis, in plant
tissue exposed to salinity, could be at several levels: (a) it
could affect permeability of the barriers of the different compartments in the cell (70); (b) it could affect the tertiary structure
of the enzynmes directly involved in protein synthesis or it could
induce dissociation of ribosomes (see literature cited in 11 and
65, 67); and (c) it may affect the expression of the genome or
interfere with transcription and translation of the relevant RNAs
and thus have a more lasting effect. There is some evidence
supporting this last possibility; the activity of enzymes isolated
from plants exposed to salinity for a prolonged period differs from
that of the relevant controls (50, 52, 71): acquired salt
tolerance in cell cultures of tobacco and pepper was maintained for
several generations on non-saline medium (72). Observation of some
submicroscopic changes in the nuclei in cells of barley roots grown
in presence of 192 mM NaCl (Fig. 7) revealed that the chromatin in
the nucleus was aggregated in strongly stained masses and not
evenly distributed as usually observed. Similar changes were
induced in nuclei of maize root cells after exposure to severe
water stress (73). It cannot be excluded that all the three,
changes in permeability, in enzyme structure and in translation and
transcription, actually occur.

A most pronounced effect of exposure to salinity is the
accumulation in the tissues of proline (38, 74 and literature cited
in 12). The rapid appearance of proline in response to stress and
its rapid disappearance after stress alleviation was taken as an
indication of adaptation. It is accepted that proline serves as a
cytoplasmic osmoticum (37). This view is supported by the fact
that most of the halophytes examined were found to contain large
amounts of proline in their amino acid pool. Exposure of plants
to salinity resulted in a marked increase in free proline content
and in some cases, but not always, also in an increase in the

total free amino acids pool in the roots. Thus, for example, the free amino acids pool, in the roots of barley grown in the presence of 192 mM NaCl, was twice as high as in the Hoagland grown controls; in roots of Alaska pea grown in 168 mM NaCl, this pool was half of that in the controls, while in roots of Laxton Progress peas (the most sensitive variety) no change in total free amino acid was observed at a salinity of 120 mM. However, their proline content was tripled, arginine was doubled and alanine increased five times (68). Proline accumulation in sorghum leaves is apparently triggered by two factors; the level of stress (there is a minimal stress below which proline accumulation does not occur) and a threshhold concentration of the monovalent cations in the tussue (R. Weimberg, pers. comm.). This is apparently not the case in barley, where monovalent ions were less effective in inducing

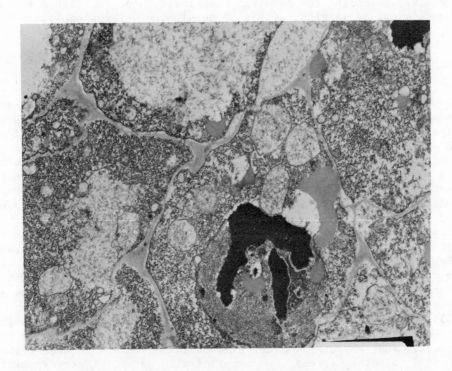

Figure 7. Electron micrograph of a cortial cell from a barley root grown in Hoagland's solution containing 192 mM NaCl. (Lerner, et al, unpub.)

proline accumulation than divalent ions (74).   It is not yet clear how stress affects the proline metabolizing mechanism, inducing such rapid sysnthesis and breakdown of this amino acid.

Proline is not the only organic osmoticum that is rapidly synthesized in the tissues of higher plants in response to salinity. The other substance is the quaternary ammonia base, glycinebetaine (21). However, the turnover of glycinebetaine is slower than that of proline. The rate of increase in the level of stress is apparently one of the important factors determining whether proline or glycinebetaine will be synthesized (21). Neither of the two substances are apparently toxic and do not cause any inhibition of enzyme activity (75, 76). Moreover, they both apparently have some remedial and/or protective effect against stress damage (21, 77, 78). The concentrations needed for that protective effect are very low, therefore their action cannot be merely osmotic.

It seems, therefore, that in plants exposed to salinity the metabolism of nitrogen containing compounds plays one of the decisive roles in the survival mechanism. It is liable to be damaged, but it is also capable of developing defense tools.

Plant hormones and salinity

There is no doubt that phytohormones are involved in the response of plants to salinity stress. Moreover, homones are apparently involved in the response of plants to other types of stress (22, 23). The content of gibberellins, IAA and ethylene change on exposure to salinity but the most marked changes occur in the content of two hormones -cytokinin and ABA(79, 80). Cytokinins are synthesized in the roots and are transported upwards. Their content decreases in the xylem exudate as the level of external salinity increases. ABA is synthesized, apparently, in the leaves and is transported downwards. ABA content, especially of the shoot, rises very quickly on exposure to salinity (as well as to drought). There may be a gradation of cytokinin/ABA ratios along the plant axis that changes under effect of salinity, but there are no experimental data which would indicate the importance of such a change. In many respects an increase in ABA concentration has an effect equivalent to that of a decrease in cytokinin concentration (23). Although both hormones are affected by drought and salinity stress, there are apparently differences in responses to these two stresses and especially after alleviation of stress. When plants exposed to water stress are supplied with water, their cytokinin content increases and the ABA content decreases. However, under salinity, when adaptation takes place, the osmotic potential of the tissue is lowered, and ABA content remains high even when the gradient for water influx is restored (23).

Both hormones may affect turgor indirectly through their effect on stomatal movement and regulation of transpiration (81, 82). ABA is known to induce stomatal closure while cytokinin causes their opening. ABA is known to affect potassium fluxes in the tissue (83) as well as translation of mRNA (84-86). Through both these mechanisms ABA can affect metabolic events in plants exposed to stress. The question is whether the changes in ABA content play a key role in these events, or they are merely a result of the general effects of stress on ABA metabolism. ABA exists in the plant in two forms: bound and free. It is not clear beyond any doubt which of these forms is the active one, or whether it is the total ABA content which is important. It is accepted that only rarely does the level of bound ABA exceed that of the free form (87). However, in pea cotyledons, germinated at 192 mM NaCl, we found that free ABA accumulated to much higher concentration than the bound form. In the control cotyledons, germinated in Hoagland's solution, the situation was reversed while the total content was much lower than in cotyledons exposed to salinity (88). In such peas, germinated in 192 mM NaCl, the increase in ABA content of the cotyledons was accompanied by increased osmolarity and increase in proline content. Technically the shoot and root could not be separated until the fourth day after germination. The free ABA of the shoots increased up to the 6th day and then decreased. Almost no ABA was detected in the roots in these experiments. The osmolarity of the shoot and the root behaved in a similar way (Fig. 8), but their proline content did not behave in the same way. Proline content of cotyledons exposed to salinity increased up to the fourth day of germination and then remained constant; in the controls it hardly changed (88).

Since in the shoot there was a correlation between internal free ABA and osmolarity, the effect of externally applied ABA was studied (Fig. 9). ABA supplied in the root medium to plants grown in Hoagland solution induced an increase in proline content of cotyledons (osmolarity could not be measured) and of proline content and osmolarity in the shoot, although there was no stress applied to the plants either osmotic or saline. The effect increased with increasing concentration of ABA. In the root osmolarity also increased in response to ABA treatment, but proline content was not affected. On the basis of these results, which are very preliminary in nature, we can suggest the hypothesis that ABA has a key role in regulating osmotic adaptation, and that the course of events may be: stress → increase in ABA content → osmotic adaptation (perhaps through excessive ion uptake) → proline synthesis (when the concentration of monovalent cations reaches a certain threshold) or some other organic solute. In the roots, either the situation is different, or, otherwise in our experiments, the concentration of the monovalent cations never reached the threshold concentration.

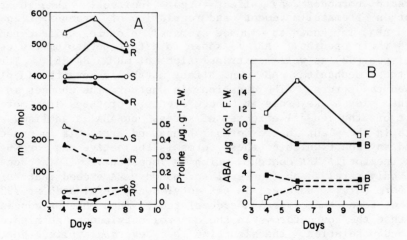

Figure 8. (a) Proline content (broken line) and osmolarity (full
          line) of roots and (▲, ●) and shoots (△, ○) of pea plants
          gown in Hoagland's solution (circles) or in presence of
          192 mM NaCl (triangles). (b) ABA content of shoots of
          the above plants. Full line, 192 mM NaCl; broken line,
          Hoagland control. F, free ABA; B, bound ABA.
       (Hasson and Poljakoff-Mayber, unpub.)

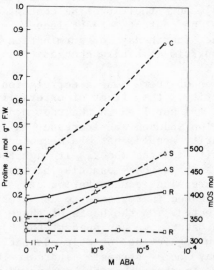

Figure 9. The effect of externally applied ABA on osmolarity (full
          line) and proline content (broken line) of pea plants
          grown in Hoagland's solution. R, root; S, shoot; C,
          cotyledon. Osmolarity measurement could not be made on
          cotyledons. ABA applied to root medium.
       (Hasson and Poljakoff-Mayber, unpub.)

Summary of ideas for discussion

The main point that I want to make here is that salinity stress is not just physiological drought but is much more complex in nature. Adaptations for survival under saline conditions are therefore not only osmotic, but can be physiological or biochemical.

Shocks due to sudden increase in salinity are rare and occur in nature only under special conditions. There is usually gradual salinization and gradual adaptation. If osmotic shocks do occur it is "downwards" by sudden desalination due to rain or irrigation. Therefore, if salinity affects turgor in the roots the change is rather slow and gradual. In the shoot, however, atmospheric drought may increase the severity of the osmotic stress, while sea water spray may increase the salinity stress. In most cases the organ that is directly and immediately exposed to salinity is the root, and the plasmalemma of the root cells is the first to be affected. It is a suitable candiate to act as a primary sensor of stress. The root is exposed to the osmotic effect of the saline soil solution as well as to the specific effect of the ions present. Changes in the structure of the plasmalemma and changes, even if slight, of turgor will be the first events to occur and will probably trigger a chain of multiple processes. Hormonal changes, which will affect the ion absorbing machinery and may affect also the activity of some enzymes, may occur next. The absorbed ions could affect the water potential and either through it, or directly, affect the structure and permeability properties of the inner membranes and organelles. The modulation of genome expression and maybe the ploidy of the cells may be affected (89, 90). This will result in changes in the structure of enzymes. The low water potential and the ions present in the cell may induce conformational changes in enzyme molecules. Pathways of metabolism may be affected as, for instance, diversion of greater part of the metabolized carbohydrate through the pentose phosphate pathway, or transfer from C_3 type of metabolism to C_4 (91); or the ability induced in Dunaliella to evolve O_2 photosynthetically in the absence of CO_2 (92).

Another result of the induced changes may be the lowering of the energy charge resulting from decrease in the ATP content and increase in ADP and AMP (55), or diversion of a larger part of respiration for "maintenance" purposes. The diversion may be needed for "repair" of the structural changes in cellular membranes, or for synthesis of various organic osmotica.

It seems that it can no longer be taken for granted that enzymes of halophytes are inhibited by high salt concentration in the same way as those of glycophytes. The findings for peroxidase and superoxide dismutase (64), as well as for other enzymes isolated from halophytes (93) render this idea as worthy of more thorough investigation. Another idea worth examining is that there is practically none, or very little, sodium in the cytoplasm. Inconsistency of the results concerning this point were already mentioned in the past but dismissed (67). If sodium ions are present in the cytoplasm in considerable amount, there is much more sense to the reported protective action of proline and glycine-betaine (75-78) which are also apparently located in the cytoplasm.

The increased activity of peroxidase induced by salinity in several plants (94) and the high resistance to salinity of this enzyme found in Halimione (64) raise the possibility that in some cases, under certain conditions, metabolic disturbances result in the formation of free radicals. In their interactions with biological membranes the free radicals may cause changes in permeability properties of membranes. Peroxidase, superoxide dis-mutase, as well as some hormones, are considered to be scavengers of free radicals. Therefore an increase in the content of these enzymes or an increase in their activity could be considered as a protective adaptation. In this connection it may be worthwhile to devote a little more attention to the effect of salinity on membrane structure and composition.

All these ideas may easily be proved wrong, but is seems to me that it is worthwhile to discuss them even if only in order to bring to our attention that adaptation to salinity is a complex process which does not depend on a single mechanism.

References

1. Levitt, J, "Responses of plants to environmental stress," Academic Press, N. Y. (1972).
2. Y. Waisel, "Biology of Halophytes," Academic Press, N. Y. (1972).
3. D. W. Rains, Ann. Rev. Plant Physiol. 23:367-388 (1972).
4. H. Greenway, J. Aust. Ins. Agric. Sci. 39:24-34 (1973)
5. A. Poljakoff-Mayber, A. and J. Gale (eds.) "Plants in Saline Environments," Ecological Studies 15, Springer Verlag, Berlin, Heidelberg, New York (1975).
6. D. H. Jennings Biol. Rev. 51:453-486 (1976).
7. J. W. Wright (ed.) "Plant Adaptation to Mineral Stress in Problem Soils," Special publication of Cornell Univ. Agr. Exp. Sta. (1976).
8. G. V. Udovenko, "Salt tolerance of cultivated plants (in Russian)", "Kolos", Leningrad (1977).

9. T. J. Flowers, P. F. Troke and A. R. Yeo, Ann. Rev. Pl. Physiol. 28:89-121 (1977).

10. J. A. Hellebust, Ann. Rev. Pl. Physiol., 27:485-505 (1977).

11. R. G. Wyn Jones, C. J. Brady and J. Speirs, "Recent Advances in the Biochemistry of Cereals." (D. L. Laidman and R. G. Wyn Jones, eds.) pp. 63-103, Academic Press, London, New York, San Francisco (1970).

12. H. Greenway and R. Munns, Ann. Rev. Pl. Physiol. 31:149-190 (1980).

13. T. C. Hsiao, Ibid. 24:519-570 (1973).

14. T. C. Hsiao, E. Fereres, E. Acevedo and D. W. Henderson, in: "Water in Plant Life," (O. L. Lange, L. Kappen and E. D. Schutze, eds.) pp. 281-305, Ecological Studies 19, Springer-Verlag, Berlin, (1976a).

15. T. C. Hsiao, E. Acevedo, E. Fereres and D. W. Henderson, Phil. Trans. R. Soc. 273:479-500 (1976b).

16. P. Matile, Ann. Rev. Pl. Physiol. 29:193-213 (1978).

17. R. A. Fisher and N. C. Turner, Ibid. 29:277-317 (1978).

18. V. Zimmerman, Ibid. 29:121-148 (1978).

19. J. D. Bewley, Ibid. 30:195-238 (1979).

20. L. Bernstein, Plant Anal. Fert. Probl. Colloq. 4:25-45 (1964).

21. R. G. Wyn Jones and R. Storey, Aust. J. Pl. Physiol. 5:817-829 (1978).

22. Y. Vaadia, Phil. Trans. R. Soc. Lond. B 273:513-522 (1976).

23. S. T. C. Wright, in: "Phytohormones and Related Compounds: A Comprehensive Treatise," Vot. II. (D. S. Lethan, P. B. Goodwin and T. J. V. Higgins, eds.) pp. 495-536, Elsevier, North Holland Biomedical Press (1978).

24. A. Poljakoff-Mayber, see Ref. 5, pp. 97-117.

25. R. L. Mott and E. C. Steward, Ann. Bot. 36:915-937 (1972).

26. S. Z. Hyder and H. Greenway, Plant and Soil 23:258-260 (1965).

27. G. J. Hoffman, J. Shalhevet and A. Meiri, Physiol. Plant 48:463-469 (1980).

28. D. W. Lawlor, J. Exp. Bot. 20:895-911 (1961).

29. R. F. Meyer and J. S. Boyer, Planta 108:77-87 (1972).

30. E. Acevedo, E. Fereres, T. C. Hsiao and D. W. Henderson, Plant Physiol. 64:476-480 (1980).

31. P. L. Steponikus, in: "Genetic Engineering of Osmorregulation," (D. W. Rains, R. C. Valentine and A. Hollaender, eds), pp. 235-258, Plenum Press, New York and London (1980).

32. D. A. Johnson, Crop Science 18:945-948 (1978).

33. J. A. Bunce, J. Exp. Bot. 28:156-161 (1977).

34. L. Erdei and P. J. C. Kuiper, Physiol. Plant. 47:95-99 (1979).

35. W. D. Jeschke and W. Stelter, Planta 128:107-112 (1976).

36. W. D. Jeschke, in: "Regulation of Cell Membrane Activity in Plants," (E. Marre and O. Cifferi, eds), Elsevier/North Holland Biomedical Press pp. 63-78 (1977).

37. G. R. Stewart and J. A. Lee, Planta (Berl.) 120:279-289 (1974).

38. T. M. Chu, D. Aspinall and L. G. Paleg, Aust. J. Plant Physiol. 3:219-228 (1976).

39. T. J. Flowers and J. L. Hall, Ann. Bot. 42:1057-1063 (1978).
40. R. G. Wyn Jones, see Ref. 31, pp. 155-170.
41. R. G. Wyn Jones, R. Storey, B. A. Leigh, N. Ahmad and
    A. Pollard, see Ref. 36, pp. 121-136.
42. H. R. Lerner, Abstracts, 13th FEBS Meeting, Jerusalem, Israel,
    August 1980.
43. H. R. Lerner, D. Ben-Bassat, L. Reinhold and A. Poljakoff-
    Mayber, Plant Physiol 61:213-217 (1978).
44. H. R. Lerner, R. Weimberg, L. Reinhold and A. Poljakoff-
    Mayber, Acta Horticulturae 89:147-149 (1979).
45. E. Neeman, M. Sc. Thesis, The Hebrew University of Jerusalem
    (in Hebrew) (1969).
46. R. Stelzer and A. Lauchli, Z. Pflanzenphysiol. 88:437-448
    (1978).
47. R. H. Nieman, Bot. Gaz. 213:279-285 (1962).
48. J. S. Boyer, Plant Physiol. 40:229-234 (1965).
49. E. Porath and A. Poljakoff-Mayber, Israel J. Bot. 13:115-121
    (1964).
50. E. Porath and A. Poljakoff-Maber, Plant and Cell Physiol. 9:
    195-203 (1968).
51. E. Hasson-Porath and A. Poljakoff-Mayber, Plant and Cell
    Physiol. 14:361-368 (1973).
52. A. Livne and N. Levin, Plant Physiol. 42:407-414 (1967).
53. A. Kalir and A. Poljakoff-Mayber, Plant Physiol. 57:167-170
    (1976).
54. D. Ephron, M. Sc. Thesis, The Hebrew University of Jerusalem
    (in Hebrew) (1966).
55. E. Hasson-Porath, E. and A. Poljakoff-Mayber, Plant Physiol.
    47:109-113 (1971).
56. T. J. Flowers, J. Exp. Bot. 25:101-110 (1974).
57. T. J. Flowers, Ibid. 23:310-321 (1972a).
58. T. J. Flowers, Phytochemistry 11:1881-1886 (1972b).
59. H. Lambers, Ph. D. Thesis, Rijksuniversiteit, Groningen,
    Holland (in English) (1979).
60. U. Luttge, W. J. Cram and G. G. Laties, Z. Pflanzenphysiol
    64:418-426 (1974).
61. H. Greenway and C. B. Osmond, Plant Physiol. 49:256-259
    (1972).
62. R. Weimberg, Ibid. 43:622-628 (1968).
63. T. J. Flowers, M. E. Ward and J. L. Hall, Phil. Trans. Royal
    Soc. London B 273:523-540 (1976).
64. A. Kalir and A. Poljakoff-Mayber, Ann. Bot. 46 (in press).
65. M. Hellal, K. Kock and K. Mengel, Physiol. Plant. 35:310-313
    (1975).
66. I. Kahane and A. Poljakoff-Mayber, Plant Physiol. 43:1115-1119
    (1968).
67. J. L. Hall and T. J. Flowers, Planta 110:361-368 (1973.
68. N. Bar-Nun and A. Poljakoff-Mayber, Ann. Bot. 44:309-314
    (1979).

69. D. Ben-Bassat,   M. Sc. Thesis, Hebrew University, Jerusalem, (in Hebrew) (1967).
70. A. Oaks, Plant Physiol. 40:142-149 (1965).
71. S. P. Treichel, S. O. Kirst and I. J. von Widlert, Z. Pflanzenphysiol 71:437-439 (1974).
72. P. J. Dix and H. E. Street, Plant Sci. Lett. 5:231-237 (1975).
73. I. Nir, S. Klein and A. Poljakoff-Mayber, Aust. J. Biol. Sci. 23:489-491 (1970).
74. T. M. Chu, D, Aspinal and L. G. Paleg, Ibid. 3:503-511 (1976b).
75. A. W. Larkum and G. R. Wyn Jones, Planta 145:393-394 (1979).
76. A. Pollard and R. G. Wyn Jones, Planta 144:291-298 (1979).
77. C. Hubac, and D. Guerrier, Ecol. Plant 7:147-165 (1972).
78. L. N. Csonka, in: "Salmonella typhimunium", see Ref. 31, pp. 35-52.
79. Y. Mizrahi, A. Blumenfeld and A. E. Richmond, Plant Physiol. 46:169-171 (1970).
80. Y. Mizrahi, A. Blumenfeld, S. Bittner and A. E. Richmond Ibid. 48:752-755 (1971).
81. M. Tal, D. Imber and C. Itai, Ibid. 46:367-372 (1970).
82. M. Tal and D. Imber, Ibid. 47:849-850 (1971).
83. S. Cocucci and M. Cocucci, Plant Sci. Lett. 10:85-95 (1977).
84. D. C. Walton, in: "The Physiology and Biochemistry of Seed Dormancy and Germination," (A.A. Khan, ed.) pp. 145-156, Elsevier/North Holland Press (1977).
85. J. H. M. Bex, Acta Bot. Neerl. 21:203-216 (1972).
86. M. G. Galli, P. Miracca and E. Sparval, Pl. Sci. Lett. 14: 105-111 (1979).
87. R. W. P. Hiron and S. T. C. Wright, J. Exp. Bot. 24:769-781 (1973).
88. E. Hasson and A. Poljakoff-Mayber, Isr. J. Bot. 29:98-104 (1980).
89. B. P. Strogonov, (1962) "Physiological bases of plants," Israel Sci. Trans., Jerusalem (1964).
90. A. L. Kurasanov and P. A. Genkel, (eds.) in: "Structure and function of plant cells under salinity: New approaches to the study of salt tolerance" (in Russian), "Nauka" Moscow (1970).
91. A. Shomer-Ilan and Y. Waisel, Physiol. Plant. 29:190-193 (1973).
92. A. Kaplan, U. Schreiber and M. Avron, Plant Physiol 65:810-813 (1980).
93. A. Shomer-Ilan, Y. B. Samish, T. Kipnis, D. Elmer and Y. Waisel, Plant and Soil 53:477-486 (1979).
94. H. C. Stevens, M. Calvar, K. Lee, B. Z. Siegel and S. M. Siegel, Phytochemistry 17:1521-1522 (1978).
94. H. C. Stevens, M. Calvar, K. Lee, B. Z. Siegel and S. M. Siegel, Phytochemistry 17:1521-1522 (1978).

GENETICS OF SALT TOLERANCE: NEW CHALLENGES

M. C. Shannon

U. S. Salinity Laboratory USDA-SEA-AR
4500 Glenwood Drive
Riverside, CA 92501 USA

SUMMARY

The contribution of genetics to the understanding of salt tolerance in higher plants is just beginning to be made. The objectives of this review are: (1) to evaluate the few genetic studies that have been performed; (2) to indicate the type of studies now in progress; and (3) to suggest new approaches in which the genetic method can aid in understanding the physiological basis of salt tolerance.

Ion exclusion mechanisms and resistance to pollen sterility are two characters that result in increased salt tolerance and are under gene control Ion exclusion mechanisms have been identified in grasses (Agrostis, Agropyron), beans (Phaseolus, Glycine), grapes (Vitis), and citrus (Citrus). Resistance to salt-stress-induced pollen sterility has been expressed in rice (Oryza).

Current studies on the genetics of salt tolerance and salt tolerant plants are limited in number and scope. Ecotypes and cultivars with increased tolerance are being tested to identify inheritance and dominance patterns. These studies are being conducted without a precise understanding of the mechanisms of the expressed tolerance.

Major areas in which the use of genetic methods are needed include: (1) inheritance studies in new lines and ecotypes isolated for tolerance; (2) genetic analysis of the progeny of regenerated plants isolated for salt tolerance in tissue culture; (3) heritability of salt tolerance from wild relatives backcrossed to their agronomic counterparts; and (4) the development of isolines for particular   physiological characters that   may contribute to salt tolerance.

INTRODUCTION

Salinity and sodicity limit crop production in many areas of the world. The practices of adding soil amendments and leaching to ameliorate salt and sodic effects are profitable only when water and cheap and abundant chemicals are available.   In many parts of the world, this approach is not possible; even in the more developed countries, the rising cost of energy and nonsaline water has put new restraints on the economical use of this technology. A promising approach to improve crop production on saline and sodic soils is to breed plants specifically to fit these environments.

The adaptation of plants to saline soils is an exiciting prospect, but presents plant breeders and geneticists with some formidable challenges. Saline soils cover a substantial portion of our earth's surface. Estimates vary from 400 to 950 x 10^6 hectares (49, 51).   Semi-arid areas in which only saline waters are available add another sizeable portion of land that could be used for crop production. By extending intensive agricultural production into such areas, it is conceivable that a second Green Revolution could be launched; however, the immediate goals are simply to increase total food production by breeding crops that can better tolerate marginally saline areas and produce higher yields on them.   We must also breed and develop crop plants adapted to lands not presently suited to agricultural use.   This may include plants that are capable of using brackish water as their only water source.

While the complexity of relations between energy and agriculture has consequences outside the scope of our present discussion, it has also an immediate importance because it presents many new possibilities.   Plant biomass is seen as a feasible alternative to some fossil fuel end products.   Competition between food and fuel crops for our limited resources of land and water must be avoided. The rapid development of salt tolerant plants that may be used as fuel or petroleum by-product substitutes is urgently needed.

The prerequisites for genetic or breeding studies are the existence of sufficient heritable variation in a particular character and a means by which genetic information can be transferred in a stable form.   Thus, it is necessary to define and measure salt tolerance in some manner such that its variation can be assessed. Also, measurable differences must exist between individuals that can produce viable off-spring so that the stable transfer of salt tolerance can be demonstrated.

The study of the genetics of salt tolerance has proceeded slowly. The reasons for this are: (1) the specific physiological effects of salinity have not been determined; (2) varietal differences have only recently been detected; (3) plant response to saline stress depends upon other interacting environmental factors such as soil water, temperature, and humidity; and (4) salt tolerance changes with plant development. These aspects will be discussed briefly in relation to past, present, and future research on the genetics of salt tolerance.

History

The effects of saline and sodic soils on plant growth have been a focus of research for nearly one hundred years (48). A comprehensive bibliography concerning the effect of salinity on plants has recently been compiled by Francois and Maas (32) and consists of over 2000 references. Publications dealing with the physiological and biochemical effects of salt on plant tissues and organelles were not included in this index, for if they had been, the volume of information would have been too large to handle. Biochemical and physiological aspects of salt effects on plants has been covered periodically in several reviews (13, 18, 31, 36, 44, 47, 48, 59).

The theory has evolved that salinity causes a combination of osmotic and ionic effects upon plants. The osmotic effects interfere with the plant's ability to extract water from the soil and maintain internal water balance; whereas, the ionic effects may interfere with cytoplasmic solute balance or in some cases disturb membrane function and cause specific ion toxicities (10, 12, 13). Plants cope with salinity in different ways and some genera and species are distinctly more tolerant to it than others. Maas and Hoffman (46) have compiled a list of salt tolerance values and comparisons for 76 agronomic and horticultural species.

As early as 1941, Lyon (45) had suggested the possibility of selecting and breeding for salt tolerance, despite the fact that few reports had substantiated the concept that differences were noted in the germination and yield capacity of certain barley and wheat cultivars at high salinites, (6, 8). Later however, cultivar tests with lettuce (Lactuca sativa), onion (Allium cepa), carrot (Daucus carota) and green bean (Phaseolus vulgaris) indicated no differences (9, 14-16). This led Bernstein (11) to conclude, "For most crops, the relative uniformity in salt tolerance among varieties suggests little likelihood of improving salt tolerance by any combination of the commonly available varieties." He suggested the use of "wild" germ-plasm resources to introduce the necessary variability into such species.

Throughout the 1960's and the early 1970's distinct salt
tolerance differences among species and ecotypes were reported.
Not surprisingly, the occurrence of salt tolerance differences
seemed to be inversely proportioal to domestication and directly
proportional to the number of cultivars tested. Dewey (24, 25)
noted enough genetic variation among 25 strains of Agropyron to
suggest a possible breeding program; however, subsequent work in
this area was not conducted. In 1963, Vose (62) reviewed his
evidence of varietal differences in the selective uptake of
nutrient ions, and Epstein and Jeffries (30) summarized the
evidence that linked selective ion transport to specific genes. In
the following decade, differences in yield and vegetative growth
were found among cultivars of barley (Hordeum vulgare), pea (Pisum
sativum), corn (Zea mays) and sugarbeet (Beta vulgaris) (42, 43,
53, 58). Also, ecotypic differences in salt tolerance among
grasses were found in Festuca rubrum, Agrostis stolonifera and
Cynodon dactylon (38, 39, 61). By the mid-1970's salt tolerance
differences among cultivars had been described for over 30 agricultural species (57).. Unfortunately, in most studies reporting salt
tolerance differences, the physiological bases for these differences were not investigated and no genetic or breeding experiments
were conducted. Nevertheless, the fundamental requirements for
genetic research had now been fulfilled. Sufficient variation had
been shown between genotypes that were interfertile and ways to
measure salt tolerance had been developed. The fundamental
concepts of ion transport had been correlated with both salt
tolerance and control through gene expression.

Plant Responses to Salinity

The effects of moderate salinity include a slowing of all
metabolic proceses contributing to growth without apparent morphological changes (28). Physiological and biochemical processes of
growing cells are fundamentally unaltered by moderate salinities;
i.e., specific enzymes, total protein, and nucleic acids are
reduced in balance with a slower growth rate (47). Only under more
severe stresses do other symptoms such as leaf injury, morphological changes and death occur.

Basic biochemical differences between halophytes and glycophytes are not apparent. For instance, most enzymes isolated from
halophytes and glycophytes do not react differently in the presence
of salinity in vitro systems (37 , 63). Development in salt-
stressed plants follows the normal sequence of ontogeny and thus
the expression of the genetic code, except for its rate, is
unaltered Nieman and Maas (52) have suggested that the supply of
metabolic energy may be a basic limiting step in the reduction of
plant growth. They suggest that salinity induces increases in the

amounts of osmotic and ionic work necessary for normal cellular maintenance. Thus, the available expendable energy must be diverted from the normal growth processes in order for the plant to survive. In summary, after almost a century of research, the primary site of salinity stress on plant metabolism has not been identified and hopes of finding a simple one gene relationship with total salt tolerance seem very much reduced.

Tolerant species or cultivars may differ in the mechanisms by which they withstand high concentrations of salt. Levitt (44) has defined salt tolerance as ion accumulation in the absence of negative effects on growth. Other terms that have been used are salt resistance and salt avoidance. Salt avoidance is the strategy that a plant uses to escape salinity effects. Examples of salt avoidance mechanisms include: delayed germination or maturity until more favorable conditions prevail; the exclusion of salt at the rootzone; or compartmentation and secretion by specialized glands or organelles. The terms salt tolerance or salt resistance are often used interchangeably to define true cytoplasmic resistance to the effect of salinity or in conjunction with salt avoidance to describe all mchanisms that may give the plant a selective advantage to saline stress.

Ion Exclusion and Accumulation

The mechanism of ion exclusion has been one of the most frequently reported differences between salt sensitive and salt tolerant cultivars. Ion selectivity may occur at the external plasmalemma, tonoplasts or internal plasmalemma. Transport systems are selective for particular ion species and may also differ between tissues. However, the integrity of ion transport systems or the membrane matrices that maintain their function and orientation may be impaired by salinity at some specific threshold concentration. There is good evidence that genetic controls regulate structure and function at several different levels.

Berstrom (21) found that certain sugarbeet (Beta vulgaris) strains translocated Na from root to the leaves at a greater rate than other strains. Low Na levels in the root were found to be a dominant characteristic and K levels likewise were under genetic control (26). In soybean (Glycine max), differences between salt sensitive and salt tolerant cultivars were attributed to differences in Cl accumulation in the shoot (2). Further study showed that in 8 crosses of parents that were dissimilar in Cl accumulation, F_2 plants segregated in ratios of 3 Cl excluders to 1 Cl includers (1). In F_2 progeny from these crosses, the F_2 excluder plants segregated in ratios of 1 excluder to 2 segregating; whereas, the F_2 includer plants bred true. Abel

concluded that the factor to exclude or include Cl in soybean
leaves and stems was controlled by a single gene pair, and that the
dominate gene was the Cl excluder.   The Cl exclusion mechanism had
been reported earlier by Cooper and his colleagues (22, 23) as a
property of certain Citrus rootstocks and had coincidently and
subsequently been observed in several other woody species (7, 17,
19, 29).   Furr and Ream (33, 54) reported that the inheritance of
salt tolerance in Citrus as measured by Cl exclusion was not
simple, but was quantitative in nature.   Only a few of the progeny
of tolerant x sensitive crosses were highly tolerant.

     The correlation between salt tolerance and ion exclusion
seems to be a strong one in most glycophytes.   As already noted,
several investigators have reported distinct ecotypic differences
in salt tolerances among several grass species (38, 39, 61).   These
differences were found to be related to Na and Cl exclusion
mechanisms between the root and leaf tissues.   More recently,
differences in the Na and Cl transport to leaf tissues have been
shown to be related to salt tolerance differences among 14 strains
of Agropyron elongatum (56).

     Greenway (34) noted differences in Na and Cl uptake between
two barley cultivars.   At high salinities 'C.P.I. 11083' excluded
both Na and Cl from leaves and inflorescences more effectively than
'Chevron.'   On the other hand, Greenway's work with halophytes led
him to conclude that the mechanism of ion accumulation was a much
more effective salt tolerance mechanism than was ion exclusion.

     Halophytes demonstrate true salt tolerance, they accumulate
salt and, in many instances, actually have a requirement for it in
order to obtain maximum growth.   They are adapted to coastal areas
where waters are usually saline and to inland deserts that have
high salinities and a scarcity of rainfall.   Glycophytes have
evolved in a divergent manner and are adapted to areas with
frequent rains and higher water quality.   Under such situations,
salt concentration and water availability vary inversely and
certain salt avoidance mechanisms have important advantages (31).

     The adaptive feature of the ion accumulation mechanism is
undoubtedly linked to osmotic adjustment and a plant's ability to
extract water from a surrounding medium of low water potential.
The accumulation of salts and the synthesis of organic osmotic
substances reduce the osmotic potential differences between plant
and soil water.   Glycophytes that exclude salts must rely more
extensively on the synthesis of osmotica to facilitate the uptake
of water into the plant.   But at high salt concentrations, ion
exclusion somehow fails.   Salts rush into sensitive plant tissues
and death soon results.   The tolerance of many salt excluders has
been related to differences in the threshold level.

Differences between salt sensitive and salt tolerant tomato species have been attributed to differences in ion exclusion and accumulation strategies (55). This work has important implications in breeding and genetics studies on salt tolerance. Lycopersicon cheesmanii a wild relative of the commercial tomato (L. esculentum) is extremely salt tolerant and accumulates salt in response to increasing salinity. The commercial cultivar is a salt excluder. Since these two species are interfertile, they present an opportunity to make dramatic increases in salt tolerance in tomato and to study differences between the exclusion and accumulation mechanisms.

Environment

In the selection for salt tolerance, non-heritable variance is especially important. The non-heritable variance of an individual or family may reflect differences of the micro-environment in which different members of the family are raised or it may reflect accidents and chances of development.

Plant phenotype is a product of genotype and environment. Salinity and environmental effects interact in several ways (47); thus, genetic relationships are complicated by the fact that genotype - environmental reactions are actually part of the salinity problem. Salinity imposes one environmental restraint upon plant growth, but little is known about the genes that affect salt tolerance. We do know, however, that the longer the chain of events between the action of the genes and their final expression in the phenotype, the greater the complexity of effects that can arise from a single alteration. As long as we can only measure the effects of salinity as a reduction in growth, we will not have a simple genetic system with which to work. Furthermore, since growth or reduction of growth is a quantitative character, measurements of the salinity effect must be made in a quantitative manner. The disciplines of applied biometrical and quantitative genetics are relatively new to the plant breeder and geneticist.

There is yet another aspect of genetics and environment that has not been seriously considered; specifically, a permanent alteration in the genome that is stimulated by the environment. These are not mutagenic effects, but rather secondary effects of the type dscribed by Durrant (27). He found that certain cultivars of flax (Linum usitatissimum), when grown in soils with certain fertilizer combinations, not only reflected the effects of the treatments in their growth, but also transmitted effects to their progeny in the next and later generations. These changes were transmitted equally by pollen and egg and thus resemble paramutation in Zea mays (20) rather than mutation induced by radiation or chemical mutagens (50).

Although it seems that the changes cited by Durrant (27) are nuclear in origin their nature and the details of their behavior in transmission to offspring are not yet known. Similar conditioning changes have been observed in <u>Pisum sativum</u> and <u>Nicotiana rustica</u> (40,41), but they have not been found in all species and are not shown by all cultivars in a particular species where they have been found.

An interesting parallel exists with salinity. Strogonov (60) reported that cotton seed derived from plants grown on saline soils displayed heritable resistance to salinity. Whether this observation was due to the selection of a simple Mendelian trait for salt tolerance or due to some type of environmental conditioning or paramutation are questions that have not been fully answered.

Genetic Fitness

A useful choice of genes is essential to study the transmission of characters. Optimally, each genotype will give rise to a distinctive, easily-identifiable phenotype. Gaps in our understanding of the true effects of salinity have limited the choice of genes with which to work and unique tolerant phenotypes can be identified only under carefully controlled conditions. There are several methods that can be used to simplify matters. Essentially, we can partition the effects that salinity exerts upon the plant or we can partition the responses of the plant to salinity. We have already discussed some non-genetic tactics for doing these things. Plants respond differently to ionic and osmotic influences of salinity, and certain species and genotypes respond differentially to specific ions. The effects of salinity can also be partitioned with respect to plant development. Salinity effects at germination, vegetative growth phases, flowering, or seed set may differ substantially.

Akbar and Yabuno (3-5), found that four rice varieties responded quite differently to salinization during flowering. Among the varieties 'Bluebonnet','IR8', 'Jonah 349' and 'Magnolia', three types of sterility caused by salinization could be classified. F_1 crosses between the relatively salt resistant 'Jonah 349' and the salt sensitive 'Magnolia' varieties were highly resistant to salinity. The F_2 populations resulted in some salt resistant combinations and salt resistant progenies were selected from the F_3 and F_4 populations. The mode of inheritance of the delayed type panicle sterility induced by salinity was investigated with the F_2 of the backross populations derived from 'Magnolia' x 'Jonah 349' crosses. Resistance to this type of sterility was found to be a

dominant character and presumably was controlled by a small number
of genes.   At least three pairs of genes seemed to be involved.
Related experiments eveluted salt tolerance of rice varieties
during the early developmental stages such as germination, seedling
and transplanting age. Effect of salinization was found to be much
more severe at the young seedling and transplanting stages than at
the germination stage, but no correlation for salinity resistance
was found among the various developmental stages or with the pollen
sterility factor noted at flowering.

        Fitness to salinity might also include different combinations
of avoidance, exclusion and tolerance mechanisms depending upon the
specific agronomic and cultural practices that would be followed.

        Two genetic methods that can be used to improve measurement
of fitness to salinity are progeny testing and the use of inbred
lines.   Unfortunately, these methods are both laborious and time
consuming. Progeny tests can be used to circumvent the need for
refined salinity and environmental control and the use of inbreds
can be used to evaluate the contribution of particular gene
differences against uniform genetic backgrounds.   This is often
referred to as the isoline technique. Since the effect of salinity
on physiological characteristics has not yet been distinguished, we
must measure growth, a continuous variable.   Continuous variations
in growth are due to genes which we cannot expect to be easily
recognizable as individuals upon segregation.   We also cannot
expect to overcome the intrinsic difficulty of the situation by
using somatic analysis, a technique that separates growth into
components.   Despite the limitations there are still imaginative
ways in which these two genetic techniques can be used to answer
some of the fundamental questions concerning salinity effects on
plants.

        Under artificial selection, yield and quality are substituted
for survival of the species as major selection criteria.   Yield
criteria may be measured on a relative or absolute scale.   This is
another way to partition plant responses.   Relative salt tolerance
is defined as growth (yield) under saline conditions relative to
growth under nonsaline conditions.   Judged by such a criterion, a
slow growing plant may rate well if salinity does not substantially
reduce its growth further.   Absolute salt tolerance is directly
related to the maximum growth or yield potential of a given plant
under high salinity regardless of its growth rate under nonsaline
conditions.   Under salinity stress, plant growth may decrease
substantially without affecting its evaluation for absolute tolerance.   For example, a vigorously growing plant that is drastically
affected by salinity would have low relative salt tolerance, but if
it still yields more than a slow-growing plant with high relative
tolerance, it would be judged as having a higher absolute

tolerance. Although absolute yield under high salinity is of
greater importance than relative yield both must be considered. If
a cultivar is relatively tolerant but grows slowly or has poor
quality characteristics, its high relative tolerance should not be
ignored if that characteristic can be transferred to high yielding
cultivars. Relative tolerance may also contribute to the overall
fitness of the species to produce well in a range of saline and
nonsaline environments – a character referred to as environmental
plasticity. Such considerations should inspire and promote significant research efforts on the genetic mechanisms of salt
tolerance.

References

1. G. H. Abel, Crop Sci. 9:697-698 (1969).
2. G. H. Abel and A. J. McKenzie, Ibid 4:157-161 (1964).
3. M. Akbar and T. Yabuno, Japan J. Breed. 24:176-181 (1974).
4. M. Akbar and T. Yabuno, Ibid 25:215-220 (1975).
5. M. Akbar and T. Yabuno, Ibid 27:237-240 (1977).
6. A. D. Ayers, Agron. J. 45:68-71 (1953).
7. A. D. Ayers, D. G. Aldrich and J. J. Cooney, Calif. Avocado
    Soc. Yearbook, 174-178 (1951).
8. A. D. Ayers, J. W. Brown and C. H. Wadleigh, Agron. J. 44:
    307-310 (1952).
9. A. D. Ayers, C. H. Wadleigh and L. Bernstein, Proc. Amer. Soc.
    Hort. Sci., 57:237-242 (1951).
10. L. Bernstein, Amer. J. Bot. 48:909-918 (1961).
11. L. Bernstein, Desalination Res. Conf., Nat. Acad. Sci., Nat.
    Res. Council Publ. 942:273-283 (1961).
12. L. Bernstein, Amer. J. Bot. 50:360-370 (1963).
13. L. Bernstein, Ann. Rev. of Phytopathology 13:295-312 (1975).
14. L. Bernstein and A. D. Ayers, Proc. Amer. Soc. Hort. Sci. 57:
    243-248 (1951).
15. L. Bernstein and A. D. Ayers, Ibid 61:360-366 (1953).
16. L. Bernstein and A. D. Ayers, Ibid 62:367-370 (1953).
17. L. Bernstein, J. W. Brown and H. E. Hayward, Ibid 68:86-95
    (1956).
18. L. Bernstein and H. E. Hayward, Annu. Rev. Plant Physiol. 9:
    25-46 (1958).
19. L. Bernstein, C. F. Ehlig and R. A. Clark, J. Amer. Soc. Hort.
    Sci. 94:584-590 (1969).
20. R. A. Brink, Quart. Rev. Biol. 35:120-137 (1960).
21. H. Berstrom, Ann. Agr. Col. 5:89-104, Sweden (1938).
22. W. C. Cooper and B. W. Gorton, Proc. Am. Soc. Hort. Sci. 59:
    143-146 (1952).
23. W. C. Cooper, B. S. Gorton and C. Edwards, Proc. Rio Grande
    Valley Hort. Inst. Proc. 5:46-52 (1951).
24. D. R. Dewey, Agron J. 52:631-635 (1960).

25. D. R. Dewey, Crop Sci. 2:403-407 (1962).

26. J. Dudley and L. Powers, J. Am. Soc. Sugar Beet Technol. 11: 97-127 (1960).

27. A. Durrant, Nature 181:928-929 (1958).

28. F. M. Eaton, J. Agr. Res. 64:357-399 (1942).

29. C. F. Ehlig, Proc. Amer. Soc. Hort. Sci. 76:323-331 (1960).

30. E. Epstein and R. L. Jeffries, Ann. Rev. Plant Physiol 15: 169-184 (1964).

31. T. J. Flowers, P. F. Troke and A. R. Yeo, Ibid 28:89-121 (1978)

32. L. E. Francois and E. V. Maas, "Plant Responses to Salinity: Indexed Bibliography," USDA Agric Rev. and Manual Ser. (1978).

33. J. R. Furr and C. L. Ream, First Intl. Citrus Symposium 1:373-380 (1969).

34. H. Greenway, Aust. J. Biol. Sci. 18:763-779 (1965).

35. H. Greenway, J. Aust. Inst. Agric. Sci. 39:24-34 (1973).

36. H. Greenway and R. Munns, Ann. Rev. Plant Physiol. 31: 149-190 (1980).

37. H. Greenway and C. B. Osmond, Plant Physiol. 49:256-259 (1972).

38. U. Gupta and P. S. Ramakrishnan, Proc. Indian Acad. Sci. 86B: 275-280 (1977).

39. N. J. Hannon and H. N. Barber, Search 3:259-260 (1972).

40. H. R. Highkin, Proc. Xth Int. Cong. Genets., Montreal, 2: 120 (1958).

41. J. Hill, Genetics 55:735-754 (1967).

42. E. R. R. Iyengar and J. B. Pandya, Sand Dune Res. Soc., Japan, 24:45-52 (1977).

43. R. C. Jaeswal, K. Singh and M. L. Pandita, Haryana J. Hort. Sci. 4:51-59 (1972).

44. J. Levitt, in: "Responses of plants to environmental stresses," (T. T. Koslowsky, ed.), Academic Press, N. Y. (1972).

45. C. B. Lyon, Bot. Gaz. 103:107-122 (1941).

46. E. V. Maas and G. J. Hoffman, J. Irrig. and Drainage Div., ASCE 103(IR2):115-134 (1977).

47. E. V. Maas and R. H. Nieman, in: "Crop tolerance to suboptimal land conditions," (G. A. Jung, ed.), pp. 277-279, ASA Spec. Publ. (1978).

48. O. C. Magistad, Bot. Rev. 11:181-230 (1945).

49. F. I. Massoud, in: "Salinity and alkalinity as soil degradation hazards," FAO/UNEP expert consultation on soil degradation, FAO, Rome (1974).

50. K. Mather, Nature 190:404-406 (1961).

51. F. N. Ponnamperuma, in: "Plant response to salinity and water stress," (W. J. S. Downton and M. G. Pitman, eds.), Assoc. for Sci. Cooperation in Asia, Sydney, Australia, (1977).

52. R. H. Nieman and E. V. Maas, Sixth Intl. Biophysics Congress, Sept. 3-9, Kyoto, Japan, p. 121 (1978).

53. M. Rai, Indian J. Plant Physiol. 20:100-104 (1977).

54. G. L. Ream and J. R. Furr, J. Am. Soc. Hort. Sci. 101:265-267 (1976).

55. D. W. Rush and E. Epstein, Plant Physiol. 57:162-166 (1976).

56. M. C. Shannon, Agron. J. 70:719-722 (1978).

57. M. C. Shannon and M. Akbar, in: "Breeding plants for salt tolerance," Workshop on membrane biophysics and salt tolerance in plants, Faesalabad, Pakistan, (1978).

58. K. R. Stino, M. Abdel-Aziz, A. Fattah, A. S. Abdel-Salam, W. A. Warid, I. A. Elmofty and M.M. Abdel-Gawwad, Desert Inst. Bull. A.R.E. 22:167-174 (1972).

59. B. P. Strogonov, in: "Physiological basis of salt tolerance of plants," Jerusalem, Israel, Program for Sci. Transl. p. 279 (1962).

60. B. P. Strogonov, E. F. Ivanitskaya and I. P. Chernyadeva, Fiziol. Rost. 3:319-327 (1956).

61. B. L. Tiku and R. W. Snaydon, Plant and Soil 35:421-431 (1970).

62. P. B. Vose, Herbage Abstracts 33:1-13 (1963).

63. R. Weimberg, Plant Physiol. 46:466-470 (1970).

OSMOREGULATION BY ORGANISMS EXPOSED TO SALINE STRESS:

PHYSIOLOGICAL MECHANISMS AND GENETIC MANIPULATION

D. W. Rains, L. Csonka, D. LeRudulier, T. P. Croughan,
S. S. Yang, S. J. Stavarek and R. C. Valentine

Plant Growth Laboratory
    and
Department of Agronomy and Range Science
University of California
Davis, CA 95616 USA

INTRODUCTION

Organisms exposed to saline environments have evolved a number
of processes which enhance tolerance to this stress. Elevated
levels of salinity present an environment high in inimical chemicals and low in water potential. The organisms must tolerate these
toxic ions and associated nutrient imbalances. This coupled with
reduced water availability limits this environment to organisms
capable of tolerating these two conditions (1).

Tolerance is achieved when the organisms have the capacity to
grow and reproduce in this environment. Agricultural importance is
achieved when these organisms are productive and economically
successful in saline areas.

The following discussion is developed in two parts. The first
part provides information on genetic engineering procedures used to
improve tolerance of organisms to salinity. These "engineered"
organisms are essential tools in evaluating the biochemical and
physiological bases for osmoregulaton and other mechanisms related
to tolerance (2).

A second area to be addressed is the use of plant cell culture
techniques to select for saline tolerant cell lines and the use of

283

these cell lines to provide information on osmoregulation in higher plants. The selection of saline tolerant cell lines may also provide germplasm for breeding programs focused on improving plant performance in saline environments.

The ultimate goal is to obtain organisms which are adapted to saline areas and to enhance exploitation of these areas for production of food, feed, fiber and chemical feedstocks.

Saline Influences on Organisms and Use of Organic Solutes and Genetic Engineering to Alter Stress Tolerance.

Organisms exposed to an increase in osmolarity in their growth media respond by a corresponding elevation of the osmotic tension in the cell interior. Bacteria may accomplish this by a build-up of the internal concentration of proline, or in some cases, glutamate, or γ-amino butyrate (3), substances which prevent harmful effects of water stress. That this kind of response is of physiological significance was indicated by the observations that proline stimulated the growth of Salmonella orianenburg in high salt media. The nitrogen-fixing bacterium, Klebsiella pneumoniae, as well as Escherichia coli and S. typhimurium show similar response to exogenous L-proline (4).

The discovery of an osmotic stress tolerance plasmid (3) has raised the exciting possibility of genetic engineering of hardy strains of Rhizobium resistant to drought and salinity. This work also provides important insight into the mechanism of drought and salt tolerance in higher plants, a rapidly emerging area of plant research. A brief summary of the construction and properties of a transmissible plasmid for osmotic tolerance against the dehydrating effects of high concentrations of salt and drought is provided below (see Fig. 1).

Stress tolerant mutants of S. typhimurium were isolated which overproduce L-proline. The selection was for strains that grow in the presence of normally toxic concentrations of the L-proline analogue, L-azetidine-2-carboxylate. (Rationale: mutants which excrete L-proline dilute out and thus antagonize the analogue.) Among the mutants resistant to L-azetidine-2-carboxylate, we found some which grew faster than the parental strain on media containing inhibitory concentrations of NaCl. The most salt-tolerant strain obtained thus far has a doubling time of 4 hr in minimal medium containing 0.65 M NaCl, as compared to the parental strain, which grows with an 8 hr doubling time. The mutation conferring L-azetidine-2-carboxylate resistance and salt tolerance was closely linked to the locus of the first enzyme of the L-proline biosynthetic pathway, proB, and it was tranferred via an F' factor

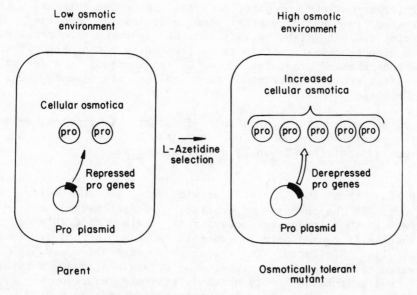

Figure 1. Genetic engineering of an osmotic tolerance plasmid.
           The plasmid carries genes for proline biosynthesis;
           the depressed plasmid has sustained a "regulatory"
           mutation which switches on proline synthesis to such
           a high level that the proline molecules function as
           a sort of "osmotic antifreeze" permitting growth in
           high osmolarity. It is interesting to speculate that
           a similar mechanism may operate for other forms of
           osmotic stress including drought, high sugars (xero-
           tolerance) and ethanol tolerance, a hypothesis cur-
           rently being tested.

(a plasmid that is transmissible among <u>Enterobacteriaceae</u>) to other strains of <u>S</u>. <u>typhimurium</u>, which as a result acquired the L-azetidine-2-carboxylate-resistant, salt-tolerant phenotype of the donor. These results corroborate the hypothesis that L-proline has a central role in osmoregulation, and suggest a method of isolation of salt-resistant mutants of other bacterial species or higher plants.

One of our goals is to develop a procedure to obtain osmotolerant mutants of other bacteria, including <u>Rhizobia</u>. As a first step in testing whether the selection of proline over-producing mutants might be a practical approach, we transferred into a nitrogen fixing strain of <u>K</u>. <u>pneumoniae</u> one of the Fls which conferred enhanced osmotolerance on <u>S</u>. <u>typhimurium</u>, and determined the effect proline over-production had on nitrogenase activity under osmotic stress.

First, we examined the effect of osmotic inhibition, achieved by the addition of NaCl, on the growth rate (Table 1) and nitrogenase activity (Table 2) of <u>K</u>. <u>pneumoniae</u>.

Nitrogenase activity was much more sensistive to osmotic inhibition than the overall growth rate of the cells. For instance, in the absence of proline, 0.4 NaCl caused a ten-fold decrease in the nitrogenase activity of strain M5A1 (Table 2, col. 1), whereas it caused only a two-fold decrease in the growth rate (Table 1). Similarly, in the absence of proline, 0.6 M NaCl decreased the nitrogenase activity of strain M5A1 over a hundred-fold, while it caused only about a six-fold reduction in growth rate. Analogous results were obtained with strain KY2 (pro-3/F'$_{128}$-proB$^+$A$^+$).

The stimulatory effect of proline on the growth rate is manifested only under conditions of extreme osmotic inhibition ($\leqslant$ 0.6 M NaCl with <u>K</u>. pneumoniae, unpub. results). However, proline exerted a much greater stimulatory effect on nitrogenase at lower osmolarities. Thus, in the presence of 0.4 M NaCl, 0.5 mM proline caused approximately a four-fold enhancement of the nitrogenase activity of strain M5A1, and approximately fifty-fold enhancement in the presence of 0.5 M NaCl (Table 2, col. 1 and 2).

The effect of the mutation resulting in proline over-production in strain KYL (F'$_{128}$ pro-$\overline{74}$) was similar to that seen when proline was supplied exogenously, in that the growth rate was stimulted only under extreme osmotic inhibition (compare strains KY1 and KY2, without proline, with 0.6 M NaCl; Table 2, col. 3 and 4). However, the mutation had a much more pronounced stimulatory effect on nitrogenase activity: in the absence of proline, at 0.4 and 0.5 M NaCl, the nitrogenase activity of strain KY1 was at

TABLE 1. The effect of exogenously added proline and of proline over-production on the growth rate of K. pneumoniae under conditions of osmotic inhibition.*

NaCl (M)	Growth Rate (generation/h)				
	M5A1 (wild type)		KY1** (F' pro-74)	KY2 (F' proA⁺A⁺)	
	-proline	+0.5 mM proline	-proline	-proline	+0.5 mM proline
0.0	0.33	0.33	0.31	0.33	0.33
0.3	0.25	0.26	0.24	0.25	0.26
0.4	0.17	0.24	0.17	0.17	0.24
0.5	0.15	0.16	0.14	0.14	0.17
0.6	0.054	0.096	0.079	0.033	0.15

*Growth was under anaerobic conditions at room temperature in sucrose minimal medium supplemented with 0.68 mM L-glutamine. See LeRudulier et al. (37) for details of media and mutant characteristics and pedigree.

**Strain KY1 is a proline overproducing mutant selected for tolerance to high osmotic environments (carries proline-overproducing plasmid). Strain KY2 is the osmotically sensitive parental strain of KY1.

TABLE 2. The effect of exogenously added proline and of proline over-production on nitrogenase activity under conditions of osmotic inhibition.*

NaCl (M)	Nitrogenase Activity (μmoles ethylene produced per hour per mg protein)				
	M5A1 (wild type)		KY1 (F' pro-74)	KY2 (F' proB⁺A⁺)	
	-proline	+0.5 mM proline	-proline	-proline	+0.5 mM proline
0.0	2.55	2.66	2.64	2.71	2.59
0.3	0.42	1.26	1.55	0.60	1.31
0.4	0.25	0.96	1.53	0.14	0.74
0.5	0.02	0.94	0.42	0.04	0.73
0.6	0.02	0.12	0.26	0.01	0.19

*See Table 1 and reference (37) for details on methods and procedures.

least ten times greater than that of strain KY2.   At 0.6 M NaCl,
the stimulatory effect of the mutation was greater than twenty-
five-fold.

In continuing the discussion of some of the key features of
the mechanisms of osmoregulation, Measures (5) found that many
varieties of a bacteria respond to osmotic stress by accumulating
high internal pools of glutamate which he proposes to play an
important role as a counterion for K^+.   However, since these
observations were obtained using a rich broth medium containing
large amounts of preformed glutamate, it is not possible from these
data to distinguish between accumulation of exogenous glutamate
versus de novo biosynthesis.   We have answered this question by
growing cells in defined medium containing no exogenous glutamate
and following glutamate pools in response to osmotic stress.   The
major point here is that glutamate levels increase markedly during
osmotic stress to levels approximately equivalent to that of K^+ and
on a time scale essentially synchronized with K^+ accumulation (4).

It is interesting to speculate that the sophisticated control
system(s) which have been evolved for modulating glutamate and
glutamine biosynthesis in bacteria may play a vital role in
osmoregulation.   Indeed, the control systems for the glutamine-
dependent route of glutamate biosynthesis (glutamate synthase-
glutamine synthetase) is among the most elaborate yet discovered in
bacteria (also note that the glutamate dehydrogenase route of
glutamate biosynthesis may respond to osmotic conditions; see 6).
The network of control of the glutamine-glutamate pathway includes
not only the now classic examples of induction and repression at
the genetic level, and feedback or end product inhibition but also
covalent modification (adenylylation) occurring at specific sites
of glutamine synthetase (7).   Covalent modification of glutamine
synthetase resulting in raising or lowering in catalytic activity
for glutamine biosynthesis and consequently glutamate production is
mediated in turn by an elaborate enzyme cascade that is triggered
by signals from the environment.   In this manner the cell's demand
for glutamate is intimately linked to changes in the environment.
It seems clear that the elegant glutamine synthetase cascade is
designed for and is capable of rapid amplification of environmental
signals into biochemical language.   Experiments are in progress to
determine whether modulation of glutamate and glutamine biosynthesis represent a part of the trigger or sensing mechanism of
osmoregulation.   The "mutant approach" is being utilized to
determine which step(s) is essential (6) during osmotic stress.
Exogenous proline was provided in the rich medium used to grow all
the cultures.   Under these circumstances, specifically proline was
found to accumulate to high levels: however, we have found that
when several of the same species used by Measures are grown in
chemically defined medium (salts and a single carbon source) the

picture is dramatically changed with proline levels being very low
and glutamine and glutamate elevated (4).   This finding has
influenced our interpretation of events occurring during osmotic
adaptation. Glutamine and glutamate synthesis is discued as being
the major source of glutamine and glutamate (some exogenous
glutamate may also be available?);   in contrast, the exogenously
supplied proline appears to be the major source of proline during
osmotic stress. Undoubtedly this scheme is oversimplified and will
require further changes as more knowledge of the mechanism is
elucidated.

It should also be pointed out that these microbial systems
should not be regarded as merely models for higher plants; indeed
osmotic tolerance is an essential trait in the utilization (fermentation) of biomass yielding ethanol and numerous other essential
fuels and chemicals.   In short, the cardinal importance of osmoregulation as an essential cellular process provides added incen-
tive for research in this area, knowledge which may have applications for selection and breeding of stress tolerant plants.

Osmoregulation in Higher Plants, Organic Solutes

Plants use organic solutes and inorganic solutes as compatible
osmotica. The accumulation of compatible solutes provide the means
for maintaining favorable water balance and cell turgor during
stress. Plants exposed to salinity respond by using a number of
mechanisms to maintain a favorable water balance (e.g., osmotic
balance).   A number of these processes and substances used by
plants to osmoregulate are similar with those described for
microbial systems (8).

A body of evidence has been accumulated supporting the role of
organic solutes as compatible osmotica (8, 9). The fact that there
is osmotic distribution within cells suggests that electrolyte
concentrations may be relatively low in cytoplasmic compartments
and the vacuole may accumulte substantial quantities of inorganic
ions (10). The reverse distribution has been indicated for organic
solutes (11).   It is also pertinent that the activity of many
enzymes is disrupted by inorganic solutes (12-14), while organic
solutes are either non-disruptive or only slightly so at high
osmotic concentrations (15).

These observations suggest that a successful means of osmoregulation in plants exposed to salinity is via organic solute
metabolism. Jefferies (8) has tabulated a list of halophytic plant
species that accumulate a number of organic solutes while growing
in saline habitats.   The solutes include proline, methylated
quaternary compounds and sorbitol. Other important organic solutes

are betaine, glycinebetaine and glycerol (9).

The role of organic solutes in plants exposed to salt stress has been discussed in considerable detail and there is little agreement about either their role or significance. It has been suggested that the accumulation of these compounds may be the result of lesions caused by salt injury (16). The accumulation of proline and glycinebetaine, however, has been associated with increased salt tolerance in some plant species, but this is not a universal observation. Wyn Jones (9) suggests that accumulation of glycinebetaine does help a salt stressed plant to adapt but the accumulation of glycinebetaine is not absolutely required for salt tolerance in higher plants.

The role proline plays in osmoregulation remains a moot question. Proline does accumulate in plants subjected to stress, but the response appears to be related to extreme osmotic shock or possible salt toxicity (17). Jefferies (8) has discussed the importance of nitrogenous compounds in the survival and productivity of halophytic species. Nitrogenous compounds such as proline could effectively regulate the storage of essential nitrogen. Proline is a good candidate because it is osmotically active, compatible with cytoplasmic constituents and can be easily converted to glutamate. This is significant because glutamate is an amino acid central to the regulation and synthesis of other essential amino acids and as discussed in this article implicated in osmoregulation in bacteria. A plant exposed to salinity could use proline for both nitrogen reserves and osmoregulation.

Although compatible organic solutes are not correlated ubiquitously with enhanced salt tolerance in plants, such a characteristic may be useful as a selection criterion for breeding plants. The presence of specific organic molecules could be used to identify plants with potential for improved performance in saline environments. A rapid chemical test would permit rapid screening of large numbers of plants for this characteristic. Currently this approach cannot be used routinely for organic solutes because of the wide discrepancy in the information on the significance of these compounds in relation to salt tolerance.

It is apparent from the previous discussion, however, that organic solutes play a central role in osmoregulation and saline tolerance of both microorganisms and higher plants.

Inorganic Solutes and Saline Stress

A number of mechanisms or processes influence the tolerance of organisms to environmental stress. In the previous discussion the effect of organic osmoticum on stress tolerance was discussed. Inorganic ions also provide osmoregulation for organisms exposed to water or salt stress.

Cells adapt by using some remarkable biochemical machinery including a potassium (K^+) accumulation system which is somehow triggered in response to increasing osmotic strength in the environment; this osmotically stimulated system appears to be already present in the cell since only a few seconds are sufficient to permit active accumulation of K^+ (probably too short a time scale to allow new rounds of protein synthesis to occur).

Accumulation of K^+ appears to be a general response to increased osmotic stress caused by both neutral organic compounds (e.g., sucrose) as well as inorganic charged molecules (e.g., NaCl).

As pointed out by Epstein and Schultz (18) and Christian and Waltho (19) inorganic ions such as K^+, commonly available in most environments, may be, energetically speaking, the cheapest form of osmoregulators. In plant systems there has been considerable emphasis on the role of ions whereas there is a much smaller literature on bacteria. Epstein and co-workers have carried out a comprehensive study on the role of potassium (K^+) in osmoregulation in Enteric bacteria with their conclusions being of considerable interest here. In their 1965 paper Epstein and Schultz say, "The ability of E. coli to maintain its internal osmotic activity equal to, or somewhat greater than that of the surrounding environment, through regulation of its cell K^+ content, is of profound functinal significance for an organism which may be subjected to a wide range of growth conditions." They further conclude that:

"1. Under a variety of conditions including those suitable for optimal growth, the osmolarity of the growth medium is a major determinant of the cell K^+ content."

"2. The growing cell responds to abrupt changes in the surrounding osmolarity with rapid changes in cell content in the direction necessary to minimize the osmotic difference."

"3. The conclusion that the bulk of the intracellular K^+ in E. coli exists in an unbound, osmotically active form, though not directly established, is strongly suggested."

Epstein and co-workers describe the interesting finding that certain classes of K^+ uptake mutants which they have analyzed extensively are diminished in their capacity to grow in medium of high osmotic strength. In other words, such mutants behave as osmotically sensitive strains, interesting indirect evidence for a key role of potassium accumulation in osmotic adaptation. This important area deserves further work.

The finding by Epstein and Schultz (18) that the increase in cellular K raises the cellular osmolarity by about half as much as the increase in medium osmolality is of particular interest since each mole of K^+ accumulated must be associated with an equivalent of intracellular anion. Thus, the increase in cellular osmolality would be equal to the increase in medium osmolality if the anions accumulated with K^+ were univalent and osmotically active.

With this as background it is interesting to speculate that the elusive anion mentioned by these workers is in fact glutamic acid harboring a negative charge at neutral pH (4).

Potassium appears to occupy a unique position in osmoregulation of organisms exposed to saline environments. In a preliminary experiment a number of Rhizobium organisms were exposed to increasing NaCl concentrations. The Rhizobium spp. were found to differ in their ionic regulation. This is illustrated by the differential levels of potasium found in the cells (Table 3). R. meliloti and R. japonicum 440 responsed similarly to NaCl. Intracellular K^+ levels increase as NaCl was increased. In contrast, R. japonicum 110 was very sensitive to NaCl and although K^+ increased in concentration the percent increase was relatively small and 0.1 M NaCl proved to be toxic to the organism. R. japonicum 110 shows little tolerance to salinity. This is of importance since the organism is a common Rhizobium species widespread in the soybean growing areas of the midwest. R. meliloti and R. japonicum 440, however, showed considerable tolerance to salinity and this correlates with the accumulation of K^+. From the previous discussion it seems reasonable to speculate that the ability of these Rhizobium to uptake K^+ in the presence of high levels of Na^+ is a potential mechanism which enhances salt tolerance. It is also encouraging to observe the apparent genetic variability in R. japonicum in response to salt stress. This suggests the potential to genetically alter these organisms and enhance salt tolerance.

Osmoregulation with Inorganic Solutes: Significance for Plants

Saline environments consist of an elevated level of inorganic ions. The ion mixture can vary extremely and have high concentrations of both essential and non-essential elements. These ions are osmotically active and can provide the plant with a ready reservoir of both compatible and incompatible solutes. In many instances the plant is faced with an extremely hostile environment, including potentially toxic ions such as Na^+ and Cl^-. Elevated salt levels disrupt enzyme and ribosomal functioning in vitro (20).

There is, however, a salt-stimulated increase in the specific activity of some enzymes from some halophytes (21). Flowers (22) assayed several enzymes from both halophytes and glycophytes for activity in the presence of salt. Only one, an ATPase from the halophyte, was stimulated by salt. The others were uniformly inhibited. Hawker (23) investigated the effect of salt on ADP-glucose from salt-tolerant and salt-sensitive species and found uniform inhibition. Poljakoff-Mayber and Meiri (24) found that salinity reduced the ability of excised pea root tips to incorporate amino acids into protein. Likewise, Hall and Flowers (25) demonstrated that protein synthesis is salt-sensitive for both halophytes and glycophytes.

Kalir and Poljakoff-Mayber (26), however found that mito-chondrial malic dehydrogenase was more sensitive to salt than was soluble MDH, and Austenfeld (27) found glycolate oxidase activity undisturbed by up to 1.0 M NaCl or KCl.

Most of the evidence cited above suggests that salt tolerance must be due in large part to a "salting-away" of ions within compartments (i.e., vacuoles), removing them from the metabolic processes, paralleled by production of osmotica to maintain favorable water relations in the cytoplasm. The evidence of Kalir and Poljakoff-Mayber (26), however, suggests actual cytoplasmic toler-ance of high levels of salt, or the ability to substitute Na^+ for K^+ in activation of some enzymes. Austenfeld (27) suggested that the apparent salt tolerance of glycolate oxidase in vitro was due to an adaptation of the enzyme to high salt levels within peroxisomes in vivo. A necessary step toward systematic under-standing of salt tolerance in plant cells would be determination of the location of organic and inorganic osmotica within the cell.

The vacuole may comprise 85-95% of total cell volume (11, 28). At this stage, it is assumed that the bulk of accumulated ions are in this compartment. It has been shown that Na^+ is pumped into the vacuole (28). Further evidence was given by Hall et al. (29). They studied the site of salt accumulation in halophytes by using Rb^+ as an electron-dense tracer and found accumulation of Rb^+ in the vacuole to the point of absence from the cytoplasm. The idea, then, is that toxic ions are sequestered inside the vacuole, and that this accumulation maintains a favorable osmotic balance between the inside of the cell and the saline external environment.

That does not explain how osmotic balance is achieved at the cytoplasm-medium interface, however. There must be accommodation to the lowered water potential in the medium through adjustment of matrix or solute potentials in the cytoplasm. By far, the larger factor is the solute potential. However, the compounds which are present in salt-treated cells and absent in cells not treated with salt should be identified. These compounds appear to act as osmotic agents maintaining a favorable cell water balance. The literature is replete with suggestions that this is, indeed, feasible.

Ion concentrations within cells are regulated by a number of processes. Ions can be excluded or uptake can be regulated. The amount and types of ions accumulated or excluded by cells in saline environments play a significant role in adapting these organisms to this stress. Exclusion can mean either not permitting ions into the cell in the first place, or pumping them out once they are in. Rains (12) has pointed out that "outpumps" would be energetically more expensive than simple exclusion at the plasmalemma. Exclusion is then seen as preferential absorption of K^+, for example, over Na^+, and salt tolerance would indicate higher than usual K^+ specificity for the transport mechanism. Nevertheless, Na^+ out-pumps have been postulated (30,31) and Pierce and Higinbotham (28) found that Na^+ was pumped out of the plasmalemma and also pumped into the vacuole of coleoptile cells of Avena sativa L. Since the tonoplast had a less specific mechanism than the plasmalemma, those researchers conferred primary selectivity on the plasmalemma. Rains (12) supports that by giving evidence that halophytes show greater affinity for K^+ at high Na^+ concentrations than do glycophytes. Exclusion of salts, either directly or by outpumps, would certainly be an energetically expensive proposition for a cell bathed in sea water.

Sodium chloride selected and non-selected alfalfa cells were grown in the presence of increasing levels of NaCl. The NaCl selected cells were found to be more tolerant of Na^+ and could grow in the presence of 1% NaCl (w/v), a level of NaCl that was toxic to the non-selected cells (41). The selected cells retained consider-ably more K^+ than the non-selected cells (Table 4). Regulation of K^+ by cells exposed to stress is a common observation and suggests that this mechanism may be common to many organisms exposed to saline environments.

The role of inorganic ions in osmoregulation of organisms exposed to stress is common to bacteria and higher plants. This ionic regulation is related to enhanced saline tolerance. Mechan-istically ion regulation would be expected to involve membrane transport processes. It seems reasonable to assume that this is under genetic control and genetic manipulation of ion regulation by organisms represents a strategy to better adapt to saline environments.

Energy Cost of Tolerating Stress

Organisms exposed to stress adapt via a number of processes. We have discussed some of these previously. Organic and inorganic

TABLE 3.   Steady state levels of potassium in Rhizobium spp.
grown under salt stress.[*]

NaCl (M)		n mole K^+/mg cell protein [***]
0.0	R. meliloti[**]	263
0.2		503
0.4		828
0.0	R. japonicum 440	575
0.2		758
0.4		900
0.0	R. japonicum 110	423
0.5		433
0.1		0 No growth

[*] Rhizobium spp. were grown on MSY medium (44) which had
modified carbon and nitrogen source. The carbon source
was Na lactate added at 1 gm/liter and the nitrogen
source was L-aspartic acid added at 1 gm/liter. The
media was adjusted to pH 6.8.

[**] Rhizobium meliloti (alfalfa) and R. japonicum 440 soybean) are fast growing Rhizobium species and were shown
to be salt resistant. R. japonicum (soybean) 110 is
slow growing and was relatively sensitive to salt (unpub. results).

[***] Intracellular K^+ was measured by the method of ref. 45.

osmoticum are accumulated to combat unfavorable water gradients.
By maintaining cellular concentrations of these substances which
exceed the external environment organisms adapt, survive and grow.
This requires expenditure of energy, either to accumulate inorganic
materials or to synthesize and maintain organic osmoticum.

TABLE 4. Potassium content of alfalfa cells grown under salt stress.*

| NaCl, conc. % (w/v) | Potassium content mg/dry wt tissue | |
	NaCl sensitive cells	NaCl selected cells
0	14.8	15.0
0.25	10.5	14.8
0.50	9.5	14.0
0.75	9.0	12.0
1.00	7.5	11.8

* Details on the characteristics of these alfalfa cells are presented in a paper by Croughan et al. (41).

TABLE 5. Energy use efficiency of salt selected (S) and non-selected (NS) alfalfa cells grown in the presence of increasing NaCl.

| NaCl (M) | Energy Use Efficiency** | |
	S	NS
0.0	0.43	0.41
0.25	0.50	0.37
0.50	0.61	0.32

** Energy use efficiency = $\dfrac{\text{dry wt. of tissue (g)}}{\text{wt. of sugar consumed (g)}}$

The cost of tolerating stress can be determined for micro-organisms by a simple accounting procedure. These heterotrophic organisms are provided a known amount of carbon substrate. The amount used to carry out various processes and to grow can be accounted for by determining biomass productions and carbon consumption. This was used by Anderson et al (32) to determine energy cost of N_2 fixation by Klebsiella. This "nodule in a test tube" was the model for the work presented below.

It was previously reported that mutant strains of K. pneumoniae were derepressed for nitrogen fixation and blocked in NH_4^+ (33, 34). Such strains provide a convenint tool for studying the physiology and energetics of N_2 fixation in vivo, where the energy consumption (as glucose) and the production of NH_4^+ can be determined directly using nongrowing cells as described before (32, 35, 36). Besides, as K. pneumoniae can evolve H_2 at a high rate via the conventional hydrogen-producing system coupled to carbohydrate fermentation, it is necessary to use mutant strains blocked in this conventional system and which produce H_2 only by the nitrogenase system. Such mutant strains which allowed investigation of nitrogenase-catalyzed H_2 evolution in vivo were isolated previously as chlorate resistant clones (35). The mutant strain N_{20} used in the present study did not take up H_2 nor evolve H_2 unless nitrogenase was present. NH_4^+ production, H_2 production and glucose consumption were determined for this strain as described by LeRudulier et al (37).

In absence of NaCl, the rates of NH_4^+ production, H_2 evolution, and glucose consumption were 1.63, 1.00 and 7.56 μmol. $h^{-1}.mg^{-1}$ protein, respectively. In presence of NaCl these rates were decreased. The evolution of H_2/NH_4^+ (Fig. 2a) remains almost constant, 0.51 to 0.61, at osmolarities below 0.4 M but increases with osmolarity above 0.4 M. Under 0.4 M NaCl stress, nitrogenase catalyzes the formation of 0.58 mole H_2 per mole NH_4^+ produced; at 0.5 M NaCl, this value increased to 0.71 mole and reached 1.0 mole at 0.6 M NaCl. The glucose/NH_4^+ ratio (Fig. 2b) increases gradually with the osmolarity of the medium. The glucose requirement varies from 4.6 (0 NaCl) to 15.0 (0.6 M NaCl) moles glucose per mole NH_4^+. To estimate the energy requirement for N_2 fixation in terms of ATP consumption, the fermentation products must be measured, since K. pneumoniae can obtain ATP from the clastic cleavage of pyruvate to acetate, in addition to that obtained through the Embden-Meyerhof-Parnas pathway to pyruvate. From the levels of acetate produced (unpub. results) the apparent ATP requirement for N_2 fixation was calculated by assuming an ATP yield of 2^+ (moles acetate/moles glucose) per mole glucose fermented (35). An apparent ATP requirement of 26 ATP/N_2 is calculated for N_2 fixation in strain N_{20} grown without NaCl. This value is considerably higher in presence of NaCl: 61 ATP/N_2 in the case of

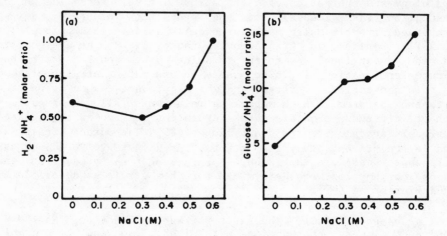

Figure 2.   The effect of NaCl concentration in glucose minimal
            medium on: (a) nitrogenase catalyzed production
            of H_2 and NH_4^+; (b) glucose requirement for pro-
            duction of NH_4^+ from N_2 in strain N-20. The culture
            was allowed to induce a maximum level of nitrogenase
            in the dialysis culture flask and then 1 ml samples
            were transferred anaerobically into flasks (10 ml)
            filled.

0.3 or 0.4 M NaCl, 67 ATP/N_2 with 0.5 M and 82 ATP/N_2 when the osmolarity of the medium is 0.6 M. If the rate of glucose consumption in the nif^-strain SK48 (unpub. results) measures "maintenance" energy requirement for the organism, then apparent ATP/N_2 value for nitrogenase reaction is about 18 without NaCl but remains still very high with NaCl, from 58 (0.3 M) to 79 (0.6 M).

At present, it is not possible to investigate the effect of proline on nitrogenase-catalyzed H_2 evolution. As a consequence of the chlorate mutation, proline has effect neither on the growth curve nor on nitrogenase activity under osmotic stress conditions. More specific mutant strains will have to be constructed to answer this question.

Stress tolerance in eukaryotic, higher plants involves a number of processes parallel to those in microorganisms. As with microorganisms these mechanisms require the diversion of energy to maintain processes enhancing tolerance to stress environments.

Quantitation of energy consumption by higher plants is complex and mostly indirect (38). Direct measurement of energy utilization by a multi-faceted organism such as a plant has been less than successful.

Plant cell culture techniques, however, do provide an opportunity to measure energy consumption directly using techniques very similar to those discussed for microbes (39). The plant cells are heterotrophic and the carbohydrate substrate used to grow cells can be added in measured amounts and the amount consumed determined directly (40).

This approach was used to measure the energy efficiency of alfalfa cells grown in the presence and absence of saline stress (40). Two lines of cells were compared. A line of alfalfa cells selected for tolerance to elevated levels of salt were compared to a line of non-selected cells. The characteristics of these two lines are detailed in previous papers (41, 42).

The energy use efficiency was determined by dividing the amount of cell material produced by the amount of glucose consumed by these cells (Table 5). In the absence of NaCl, the salt tolerant cells show a similar value for energy use efficiency as the non-selected cells. As NaCl concentrations are increased the non-tolerant line shows increased efficiency in energy use with increasing levels of NaCl. This greater efficiency of energy utilization by the salt selected plant cells could have considerable implications on the survival and productivity of plants exposed to saline enviroments. Enhanced efficiency of substrate utilization could conceivably provide adequate substrate for both osmoregulation and biomass production.

The decrease in energy use efficiency (0.41 to 0.32) as a function of increased NaCl stress by the non-selected line may indicate the diversion of energy substrates away from growth to osmoregulating processes. A reduction in growth would result in a reduction of plant productivity by this alfalfa cell line in stress situations, an interpretation consistent with the response of this cell line to salinity.

Conclusions

A number of biological processes influence or are influenced by saline stress. Some of the processes have been genetically defined and can be manipulated by traditional breeding approaches (43). Some organisms can be genetically engineered (2) to be better adapted to saline environments.

A major obstacle to the application of genetic manipulation to improve performance of these organisms in saline environments is the lack of understanding of the mechanisms responsible for enhanced salt tolerance.

In this paper we have compared and contrasted recognizable salt tolerant mechanisms of bacteria and plants and have discussed how this information might be applied to genetic alteration of desired characteristics.

References

1. E. Epstein, in: "Genetic Engineering of Osmoregulation: Impact on Plant Productivity for Food, Chemicals and Energy," (D. W. Rains, R. C. Valentine, and A. Hollaender, eds.) pp. 7-21, Plenum Press, New York (1980).
2. J. Mielenz, K. Andersen, R. Tait, and R. C. Valentine, in: "The Biosaline Concept: An Approach to the Utilization of Underexploited Resources," (A. Hollaender, ed.), PP. 361-371 Plenum Press, New York (1979).
3. L. N. Csonka, in: "Genetic Engineering of Osmoregulation," (D. W. Rains, et al., eds.) pp. 35-52, Plenum Press, N. Y. (1980).
4. L. Csonka, D. LeRudulier, S. S. Yang, A. Valentine, T. Croughan, S. J. Stavarek, D. W. Rains and R. C. Valentine, in: "The Importance of Water and Temperature Stress With Respect to Plant Growth and Yield in Humid, Temperate Climates: Future Research Needs," (P. J. Kramer, and C. R. Raper, Jr., eds.) (In Press.)
5. J. C. Measures, Nature 256:298-400 (1975).

6. S. Dendiger, L. G. Patil and J. E. Brenchley, J. Bacteriol 141: 190-198 (1980).

7. P. B. Chock, S. G. Rhee and E. R. Stadtman, Ann. Rev. Biochem. 49:813-843 (1980).

8. R. L. Jeffries, in: "Genetic Engineering of Osmoregulation: Impact on Plant Productivity for Food, Chemicals and Energy" (D. W. Rains, et al, eds.), pp. 135, Plenum Press, N. Y. (1980).

9. R. G. Wyn-Jones, ibid., pp. 155 (1980).

10. W. D. Jeschke, in: "Recent Advances in the Biochemistry of Cereals," (D. L. Laidman and R. G. Wyn Jones, eds.), pp. 48, Academic Press, N. Y. (1979).

11. T. J. Flowers, P. F. Troke and A. R. Yeo, Ann. Rev. Plant Physiol. 28:89-121 (1977).

12. D. W. Rains, Ibid., 23:367-388 (1972).

13. D. W. Rains, in: "Microbiol Conversion Systems for Food and Fodder Production and Waste Management," Kuwait Inst. for Scientific Research and Kuwait University, pp. 55-79, (1978).

14. D. W. Rains, in: "The Biosaline Concept: An Approach to the Utilizatin of Saline Environments," (A. Hollaender, ed.), pp. 47, Plenum Press, N. Y. (1979).

15. A. Pollard and R. G. Wyn Jones, Planta 144:291-294 (1979).

16. B. P. Strogonov, Adak. Nauk. SSSR, translated from Russian, Israel Prog. Sci. Transl., Jerusalem (1964).

17. A. H. Hanson and C. E. Nelson, Plant Physiol. 62:305-312 (1978).

18. W. Epstein and S. G. Schultz, J. Gen. Physiol. 49:221-234 (1965).

19. J. H. B. Christian and J. A. Waltho, J. Gen. Microbiol. 43: 345-355 (1966).

20. H. Greenway and C. B. Osmond, Plant Physiol. 49:256-259 (1972).

21. S. P. Treichel, G. O. Kirst and D. J. von Willert, Z. Pflanzenphysiol. Bd. 71:437-449 (1974).

22. T. J. Flowers, Phytochemistry 11:1881-1886 (1972).

23. J. S. Hawker, Aust. J. Plant Physiol. 1:491-501 (1973).

24. A. Poljakoff-Mayber and A. Meiri, "The Response of Plants to Changing Salinity," final Tech. Rep. Hebrew Univ., Jerusalem. Bet Dagan, Volcani Inst. Agric. Res. (1969).

25. J. L. Hall and T. J. Flowers, Planta (Berlin) 110:361-368 (1973).

26. A. Kalir and A. Poljakoff-Mayber, Plant Physiol. 55:155-162 (1975).

27. F. Austenfeld, Physiol. Plant. 36:82-87 (1976).

28. W. S. Pierce and N. Higinbotham, Plant Physiol. 46:663-673 (1970).

29. J. L. Hall, A. R. Yeo and T. J. Flowers, Z. Pflanzenphysiol. Bd. 71:(S)200-206 (1974).

30. D. H. Jennings, New Phytol. 67:899-911 (1968).
31. M. C. Pitman and H. D. W. Sadler, Proc. Natl. Acad. Sci 57:44-49 (1967).
32. K. Andersen, K. T. Shanmugam and R. C. Valentine, in: "Genetic Engineering for Nitrogen Fixation," (A Hollaender, et al. eds.) pp. 95-110, Plenum Press (1977).
33. K. T. Shanmugam and R. C. Valentine, Proc. Natl. Acad. Sci. 72:136-139 (1975).
34. K. T. Shanmugan, C. Morandi, K. Andersen and R. C. Valentine, in: "Proc. III Int. Conf. Enzyme Engineering," (K. E. Pye and H. H. Weethall, eds.) pp. 193-205, Plenum Press, N. Y. (1978).
35. K. Anderson and K. T. Shanmugam, J. Gen. Microbiol. 103: 107-122 (1977).
36. K. Anderson, K. T. Shanmugam, S. T. Lim, L. N. Csonka, R. Tait, H. Hennecke, D. B. Scott, S. S. M. Hom, J. F. Haury, A. Valentine and R. C. Valentine, TIBS 5:35-39 (1980).
37. D. LeRudulier, S. S. Yang and L. N. Csonka, in: "Genetic Engineering of Symbiotic N_2 Fixation and Conservation of Soil Nitrogen," (J. M. Lyons, et al, eds.) Plenum Press, New York (In Press).
38. F. W. T. Penning de Vries, A. H. M. Brusting and H. H. van Larrs, J. Theor. Biol. 45:339-377 (1974).
39. A. Kato and S. Nagai, European J. Appl. Microbiol. Biotechnol. 7:219-225 (1979).
40. T. P. Croughan, S. J. Stavarek and D. W. Rains, in: "Proc. of Propagation of Higher Plants through Tissue Culture: Emerging Technologies and Strategies," Environ. & Exptl. Bot. (In Press).
41. T. P. Croughan, S. J. Stavarek and D. W. Rains, Crop Sci. 18:959-963 (1978).
42. D. W. Rains, T. P. Croughan and S. J. Stavarek, in: "Genetic Engineering of Osmoregulation: Impact on Plant Productivity for Food, Chemicals and Energy," (D. W. Rains, et al, eds.) pp. 279-292, Plenum Press, N. Y. (1980).
43. E. Epstein, J. D. Norlyn, D. W. Rush, R. W. Kingsbury, D. B. Kelley, G. A. Cunningham and A. F. Wrona, Science 210: 399-404 (1980).
44. S. T. Lim, Plant Physiol. 62:609-610 (1978).
45. H. Sanui and N. Pace, Anal. Biochem. 25:330-346 (1968).

THE ROLE OF MEMBRANES AND TRANSPORT IN THE SALT TOLERANCE OF

HALOBACTERIA

Janos K. Lanyi

Dept. of Physiology and Biophysics
University of California
Irvine, CA   92717

SUMMARY

The ability of halobacteria to grow at several molar NaCl
concentrations depends on the effective removal of large amounts of
sodium ions from the cytoplasm and its replacement with potassium,
an ion more compatible with intracellular processes.   This mechanism appears to be a general one of salt tolerance, although in moderately halophylic bacteria, alga and halophytic plants, osmotic
balance is achieved with certain amino acids, sugars, polyols or a
combination of these with potassium ions.   It seems therefore that
salt tolerance includes the ability to transport sodium ions across
the cytoplasmic membrane at rates many times more than necessary
for growth at lower salt conditions.   This transport is always
energetically uphill, i.e. sodium is removed into the external
medium against a concentration difference across the membrane, and
against an opposing electrical potential difference.

According to the principles of chemiosmotic energy coupling in
biological membranes, metabolic energy in cells is generated and
stored in the form of transmembrane gradients for cations, such as
proton, sodium and potassium ions.   These gradients are generated
by membrane enzymes which catalyze primary energy conversion,
"pumps", and ion recirculation through other specific membrane
enzymes serves to couple the gradient created to energy requiring
processes.   In the halobacteria, two light-driven membrane pumps
have been investigated: bacteriorhodopsin, which extrudes protons
from the cell, and halorhodopsin, a recently discovered translocator for sodium ions.   Bacteriorhodopsin and halorhodopsin

are retinal proteins, and upon illumination of these pigments, a
series of photochemical intermediates are produced which relax
within 10-20 msec, and in an unknown way are coupled to the
vectoral uptake and release of protons or sodium ions, respectively. There is reason to believe, however, that the main pathway
of sodium transport is not halorhodopsin. Rapid sodium extrusion
from halobacterial cells is via a sodium/proton antiporter, which,
by exchanging sodium ions for protons across the membrane, will
utilize the energy of the proton gradient to drive sodium transport. As far as it is known, such exchange is a more widespread
mechanism for sodium transport in procaryotes, but primary pumps
for sodium do exist in mammalian cells, and have been suggested to
operate in plants as well.

SALT TOLERANCE AND SODIUM TRANSPORT

Large-scale biomass production is a potentially viable alternate source of energy, yet huge regions of the world endowed with
high solar flux are unsuitable for crops because of arid conditions
and/or high soil salinity. The lowered water activity is a severe
environmental constraint, which prevents the agronomic development
of these areas and threatens the productivity of a significant and
increasing portion of presently cultivated lands.

Little is known about the cellular and molecular mechanisms
involved in salt tolerance and xerotolerance in plants. Even
though halophytic plants have developed effective strategies of
dealing with high salt concentrations (1-3), studies of these
plants have been difficult and slow. In fact, the fundamental
questions of the biochemistry of salt tolerance have been raised
only with microorganisms (4,5), which have traditionally served as
models for higher organisms in modern biology. As might be expected, what is known about salt tolerance in vascular plants has
revealed considerable similarities at the cellular and molecular
levels between these systems and microorganisms.

The fundamental problem of salinity, which must be overcome by
bacterial cells, is the disruptive effect of Na^+ on ribosome structure and function and other salt sensitive enzyme systems (4,6).
While most bacterial, plant and animal cells contain mechanisms for
removing Na^+ and for transporting Cl^-, the most powerful means of
dealing with salt were naturally developed in organisms which live
in highly concentrated brines. Some of these organisms can grow at
NaCl concentrations up to saturation (about 30% w/v), and they
maintain low internal Na^+ against such external concentrations by
utilizing energy linked membrane transport systems. This strategy
is common to all halotolerant microorganisms, but the water loss
which inevitably follows the removal of salt from cells is handled
differently in the different kinds of organisms. The strategies

employed for osmoregulation are the results of adaptation to specific environmental constraints. Thus, where salinity is high (>20%
w/v) and constant, certain halobacteria evolved means to accumulate
high internal concentrations of K^+ in place of Na^+ (7), which then
serves to balance water osmotically across the cell membrane.
Although KCl is a somewhat more benign substance than NaCl, extensive modifications in the composition and physical chemistry of
proteins, ribosomes and membranes had to occur to accomodate this
salt as a solute compatible with intracellular metabolism (4,8).

Where salinity in nature is lower and variable, and rapid
adaptation to the changing conditions is a desirable characteristic, the moderate or facultatively halophilic bacteria and some
green algae developed another strategy: the accumulation of large
quantities of sugars, amino acids or other small molecules, such as
betaine, in place of NaCl and sometimes in combination with K^+, for
maintaining osmotic balance. For example, Ben-Amotz and Avron
have shown (9) that the green algae Dunaliella can reversibly
convert its large, but osmotically inactive polysaccharide reserves
to glycerol in response to an increase in salinity. Also, Avi-Dor
and Rafaeli described osmotic adaptation in moderately halophilic
bacterium with a growth optimum between 0.5-1.0 M NaCl (10).
Sodium ions could not be replaced by potassium or sucrose, but
growth at high salt concentrations also required betaine or choline
(11), the latter accumulated by active transport and then converted
to betaine (11,12). Betaine reached an intracellular concentration
of nearly 1 M, and obviously acted as an osmotically protective
agent since it allowed for growth and respiration in intact cells,
but had no protective effect against salt on isolated enzymes.
Furthermore, the stimulatory effect of betaine on respiration was
observed only when the cells were stressed with non-permeating
solutes, such as NaCl, KCl or sucrose, which would result in water
loss (13). Since most enzymes and other essential constituents can
function in the presence of high concentrations of simple molecules, such as betaine and glycerol, modification of the intracellular protein was unnecessary in these cells. On the other hand,
this mode of adaptation is energetically costly and necessitates
the development of extensive control mechanisms for sensing the
environment and regulating the cellular response. Higher plants in
general have adapted this physiological response to salt, and it is
important to note that betaine is found in some of agriculture's
most salt-tolerant species to be the osmoregulating sustance, e.g.
in cotton, barley and sugar beets (14,15).

PRINCIPLES OF MEMBRANE TRANSPORT

The common element among the above described different mechanisms is the removal of Na^+ from the cell interior. The Na^+ transport is an energy requiring process and must be integrated

into the overall scheme of ion circulation across the cell membrane. The circulation of the ions is a consequence of the electrical and concentration gradients between the cell interior and exterior. These gradients not only serve the needs of the organism for maintaining benign interior conditions, but are essential elements in the conservation of energy and its transmission from energy producing to energy consuming membrane reactions (16). Thus, the transfer of energy in such membrane systems depends on vectorial enzyme reactions, catalyzed by membrane proteins which generate or discharge ionic gradients. The gradients normally encountered are those of H^+, Na^+, and K^+. The membrane components which couple the translocation of these ions to a chemical reaction, (e.g., terminal oxidation of ATP hydrolysis) or to photochemical events (e.g., in bacteriorhodopsin) are termed primary energy transducers, or pumps (17). Others, which couple the translocation of inorganic ions or metabolites to one another (e.g., the proton/sodium-antiporter) are termed secondary energy transducers. Both kinds of membrane components will cause energetically uphill movement across the membrane, i.e., active transport.

The elctrochemical potential difference, $\Delta\tilde{\mu}_i$, for a charged species, i, between the two phases separated by the membrane is:

$$\Delta\tilde{\mu}_i = m\Delta_\psi - \Delta\mu_i = m\Delta_\psi - \frac{RT}{F} \ln [i]_{out}/[i]_{in} \quad [1]$$

where m is the net charge of i, Δ_ψ is the electrical potential difference between the two phases, $\Delta\mu_i$ is the chemical potential difference for i, $[i]_{out}$ and $[i]_{in}$ are the concentrations (activities) of i in the two phases, i.e., in the external medium and the cell interior, respectively, and R, T and F have their usual meanings. For protons equation [1] takes the familiar form:

$$\Delta\tilde{\mu}_{H^+} = \Delta_\psi - 2.3 \frac{RT}{F} \Delta pH \quad [2]$$

where $\Delta\tilde{\mu}_{H^+}$ is the electrochemical potential difference for protons (protonmotive force), expressed in mV, and pH is the pH difference across the membrane. Since the value of 2.3 RT/F is approximately 60 mV, a ten-fold concentration difference of protons ($\Delta pH = 1$) is equivalent to a membrane potential of 60 mV.

In general, the study of transport processes may be divided into determinations of (a) transmembrane fluxes and gradients under various conditions, and (b) structural and functional properties of the proteins involved in the binding and translocation of the ions. The former approach has been utilized far more often than the latter, and has provided a number of models for the mechanism and energetics of primary and secondary transport systems. For the functioning of secondary energy transducers in such models, extensive use is made of the concepts of uniport, symport and

antiport (17) to denote processes catalyzed by appropriate porters
(transport catalysts) which result in either single translocation
across the membrane, coupled translocation in the same direction,
or coupled translocation in the opposite direction, respectively.

BACTERIORHODOPSIN

During the past several years, particular attention has been
paid to the light-driven transport processes in H. halobium membranes (18,19).   Many of these are initated by a remarkably simple
ion pump, bacteriorhodopsin. This retinal-protein uses light-energy
for the active transport of protons, and consists of a single poly-
peptide chain of 248 amino acids (20), arranged in seven nearly
completely helical segments looped across the width of the membrane
(21).   The retinal is bound via a Schiff-base linkage to one of the
lysines in the protein, and it appears that this relatively simple
structure contains all that is required for the light-driven proton
translocation (19).   It is of great importance for investigations
of this system, however, that bacteriorhodopsin is normally part of
a more complex structure, and is found in two-dimensional crystalline arrays containing approx. 10^5 molecules (22).   Such a unique
organization for a membrane protein has not only permitted the easy
isolation of membranes containing exclusively bacteriorhodopsin,
but made possible revealing X-ray, electron and neutron diffraction
studies (21,22-24).   Recently, the amino acid sequence of bacteriorhodopsin has become available (20), and since the three-
dimensional crystallization of the protein has also been achieved
(25) it is likely that within a few years the structure of bacteriorhodopsin will be known to the same degree of resolution as for
many soluble proteins.

In contrast, the mechanism of function in bacteriorhodopsin is
not well understood.   Flash spectroscopic studies (19) have shown
that the chromophore (570 nm) is converted after a few picoseconds
of absorbing a photon into another (K) form, which decays within 10
msec at room temperature through a series of other spectral intermediates (L, M, N, and O).   During this time a proton is released
on the side of the membrane facing the cell exterior, followed by
proton uptake on the cytoplasmic side.   It has been proposed that
the reversible protonation of the retinal Schiff-base is the
central event in proton translocation (19,26).   Protons to and from
this group will be supplied by proton conduction either via an ice-
like matrix in the protein (27) by an Onsager mechanism or less
likely, an aqueous channel.   Initiation of the proton migration,
after absorption of a photon by the retinal, may be induced by
charge delocalization along the retinal chain or by configurational
change of the retinal relative to other groups in the protein. The
reversible protonation of a tyrosine observed during the photocycle
(28,29), the pH dependence of the photointermediates (30), as well

as a distinct pK for light-induced proton release (31) suggest
that donor and acceptor groups exist for protons other than the
Schiff-base.

Bacteriorhodopsin will transport protons against an electrochemical potential difference up to at least 200 mV and continuous
illumination of intact H. halobium cells, cell envelope vesicles
and liposomes reconstituted with purified bacteriorhodopsin will
result in the development of a pH difference and/or electrical
potential across the membranes. The energy conserved in this gradient of protons has been thought to drive ATP synthesis via a
proton translocating ATPase (32-34), to generate a sodium gradient
via a sodium/proton antiport system (35,36), and indirectly through
the gradient of sodium to drive various sodium/amino acid symport
systems (37). Thus, bacteriorhodopsin represents a source of
metabolically useful energy, alternative to the respiratory chain
in H. halobium.

SODIUM PROTON ANTIPORTER

Since the gradient of protons across the Halobacterium membrane
represents conserved energy, the action of light on bacteriorhodopsin is expected to be capable of energizing all transport processes
coupled to proton movement. We have found (35) that when H. halobium cell envelope vesicles or whole cells were illuminated Na^+
efflux occurred and high Na^+ concentration gradients (out $\gg$ in)
developed. Since under these conditions Na^+ moved against its
electrical potential (inside negative in these systems), as well as
against its chemical potential (see Eq. 1), the transport was considered to be active. We also observed (35) that when both NaCl
and KCl were included inside the envelope vesicles (e.g. 1.5M NaCl
plus 1.5M KCl), the pH changes measured during illumination in the
vesicle suspension were not simple as when only KCl was present.
Depending on the amount of NaCl load inside the vesicles, there was
an initial period of smaller pH change, followed by an increase to
the value found in the absence of NaCl. The lengths of these
initial periods during the illumination roughly coincided with the
time required to deplete the vesicles of ^{22}Na. These observations
suggested to us that the efflux of Na^+ was accompanied by an influx
of H^+, which decreased the size of the pH gradient. The simplest
model which accounted for the results was coupled H^+/Na^+ exchange,
i.e. that an antiporter existed for these ions. When the experiment was performed at pH values higher than 6.5, the Na^+ dependent
influx of H^+ during illumination became more pronounced. At a pH
>7, the initial acidification of the medium was followed by a rise
in pH, leading to a transient reversal of the pH gradient. The
reversal strongly suggested that the antiport could not proceed
with a stoichiometry of 1:1, because such a process would be driven
only by ΔpH and would be abolished when the pH difference

approached zero.   On the other hand, at higher H^+/Na^+ stoichiome-
tries, the exchange would be driven also by the electrical potential since it would include net charge translocation, i.e. the
transport would be electrogenic.   The $\Delta\psi$ in this model would
balance the system at a reversed pH gradient.   Results consistent
with this model of sodium transport were obtained also by Eisenbach
et al (36), and later Caplan et al (38) showed that the light
driven Na^+ extrusion could be achieved even when the ΔpH was
artificially reversed (acid inside), a strong argument for the
electrogenicity of the antiport.

The bacteriorhodopsin-sodium/proton antiport model for the
circulation of sodium ions in H. halobium envelope vesicles was
consistent with the observed effects of ionophores in this system.
In many of these studies, the appearance of sodium gradients (out
$\gg$ in) was followed by measuring the uptake of [^{3}H] glutamate which
is driven by these gradients.   It was found (39) that uncouplers
(proton conductors) inhibited the transport considerably when added
before the illumination, but had much less effect when added after
a period of exposure to light.   Thus, it was concluded that the
sodium gradient was created by a mechanism involving proton circulation.

We determined the H^+/Na^+ exchange stoichiometry in H. halobium
membranes by an indirect method, utilizing the Na^+ gradient dependent accumulation of [^{3}H] serine to measure $\Delta\mu_{Na}+$ in the steady-
state (40).   A plot of $\Delta\mu_{Na}+$ vs. proton electrochemical potential
difference, produced at different light-intensities, yielded a
straight line between 0 and 100 with mV with a slope of 1.8,
suggesting a stoichiometry of 2 H^+/Na^+.   We also determined the
rate of Na^+ efflux at different magnitudes of $\Delta\mu_H+$ and found that
the transport rate was very low (but not zero) at $\Delta\mu_H+$ more positive than -130 to -150 mV, but at H^+ gradients more negative than
these values, the rate of Na^+ extrusion was a steep linear function
of $\Delta\mu_H+$.   After studying this effect under conditions where (a) the
counter-ion to Na^+ movement was chosen to be either K^+ of Cl^-, and
(b) the electrical and chemical components of the protonmotive
force were varied relative to one another, we concluded that the
sodium/proton antiporter in H. halobium vesicles if "gated" at a
critical electrochemical potential dfference for protons.   Little
can be said at this time about the molecular mechanism of the regulation of this transport catalyst.   On the other hand, the physiological consequences of the observed gating of Na^+ transport are
far-reaching. The threshold potential for the gating is at or just
below the promotive force measured in resting (non-energized but
not starved) H. halobium cells.   If the gating observed in the
vesicles functions in the same way in intact cells, energizing the
cells by illumination or aerobic respiration will raise $\Delta\mu_H+$ above
the threshold, to the extent necessary for the rapid extrusion of

Na^+.   Lowering the protonmotive force by starving the cells will
not result in massive Na^+ influx by reversal of the antiporter
function, however, since this porter functions much less effectively below the threshold protonmotive force. Thus, the physiolog-
ically important process, i.e., the transfer of energy from the H^+
gradient to the Na^+ gradient is encouraged, while the reverse is
prevented.

HALORHODOPSIN

     Recently, the observatin of light-driven sodium transport
independent of proton gradients in vesicles prepared from various
H. halobium strains prompted Lindley and Mac Donald (41) to propose
the existence of another pump, a light-driven sodium translocator
in these membranes.   This pump was later named halorhodopsin (Y.
Mukohata, pers. comm.) Lindley and MacDonald found that membrane
vesicles from a bacteriorhodopsin deficient red pigmented strain of
H. halobium R_1mR, described by Matsuno-Yagi and Mukohata (42), lost
^{22}Na during illumination more rapidly than in the dark. The sodium
transport was insensitive to proton conductors.   A membrane potential of about -90 mV developed during illumination, which could be
abolished with valinomycin and K^+ but not with proton conductors.
From these results, Lindley and MacDonald argued that in this system the membrane potential was caused by the extrusion of sodium
ions rather than protons. Protons, in fact, were not extruded, but
taken up by the vesicles during the illumination, as detected by pH
increase in the medium and labeled N-methylmorphine accumulation.
The proton uptake was dependent on the presence of sodium, and was
diminished by valinomycin and K^+.   Proton movements thus appeared
to be secondary, and driven by the electrical potential created by
sodium extrusion.

     MacDonald and coworkers (43) obtained results consistent with
a light driven sodium pump also in vesicles prepared from
bateriorhodopsin-containing H. halobium R_1 strain.   Unlike previously studied vesicles from this strain, which release protons
during illumination (44), a small fraction of vesicles collected at
high centrifugal forces showed light-induced proton uptake. The pH
rise in the medium during illumination was enhanced by uncouplers
and inhibited by gramicidin or valinomycin and K^+.

     We have been able to confirm these results suggestive of a
primary sodium pump, in membrane vesicles prepared from various H.
halobium strains, particularly from those lacking bacteriorhodopsin.   Electrical and pH gradients were measured quantatively
during illumination, in parallel experiments with fluorescent dyes,
dio-C_5 and 9-aminoacridine, respectively (45).   Vesicles from the
bacteriorhodopsin deficient R_1mR strain developed -140 to -150 mV
inside negative electrical potentials upon illumination, balanced

by acid inside pH difference, so as to make the net protonmotive force in the steady-state near zero. This result is consistent with the absence of active transport for protons. Uncouplers had no effect on the magnitude of these gradients, but accelerated proton uptake. Slowly sedimenting vesicles from the bacteriorhodopsin-containing R_1 strain developed inside negative membrane potentials upon illumination, which was about -100 mV in excess of the opposing pH gradient. With uncouplers present, the light-induced electrical potential decreased and the pH gradient increased, and the net protonmotive force approached zero. We suggested, therefore, that in these vesicles pumps are present for both protons (bacteriohodopsin) and sodium (holorhodopsin), and that these both function in the outward direction (45). Presumably, this orientation is the same in the cytoplasmic membrane of intact H. halobium cells as well.

MacDonald's observations, together with our results discussed here, strongly suggested the existence of a light-driven sodium pump in H. halobium membranes, but the presence of a pigment associated with this pump, and its relationship to bacteriorhodopsin, had still to be established by spectroscopic means. We obtained action spectra for the creation of electrial and pH gradients during illumination (45), which exhibited maxima at 585-590 nm, removed from the 568 nm absorption band of bacteriorhodopsin. More recently Greene, et al (46) published action spectra for proton uptake in R_1mR vesicles, obtained with tunable laser illumination. Their result (590 nm maximum) confirms the red-shift from the absorption band of bacteriorhodopsin.

For more accurate spectroscopic work interference from the red carotenoids, normally present in H. halobium, was eliminated by the use of a "colorless" strain isolated by Weber, labeled ET-15. This strain lacks bacteriorhodopsin as well, to a detection limit of a few molecules per cell (47), but shows the passive light-dependent proton movements which had suggested the existence of halorhodopsin. We found (48) that difference spectra between vesicles prepared from the ET-15 strain and vesicles from a retinal-deficient, light-unresponsive H. halobium strain contained an absorption band near 590 nm.

A useful method for obtaining spectroscopic information about halorhodopsin proved to be bleaching and reconstitution with added retinal. These experiments were patterned after methods worked out for bacteriorhodopsin. The latter is known to be bleached after prolonged (3-6 hrs) illumination in the presence of hydroxylamine (49), a treatment which cleaves the lysine Schiff-base to form the retinal oxime. After removal of excess hydroxylamine, the chromophore absorption band of bacteriorhodopsin is quantitatively restored with added retinal. Difference spectra of bleached H.

halobium ET-15 vesicles, with and without added trans-retinal, exhibited a new 588 nm absorption band which grew during a 2-3 hr incubation time at the expense of the 385 nm absorption band of free retinal. With unbleached but hydroxylamine-treated vesicles, the retinal band remained undiminished and the 588 nm band did not arise (48). From such experiments, an approximate extinction coefficient of 48,000 M^{-1} cm^{-1} (retinal equivalents) was calculated for the pigment. Using this value, the amount of halorhodopsin in ET-15 membranes was estimated to be 0.13 nmol/mg protein, or about 5% of the bacteriorhodopsin content of typical vesicle preparations from the R_1 strain.

Partial bleaching of ET-15 vesicles, followed by retinal reconstitution as well as measurement of absorbance changes at 588 nm and light-induced proton uptake, established a linear relationship between the spectroscopic and transport properties. This finding, and the agreement between the action spectra obtained earlier and the absorption spectra, stongly suggested that the 588 nm absorption band detected is associated with the sodium pump. Absolute absorption spectra of bleached and reconstituted ET-15 membranes, free of light-scattering effects, exhibited no particularly strong absorption bands near 590 nm, besides the 588 nm pigment.

Similarly to bacteriorhodopsin, halorhodopsin exhibits a photocycle, but the absorption maxima of the photointermediates and their rate of production and decay are different in the time pigments (47). In both cases, however, the overall recovery time of the pigment after the flash-bleaching is about 10 msec.

The above results indicate that a retinal protein, halorhodopsin, with an absorption band at 588 nm, is implicated in the light-driven sodium pump. While bacteriorhodopsin and halorhodopsin appear to be quite similar, they show differences in transport function, spectroscopic properties and stability. It is tempting to speculate that the two pumps, for protons and for sodium ions, might function analogously, but with different cation specificities. However, given the proposed central role of the protonated Schiff-base in the transport by bacteriorhodopsin, it would be difficult to envision analogous mechanisms for proton and sodium translocation. Without excluding this possibility as yet, we are considering also more complex models for halorhodopsin, which contain a bacteriorhodopsin-like proton translocator intimately coupled to a sodium/proton exchanger. Little direct information is presently available to decide in favor of such a model.

The relative importance of the bacteriorhodopsin-sodium/proton antiporter mechanism and halorhodopsin in light-dependent sodium transport needs still to be established. Our results (50) suggest that the former accounts for virtually all of the sodium flux in bacteriorhodopsin-containing halobacteria. Thus, in $\underline{H}$. halobium membrane vesicles lacking functional halorhodopsin light-dependent Na^+ transport was as rapid as in vesicles containing this pigment, and this transport was entirely uncoupler-sensitive. Such proton-circulation dependent sodium transport could be driven also by artificial electron donors in the dark, even in strains lacking retinal protein altogether.

References

1. P. J. Mudie, in: "Ecology of Halophytes" (R. J. Reinold and W. H. Queen, eds.) Academic Press, New York, 565-597 (1974)
2. J. A. Hellebust, Annu. Rev. Plant Physiol. 27:485-505 (1976)
3. T. J. Flowers, P. F. Troke and A. R. Yeo, Annu. Rev. Plant Physiol. 28:89-121 (1977).
4. J. K. Lanyi, Bacteriol. Rev. 38:372-390 (1974).
5. J. K. Lanyi, in: "Strategies of Microbiol. Life in Extreme Environments" (M. Shilo, ed.), Dahlem Konferenzen, Verlag Chemie, Weinheim, 125-135 (1979).
6. S. T. Bayley, CRC Crit. Rev. Microbiol. 6:151-205 (1978).
7. H. Larsen, Advan. Microbiol. Physiol. 1:97-132 (1976).
8. A. D. Brown, Bacteriol. Rev. 40:803-846 (1976).
9. A. Ben-Amotz and M. Avron, Plant Physiol. 51:875-878 (1973).
10. D. Rafaeli-Eshkol and Y. Avi-Dor, Israel J. Chem. 5:99 (1967).
11. D. Rafaeli-Eshkol and Y. Avi-Dor, Biochem. J. 109:679-685 (1968).
12. D. Rafaeli-Eshkol and Y. Avi-Dor, Biochem. J. 109:687-691 (1968).
13. C. Skedy-Vinkler and Y. Avi-Dor, Biochem. J. 150:219-226 (1975)
14. R. Storey, N. Ahmad and R. G. Wyn Jones, Oecologia 27:319-332 (1977).
15. R. Storey and R. G. Wyn Jones, Phytochem. 16:447-451 (1977).
16. P. Mitchell, Theor. Exp. Biophys. 2:160-216 (1969).
17. P. Mitchell, Symp. Soc. Gen. Microbiol. 20:121-166 (1970).
18. J. K. Lanyi, Microbiol. Rev. 42:682-706 (1978).
19. W. Stoekenius, R. H. Lozier and R. A. Bogomolni, Biochim. Biophys. Acta 505:215-278 (1979).
20. Q. Ovichinnikov, A. Yu, N. A. Abdulaev, M. Yi Feigina, A. V. Kiselev and N. A. Lobanov, FEBS Lett. 100:219-224 (1979).
21. R. Henderson and P. N. T. Unwin, Nature 257:28-32 (1975).
22. A. E. Blaurock and W. Stoeckenius, Nature 233: 152-155 (1971).
23. R. Henderson, J. Mol. Biol. 93:123-138 (1975).
24. G. I. King, W. Steockinius, H. L. Crespi, and B. P. Schoenborn, J. Mol. Biol. 130:395-404 (1979).
25. H. Michel and D. Oesterhelt, Proc. Nat. Acad. Sci. 77:1283-1285 (1980).
26. A. Lewis, J. Spoonhower, R. A. Bogomolni, R. H. Lozier and

W. Steockenius, Proc. Nat. Acad. Sci. 71:4462–4466 (1974).

27. J. F. Nagle and H. K. Morowitz, Proc. Nat. Acad. Sci. 75:298–302 (1978).

28. R. A. Bogomolni, L. Stubbs and J. K. Lanyi, Biochemistry 17: 1037–1041 (1978).

29. B. Hess and D. Kuschmitz, FEBS Lett. 100:334–340 (1979).

30. R. H. Lozier and W. Niederberger, Fed. Proc. 36:1805–1809 (1977).

31. R. Renthal, Biochem. Biophys. Res. Commun. 77:155–161 (1977).

32. A. Danon and W. Stoekenius, Proc. Nat. Acad. Sci. 71:1234–1238 (1974).

33. R. Hartman and D. Oesterhelt, Eur. J. Biochem. 77:325–335 (1977).

34. G. Wagner, R. Hartmann and D. Oesterhelt, Eur. J. Biochem. 89: 169–179 (1978).

35. J. K. Lanyi and R. E. MacDonald, Biochemistry 15: 4608–4614 (1976).

36. M. Eisenbach, S. Cooper, H. Garty, R. M. Johnstone, H. Rottenbereg and S. R. Caplan, Biochmim. Biophys. Acta. 465: 599–613 (1977).

37. R. E. MacDonald, R. V. Green and J. K. Lanyi, Biochemistry 16: 3227–3235 (1977).

38. S. R. Caplan, M. Eisenbach, S. Cooper, H. Garty, A. Klemperer and E. P. Bakker, in: "Bioenergetics of Membranes" (L. Packer, G. C. Papageorgiou and A. Trebst, eds.), Elsevier/North Holland, Amsterdam, 101–114 (1977).

39. J. K. Lanyi, V. Yearwood-Drayton and R. E. MacDonald, Biochemistry 15:1595–1603 (1976).

40. J. K. lanyi and M. P. Silverman, J. Biol. Chem. 254:4750–4861 (1979).

41. E. V. Lindley and R. E. MacDonald, Biochem. Res. Commun. 88: 491–499 (1979).

42. A. Matsuno-Yagi and Y. Mukohata, Biochem. Biophys. Res. Commun. 78:237–243 (1977).

43. R. E. macDonald, R. V. Greene, R. D. Clark and E. V. Lindley, J Biol. Chem. 254:11831–11838 (1979).

44. R. Renthal and J. K. Lanyi, Biochemistry 15:2136–2143 (1976).

45. R. V. Greene and J. K. Lanyi, J. Biol. Chem. 254:10986–10994 (1979).

46. R. V. Greene, R. E. MacDonald and A. J. Perrault, J. Biol. Chem. 255:3245–3247 (1980).

47. R.A. Bogomolni, R. A. and H. J. Weber, Photochem. Photobiol.; (In press).

48. J. K. Lanyi and H. J. Weber, J. Biol. Chem. 255:243–250 (1980).

49. D. Oesterhelt, L. Schumann and H. Gruber, FEBS Lett. 44: 257–261 (1974).

50. B. F. Luisi, J. K. Lanyi and H. J. Weber, FEBS Lett. 117: 354–358 (1980).

USE OF BRACKISH AND SOLAR DESALINATED WATER IN

CLOSED SYSTEM AGRICULTURE

J. Gale

Jacob Blaustein Desert Research Institute
Sde Boqer and the Department of Botany
Hebrew University of Jerusalem, Israel

SUMMARY

Closed system desert agriculture (CSA) needs only a small supply of fresh water; about one hundredth, per farming family, of that required for open field irrigated cultivation. This may be provided by solar desalination of brackish water, which is uniquely appropriate in this context. Furthermore, the environmental conditions in CSA may be such as to allow plants to be grown with relatively saline water. These include moist soil or hydroponics, humid atmosphere, moderate temperatures and high concentrations of carbon dioxide in the ambient air. Fossil brackish water, containing 500-4000 ppm Total Soluble Salts (TSS) is frequently available in desert regions of the world, in small quantities, albeit as a non-renewable source.

About 48,000 m^3 per annum of fresh water are required by a single farmer engaged in conventional, irrigated, open-field agriculture in an arid region. In contrast to this it is calculated that in Israel (by way of example) in a region having an annual rainfall of 100 mm, only 100 m^2 of a single stage solar still could provide the fresh water requirements for 0.2 hectare of CSA, supporting the same farming family. The still would use surplus heat from CSA and would distill about 200 m^2 of brackish water (2000 ppm TSS) per annum. Total brackish water used would be 500 m^3 per farm. It would be diluted to 1000 ppm TSS with distilled water and with rainwater gathered from 0.2 hectare roof of CSA.

INTRODUCTION

The ultimate purpose of applied biosaline research is to

facilitate the production of food and saleable farm products by
people living in areas which suffer from chronic shortage of fresh
water.   These arid zone settlers should be able to maintain a
standard of living comparable to that of their urban countrymen.

Although agrotechnical methods are being developed for the use
of brackish water in the field, and varieties can be selected for
greater resistance to salt, it is preferable to use this often
limited and potentially deleterious resource as little as possible.
Closed system agriculture is a promising technology whereby this
aim of using little brackish water may be achieved.   The main
problem of desert CSA technology is however, not water, but the
dissipation of surplus energy during the daytime.

It will be argued below that the fresh water requirement for
CSA is extremely small and that this requirement may be met by
solar desalination of available brackish water.   It will also be
shown that the special environmental conditions prevalent in CSA
may increase plant tolerance to saline conditions, allowing the use
of brackish water for crop production.

Closed System Agriculture (CSA)

Controlled Environment Agriculture has been proposed and
studied for many years by Rappaport (1), Bettaque (2), De Bivort et
al (3) and others.   One such system has operated for some years at
nearby Puerto Penasco, Mexico (4).   Problems of plant growth within
such systems have been reviewed recently (5).   Consequently, only a
brief mention is made here of the general advantages and disadvantages of CSA for arid zones and this mainly in relation to the use
of brackish water.

A closed system is a greenhouse which is kept closed for as
many hours of the day as possible.   Solar heat and water are partly
recycled, reducing the use of fossil derived energy and fresh
water.   Because the greenhouse is closed, carbon-dioxide fertilization becomes practical throughout most of the day.   This increases
the efficiency of plant use of the abundant solar radiation typical
of arid regions (e.g., see 6).   The high humidity within the system
and the high levels of CO_2 (in the range of 1000-15000 ppm versus
about 330 ppm in the open air) reduce water consumption by lowering
the leaf to air vapor pressure gradient and by increasing the
diffusion resistance of the stomates (7).

Other advantages of CSA are common to all greenhouses,
although some are particularly pertinent to arid zone agriculture.
Among these are protection from wind and consequent dust/sand
abrasion, the possibility of closely controlling irrigation and
mineral nutrition and closer control of insects and pathogens.

Furthermore, CSA is a potential means of turning a short, unreliable and frequently unfavorable growing season into a year round growing season with an extremely high potential productivity (8).

There are two main disadvantages to CSA: the high cost of capital investment and the high technological level which may be required for establishing and maintaining the system. Both capital and technology are frequently lacking in developing arid regions.

CSA Climate and Plant Tolerance of Saline Water

Fossil brackish waters (500-4000 ppm TSS) are frequently available in desert aquifers (9). Their use for irrigation in the open field or for fish or algae culture, requires special techniques, which are discussed elsewhere in this workshop. At best these fossil aquifers are a limited non-renewed resource. For example, it has been estimated that in the Negev (the southern desert half of Israel) some 150×10^6 m^3/yr of brackish fossil water could be made available for some 50 years. This should be compared to the 2000×10^6 m^3/yr of fresh, renewable water, available in the north (more than 97% of which is already being used). Hence, in order to husband this resource until low cost desalinized sea water becomes a reality, it is essential to use the fossil water as sparingly and at as high a salt concentration as possible, in order to reduce the expense of desalination.

Although the climate in CSA can be brought to any desired specification, by suitable engineering, to do so at an acceptable cost is perhaps the main problem of CSA research. In fact, the climate which is most easily obtained in arid-zone CSA is, paradoxically, almost tropical: a high level of water availability at the plant roots, high air humidity, high but not extreme air temperatures, moderate to high light intensity and a year round growing season. If we add to this list of tropical conditions a high level of carbon dioxide (given CO_2 supplementation) we have, as will be shown below, a unique combination of environmental conditions under which many plants may tolerate much higher levels of salt than they would in the open field. Although this requires further verification with respect to any particular crop, species or variety, an increasing body of evidence supports this general hypothesis.

Most but not all plants show osmotic adjustment under saline conditions (e.g., see 10). Even so, many species show increased tolerance under conditions of high air humidity (11-14). In addition there is evidence of increased tolerance to salt under conditions of moderate light intensity and air temperatures. It is however extremely difficult to separate the effects of these interacting factors (15).

Recently we have found that at low levels of salinity many, but again not all, plants suffer from a shortage of assimilates (Schwarz and Gale, unpub.). This results from reduced photosynthesis caused by partial closure of stomates (which occurs despite osmotic adjustment; 10) and from an increased use of assimilates in maintenance induced respiration. A corollary of this is that plants should be more tolerant of salt under conditions of high [CO$_2$], which tends to increase the rate of photosynthesis. This was indeed found to be the case, as shown in Table 1, for Cocklebur (Xanthium Strumarium).

TABLE 1. GROWTH OF XANTHIUM STRUMARIUM SEEDLINGS AT NORMAL AND AT HIGH LEVELS OF CARBON-DIOXIDE IN SALINE CULTURE SOLUTION.

	[CO$_2$] ppm	Control Plants	Plants in Saline Solution (130mM NaCl)	Depression of Growth
Tops	320	352 ± 15	128 ± 9	64%
	2500	542 ± 24 (154%)	443 ± 16 (346%)	18%
Roots	320	163 ± 14	114 ± 12	30%
	2500	248 ± 18 (152%)	130 ± 17 (114%)	48%
Tops and Roots	320	515 ± 7	242 ± 12	53%
	2500	790 ± 28 (153%)	573 ± 18 (236%)	28%

Milligram dry weight increase during 12 days, ± s.e.. N = 6. In parentheses: High as percentage of low [CO$_2$] treatment. Plants grown in half Hoagland culture solutions. Initial dry weight (calculatd) Tops 477 ± 3, Roots 81 ± 2 mg.

Other environmental conditions: 12 hours light at 415 µE m^{-2}sec^{-1} (400-700 nm). Air temperatures: 26°C day and 18°C night. Day/Night air humidity: 65 ± 5% R.H.

The response of the shoots of plants growing under saline conditions to elevated CO_2 ppm was much higher (346%) than for the controls (154%) although this effect was not transmitted to the roots. Even so plants in salt responded 236% versus 153% for the controls. A similar response to high [CO_2] was found for beans (Phaseolus vulgaris) and saltbush (Atriplex halimus).

In conclusion, taking into account our own and other reported results of the interaction of environmental parameters and salinity, we estimated that under CSA conditions many plants could tolerate 1000 ppm TSS salt. We have used this figure for the calculations presented below. However, as noted above, this will have to be verified experimentally and the tolerated salt level adjusted, for each plant species and variety.

Combined Solar Desalination of Brackish Water and CSA

When large quantities of fresh water are required, at a steady rate of production, solar desalination is not considered to be competitive with conventional desalination systems (16,17). However, for the following reasons solar distillation seems to be particularly appropriate when used in combination with CSA in desert regions: relatively low volume requirement; low cost of land; high solar irradiance (e.g., 8500 MJ $m^{-2}y^{-1}$ in the Negev versus some 4000 MJ in Western Europe); availability of brackish water; little storage requirement for fresh water (can be fed directly to hydroponic system of CSA); can utilize energy from CSA (which also serves to dissipate surplus energy in summer); intrinsic linkage of the rate of production by the still and the fresh water demand rate of the CSA (both are largely dependent upon daily solar radiation).

The solar still may be an integral part of the closed system (2,18). However, this approach has a number of shortcomings as discussed by Ronchaine (19). At the Sde Boqer Institute for Desert Research we are investigating a system based on the liquid optical filter concept (20) in which a solar still is integrated within but is physically separated from the greenhouse (Fig. 1).

Details of the operation of a liquid filter roof greenhouse have been given before (5,20). In brief, non-photosynthetic solar energy (> 700 nm, about 50% of the solar spectrum) is absorbed by the liquid filter and transferred to the heat store. At night the circulation of the now heated filter fluid, through the roof, prevents rapid cooling of the greenhouse. The roof (and filter) cools by long wave radiation and by convection. Photosynthetic light (400-700 nm) enters the greenhouse, allowing for plant growth. Figure 1 shows how a solar still can be integrated into the system and the run-off water from the CSA added to the total fresh water acquired.

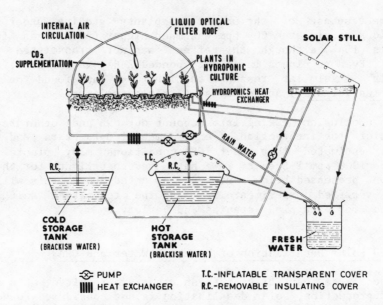

Figure 1. A closed system greenhouse combining a liquid optical
filter and solar still for utilizing brackish water for
plant cultivation in hot arid regions.

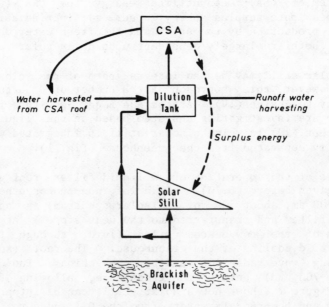

Figure 2. Water sources for CSA in arid and semi-arid regions.

Estimation of Water Budget for Arid Zone Farming Using CSA and
Brackish Water

The various possible sources of water for CSA in an arid region
or semi-arid region are shown in Figure 2.   There are a number of
possible sources of water for CSA in arid regions.   These include
run-off water from desert surface catchments (21), rainfall gathered
from the roof of the closed system (CS) and from the upper surface
of the solar still and the solar distilled water. The rain and
distilled water can.then be mixed with brackish water to a level
which plants in the CS environment can tolerate - about 1000 ppm
TSS, as suggested above, versus a maximum of about 250 ppm
acceptable for conventional open field agriculture.

An estimate of the brackish water and solar still requirements
in the semi-arid Sde Boqer region is given in Table 2. The calculation is based on the following assumptions and conditions: CS area
per farmer - 0.2 hectare; available brackish water - 2000 ppm TSS;
solar still production -   4 1 m^{-2}day^{-1}; salt level tolerated by
plants in CS - 1000 ppm TSS; rainfall - 100 mm/annum; fresh water
requirement in CSA - 0.3 m^3 m^{-2}yr^{-1} (= 600 m^3 per 0.2 h farm).

TABLE 2. ESTIMATE* OF BRACKISH WATER AND SOLAR STILL REQUIREMENTS
FOR CSA AT SDE BOQER**

100 m^2 solar still	- 150 m^3 fresh water/annum
Extra nighttime distillate with heat transfer from CSA (mainly during summer)-	50 " " " "
Rainfall harvesting from roof of CSA at 50% collection efficiency	- 100 " " " "
Total fresh water collection.	- 300 m^3 fresh water/annum
After 1:1 dilution with 2000 ppm T.S.S. brackish water	- 600 m^3 water at 1000 ppm T.S.S.

*Calculations based on assumptions and conditions given in the text.
**In Israel Negev Desert: 30°38'N, 34°38'E; elevation 400 m.
Average annual (winter only) rainfall - 100 mm.

In conclusion, and with reference to the estimate given in Table 2, we believe that using CSA and solar distillation, only a very small quantity of brackish water, about 500 m^3yr^{-1}, may be sufficient for the farming needs of a single family. This figure should be compared with 48,000 m^3 of fresh water required by a single conventional farm unit, based on 4 hectares of irrigated, intensive, mixed agriculture, in an arid region.

Although the figure used for water consumption by CS (600 $m^3/0.2$ h/yr) may be even conservatively high, other parameters require confirmation. These include the extra night time distillation obtained by using surplus heat from the CSA and the tolerance to salt of plants within the CSA environment. However, even if we budget an extra 500 m^3 fresh water, per annum, for domestic use the total fresh water required per family is still very small.

The main technological problem to be overcome is to design, construct and maintain the CSA and solar still at a cost which can be borne by the return from the agricultural produce, while allowing the farmer an adequate income. To this we may add that the technology must be such as can be serviced by the farmers for whom the CSA is designed.

ACKNOWLEDGEMENT

This review includes results and ideas of the CSA group at Sde Boqer and cooperators from the Hannover (Germany), Hebrew, Telaviv and Ben-Gurion (Israel) Universities, the Haifa Technion and the IBM Co. Science Unit (Israel).

This research is supported by a grant from the National Council for Research and Development, Israel and the KFK Karlsruhe, Germany.

REFERENCES

1. E. Rappaport, Agriculture in a Closed Space, Israel Patent No 5528 (1952).
2. R. Bettaque, El Campo 61:16-26 (1977).
3. L. H. De Bivort, T. B. Taylor and M. Fontes, National Technical Information Service, U.S. Dept. Commerce PB-279-211 (1978).
4. M. H. Jensen and H. M. Eisa, Publ. of Environmental Research Laboratory, Univ. of Arizona, Tucson, AR, 118 p. (1972).
5. J. Gale, in: "XXI Symp. British Ecological Society," J. Grace, E. D. Ford and P. G. Jarvis, eds., Blackwells (in press).
6. H. Z. Enoch, Acta Horticulturae 87:125-129 (1978).
7. P. Gaastra, Meded. Landbouwhogesch. Wageningen 59:1-68 (1959).

8. J. A. Bassham, Science 197:630-638 (1977).
9. A. S. Issar, ed., "Brackish water as a factor in development,"
    Publ. Inst. Desert Res. Sde Boqer, Ben-Gurion Univ. of the
    Negev, Israel (1975).
10. J. Gale, H. C. Kohl and R. M. Hagan, Physiol. Plant. 20:408-420
    (1967).
11. G. J. Hoffman and S. L. Rawlins, Agr. J. 63:877-880 (1971).
12. G. J. Hoffman, S. L. Rawlins, M. H. Garber and E. M. Cullen,
    Agr. J. 63:822-826 (1971).
13. J. Gale, R. Naaman and A. Poljakoff-Mayber, Australian J. Biol.
    Sci. 23:947-952 (1970).
14. J. T. Prisco and J.W. O'Leary, Plant and Soil, 39:263-276
    (1973).
15. J. Gale, in: "Plants in Saline Environments," A.
    Poljakoff-Mayber and J. Gale, eds., pp. 186-192, Springer-
    Verlag, Berlin (1975).
16. S. G. Talbert, J. A. Eibling and G. O. G. Lof, "Manual on solar
    distillation of saline water," Research Report to O.S.W. of
    U.S. Dept. Interior, Batelle Memorial Inst., Columbus, OH
    (1970).
17. M. A. Kettani, Sunworld 3:76-85 (1979).
18. F. Trombe and M. Foex, in:"U.N. Conference on New Sources of
    Energy," Paper 35/S/64 (1961).
19. J. F. M. Ronchaine, Bull. Rech. Agron. de Gembloux, Hors.
    Series 314-319 (1971).
20. J. P. Chiapale, J. Demagnez, P. Denis and P. Jourdan,
    in: "12e Colloque National de Plastiques en Agriculture,"
    pp. 87-90 (1976).
21. H. W. Lawton and P. J. Wilke, in: "Agriculture in semi-arid
    environments," A.E. Hall, G.H. Cannell and H.W. Lawton,
    eds., pp. 1-44, Springer-Verlag (1979).

SALINE TOLERANCE IN HYBRIDS OF <u>LYCOPERSICON</u> <u>ESCULENTUM</u> X

<u>SOLANUM</u> <u>PENELLII</u> AND SELECTED BREEEDING LINES

R. F. Sacher , R. C. Staples,   and R. W. Robinson

Boyce Thompson Institute for   New York State Agricultural
Plant Research   Experiment Station
Tower Road   Geneva, N. Y. USA
Ithaca, N. Y. USA

SUMMARY

The F_1 hybrid <u>Lycopersicon</u> <u>esculentum</u> cv. New Yorker x <u>Solanum</u> <u>penellii</u> was backcrossed once to <u>L. esculentum</u> and the progeny selfed to the F_9 generation. Selection was for horticultural type, earliness and cold tolerance. Salinity tolerance of the F_8 selected lines and the parent species suggested that salt tolerance is a polygenic complex and that favorable factors from both species can contribute in an additive manner. Ability to closely regulate Na^+ or Cl accumulation in the leaves was significantly correlated with plant growth performance in saline irrigation. Regulation of ion accumulation was assessed as the ratio of foliar content in salt grown plants to foliar content in control plants, defined as <u>ionic regulation index</u> (IRI). A low index indicated close regulation of accumulation of a particular ion.

$$IRI = \frac{\text{foliar ion content (\% d.w.) with 0.1 N NaCl irrigation}}{\text{foliar ion content (\% d.w.) with control irrigation}}$$

A group of plants which regulated both Na^+ and Cl contents had the highest percentage of resistant lines and the greatest degree of resistance. A group of lines which showed poor regulation of both Na^+ and Cl were all salt sensitive; however, Na^+ and Cl^- accumulated to some extent in all lines tested. The occurrence of salt sensitive lines which also had good ion regulation implies an interaction of the ability of tissues to tolerate elevated Na^+ and/or Cl levels with the ability of the plant to protect sensitive tissues by ion exclusion.   Leaf Na^+ content was found to be highly heritable (H = 0.96) and the low sodium characteristic was dominant.

INTRODUCTION

Salinity of soils and water constitutes a major detrimental factor in production on marginal areas.   Irrigation allows use of certain desert and semi-desert areas but these soils and waters are increasingly impacted by rising salinity levels.   Furthermore, expansion of production areas is now limited by salinity of soils or water supplies.   Crop responses to salinity should be understood for immediate selection goals and to provide models for future design of new or improved crop species.

Tal and his colleagues have compared a number of wild taxa of tomato with cultivars of L. esculentum. In 1971, they reported that the better osmotic adjustment of L. peruvianum was responsible for its superior performance compared with L. esculentum under saline conditions (1).   Transpiration was higher in the wild plants under low salinity, but this decreased more under high salinity than in the commercial tomato (2).   Root resistance to water flow and conductivity of the guttation fluid were higher in the wild plants. The authors concluded that the greater accumulation of ions in the wild taxa under salinity was the main cause for the differences in leaf transpiration between the wild and cultivated species.

Research by Epstein and his colleagues (3-6) has shown that several crops can be bred with wild relatives to create progeny tolerant to the salinity of sea water (35,000 ppm).   For example, barley grown in the field on sandy soil produced a yield estimated at 1580 kg/ha when irrigated with undiluted sea water.

Tomato fruit has been produced by Rush abnd Epstein (4) using a cross between commercial tomato, Lycopersicon esculentum, and the wild relative, L. cheesmanii, when irrigated with 70% seawater.   In a study of L. cheesmanii and L. esculentum, Rush and Epstein (4) showed that while sodium was steadily accumulated in leaves of L. cheesmannii, sodium levels in L. esculentum rose only slightly as the sodium chloride in the external solution was increased to 0.5 M.   The salt levels then rose rapidly in leaves of commercial tomato and the plants died.   Accumulation continued in L. cheesmanii at least to a concentration of 70% sea water, which also suggested to these workers that L. cheesmanii had responded to salinity like many haolphytes do.

In this paper we review some of our work on salinity tolerance using a cross between tomato and the drought tolerant Solanum pennellii.   We show that salinity tolerance among these progeny

depends largely upon protection from intolerable concentrations of
salt, and we present data which suggests that this is due to a combination of regulation of sodium exclusion coupled with specific
ion tolerance of tissues to moderate increases in salt levels.

MATERIALS AND METHODS

Breeding. Lycopersicon esculentum Mill cv. New Yorker by
Solanum pennellii Correll P.I. 246502, hybrids were made at the New
York Agricultural Experiment Station, Geneva, NY.   The F_2 generation proved to contain no individuals with even rudimentary aspects
of desirable horticultural type.   Subsequently, the F_1 was backcrossed to L. esculentum and selected progeny selfed to the F_9
generation. Primary selection criteria were frost tolerance, early
yield and good horticultural type.   Frost tolerance was evaluated
by subjecting plants to -8 C for four hours and selecting
survivors.

Selection for salt tolerance. Breeding lines beginning at the
F_8 generation were evaluated for salinity tolerance.   Plants of
each line were germinated and grown for 2 weeks in a mix of milled
peat moss and vermiculite (1:2, v/v).   They were then transplanted
to pots containing washed 40 mesh quartz sand and irrigated by an
emitter tube drip system (Chapin, Watertown, NY).   Solutions were
not recycled.   Initial irrigation to all plants was with control
solution consisting of half-strengh Hoagland's (Peters Hydrosol, W.
R. Grace, Inc., Allentown, PA) solution in tap water.

Greenhouse experiments were in completely randomized designs.
Plants were replicated 6 times in each line and salt treatment.
New Yorker tomato plants were grown simultaneously with each breeding line group to provide an internal standard of known salinity
response.   After two weeks of control irrigation (4 weeks total
age) the solution supplied to test plants was modified by addition
of NaCl to 0.1 (low) or 0.2 (high) concentration.   Total salts in
each treatment were: control, 693 ppm; low salt, 6,543; and high
salt, 12,393 ppm.   Control solution had 0.04 ppm Na and 3.62 ppm
Cl.

Plant analysis.   After growth on salt for two weeks, plants
were harvested, oven dried and analyzed as composite samples for
Na^+ by flame photometry and individually for Cl by use of a selective ion electrode (Orion Research Inc., Cambridge, MA).

Sodium or Cl content of dried leaves are reported as percentage of dry weight and as ionic regulation index (IRI) calculated as
percentage of dry weight in salt grown plants divided by percentage
of dry weight in control plants.

RESULTS

Tolerance to Salt

Growth in salt. Breeding lines derived from the cross between L. esculentum and S. pennellii (see Materials and Methods) were tested for tolerance to NaCl. Six of the 31 lines tested tolerated 100 mM NaCl and produced at least 80% or better of the dry weight produced in the absence of salt. Some of this data is given in Table 1 to show the ranges of values obtained.

Leaves from our hybrids showed marked increases in content of Na^+ and Cl when grown on salinized sand cultures (Table 2). The relative amount of salt which accumulated (IRI) in the leaves was negatively correlated with plant performance. A negative correlation was found for both the Na^+ ion (Fig. 1A) and the Cl ion (Fig. 1B). The correlations also held for plants grown in 200 molar added salt (data not shown). These results suggest that tomatoes can regulate the accumulation of ions in their tops. Even though salts accumulated in the tops of all lines, those which were the most successful in terms of relative performance on saline irrigation were able to regulate ion uptake and maintain a condition closer to the control status than did the sensitive lines. Regulation of Na content accounted for a greater amount of variation in sensitivity than did Cl regulation (r^2 Na = 0.56, r^2 Cl = 0.09).

When IRI values obtained for each breeding line were classified as high or low for each ion there were four possible combined ion regulation classes. The occurrence of resistant or sensitive individuals ion each such class is reported in the histogram of Figure 2. All of the lines which were saline tolerant regulated the relative increase in either Na^+, Cl or both. All of the lines which were poor regulators of both Na^+ and Cl were sensitive to salt. Regulation of the accumulation of at least one of the salt ions in the leaves was necessary but not sufficient to impart salt tolerance as evidenced by the fact that even in the nine lines which effectively regulated both Na^+ and Cl there were four which were sensitive to salt.

Sodium ion regulation. The absolute amount of foliar Na^+ and Cl^- increased with salination of irrigation water to some extent in all lines regardless of sensitivity to salt (Table 2). Examination of the Na^+ content after 0.1 N salination shows that there must be a component of specific ion tolerance as well as ion exclusion involved in salt resistance of these lines. Two lines, 79-90 and 79-91, had Na^+ contents when salt stressed which were not greatly different from the highest Na^+ content which had been found in the control plants; and resistance apparently was due to Na^+ exclusion. Two other lines, 79-84 and 79-96, had identically high Na^+

TABLE 1.   Growth of selected lines of tomatoes on sand irrigated
            with nutrient solution salinized with 0.1N NaCl.   Each
            value is an average of six plants.

Tomato Line[1]	Relative Plant dry weight	IRI[2] Na$^+$	Cl$^-$
79-96	103.1	4.2	3.6
79-89	93.6	3.8	2.6
79-90	93.4	3.5	2.2
79-91	90.6	7.0	2.8
79-86	80.8	4.5	1.8
79-71	44.3	5.1	3.3
79-84	38.0	10.9	3.9
79-85	31.3	4.6	3.3
79-68	28.3	8.8	3.6
New Yorker	68.6 $\pm$ 6.4	6.1 $\pm$ 1.6	3.2 $\pm$ 0.7

[1]Selected breeding lines of [L. esculentum cv. New Yorker
  x S. pennellii BC$_1$]F$_8$.
[2]Ionic regulation index.

TABLE 2.   Accumulation of salt by selected lines of tomato after
            growing on sand irrigated with half strength Haogland's
            solution either without further addition (Control) or
            with 100 mmolar NaCl.

Tomato Line	Performance[1] Rating	Sodium (% dry weight) Control	0.1 N NaCl	Chloride (% dry weight) Control	0.1 N NaCl
79-76	R	0.26	2.1	0.89	3.0
79-84	S	0.32	3.5	1.50	5.8
79-85	S	0.50	2.3	1.60	5.2
79-86	R	0.49	2.1	1.30	4.8
79-87	S	0.40	2.3	0.90	6.8
79-88	R	0.50	2.3	2.10	6.7
79-89	R	0.58	2.2	2.50	6.4
79-90	R	0.34	1.2	2.40	5.2
79-91	R	0.23	1.6	2.60	7.4
79-92	S	0.22	2.1	0.79	3.0
79-93	S	0.48	2.5	0.74	3.2
79-94	R	0.88	2.5	0.88	3.0
79-96	R	0.81	3.5	0.76	2.7
79-97	R	0.43	2.3	0.72	3.1

[1]Performance rating; (R), resistant (dry weight on salt was 80% or
better of that without salt); and (S), sensitive.

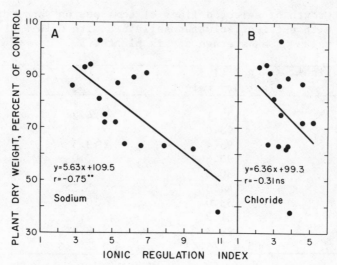

Figure 1. Regression of dry weight --of plants grown with 0.1 N NACl expressed as percent of dry weight of control plants --on sodium (A) and chloride (B) ionic regulation indices (IRI, see text). Plants were advanced selections from _L. esculentum_ x _S. pennellii_. ** indicates regression significant at 0.01 level.

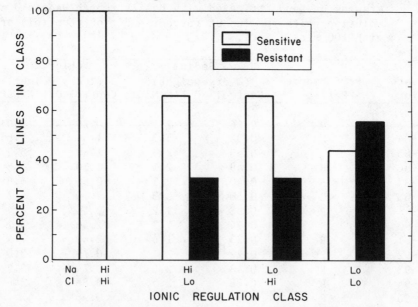

Figure 2. Percentage of tomato breeding lines in populations classified by ionic regulation indices (IRI) as follows: Hi Na^+ IRI > 6.1; Hi Cl IRI > 3.2. Resistant lines, when grown with 0.1 N NaCl irrigation, maintain plant weight greater than 80% of control plant weight.

contents, yet the growth of 79-84 was very salt-sensitive while 79-96 was very salt tolerant. Plant dry weight of 79-96 when grown on 0.1N saline irrigation was 103% of control plant weight while dry weight of 79-84 on 0.1N saline irrigation was only 38% of control plant dry weight (Table 1).

Comparison of 79-84 and 79-96 thus suggests that 79-96 tolerates elevated sodium levels whereas 79-84 does not. The cause of tissue sensitivity in these lines cannot be determined from these data, but it can be seen that some of these lines are able to tolerate high foliar sodium without adverse effects upon growth while others cannot.

Chloride ion regulation. Similarly there are lines which accumulate high amounts of Cl (Table 2) in the leaves yet are salt resistant (79-88, -89, -91) implying tolerance to Cl in these lines. While there are Cl excluding lines which are salt resistant (79-86, -90, -94, -96, -97, -100), there are also chloride excluding lines which are sensitive (79-84, -85, -87, -92, -93). This demonstrates that capacity to exclude Cl cannot be the sole cause of chloride tolerance.

Consequently, chloride may be toxic to those lines (cf. 79-84, -85) which have high leaf Cl and which are salt sensitive. When the resistant and sensitive groups were averaged (Table 3) it was found that foliar contents of Na^+ and Cl after salt treatment were not significantly different at the 0.05 level by t test. The difference between mean values for Cl IRI was also not statistically significant. Mean IRI values for sodium were significantly different at the 0.05 level with the resistant group having closer control of sodium accumulation than the sensitive group.

TABLE 3.   Resistant and sensitive line group means for Na^+ and Cl^- leaf content and Na^+ and Cl^- Ionic Regulation Index (IRI, see text).

	Na^+ % d.w.	Cl^- % d.w.	Na^+ IRI*	Cl^- IRI
Resistant	2.23	4.57	4.46	3.32
Sensitive	2.36	3.89	6.85	3.58

* Difference between means statistically significant at 0.05 level by t test.

Inheritance of Salt Tolerance.

Exclusion and close regulation of sodium were inherited in a dominant manner in the F_1 and backcross to L. esculentum (Table 4) and the exclusion (low leaf Na^+) characteristic. is apparently associated with close sodium regulation (low IRI). Chloride exclusion was dominant in the F_1 while chloride regulation apparently was recessive. The backcross data indicate that a simple dominance mechanism does not describe the inheritance of either parameter or response to chloride.

Preliminary frequency analyses of F_2 and BC_1 populations for leaf sodium content (Fig. 3a, b) suggested that a small number of genes — perhaps only one — were involved in sodium accumulation. Heritability of leaf sodium was calculated to be 0.96 based upon variances of the F_1 and F_2 populations. Chloride content in F_2 populations (Fig. 4) did not suggest more than the likelihood of a quantitative inheritance mechanism. The capacity for growth with 0.1 N additional salt was conferred largely from S. pennellii (Table 4) and was only partially explained by differences in ion accumulation and/or regulation.

Heavy selection pressure has resulted in the production of several advanced breeding lines with good horticultural plant type and potentially economic fruit types including recovery of the parental type and quality. Superior advanced lines were entered in competitive variety trials in 1980 and were found to exceed their New Yorker tomato parent in early yield (Robinson, unpub.)

TABLE 4.   Sodium and Cl contents and ionic regulation indices (IRI, see text) in leaves of tomato plants grown on 0.1 N NACl salinated nutrient solution. IRI and weight comparisons were relative to control plants on nutrient solutions.

	Na^+ % d.w.	Na^+ IRI	Cl^- % d.w.	Cl^- IRI	Plant dry weight % of control
L. esculentum	2.5	4.7	6.4	2.9	67
S. pennellii	5.3	10.4	2.4	3.5	103
F_1	2.5	5.0	2.1	3.3	134
BC_1	2.0	3.8	2.1	3.8	66

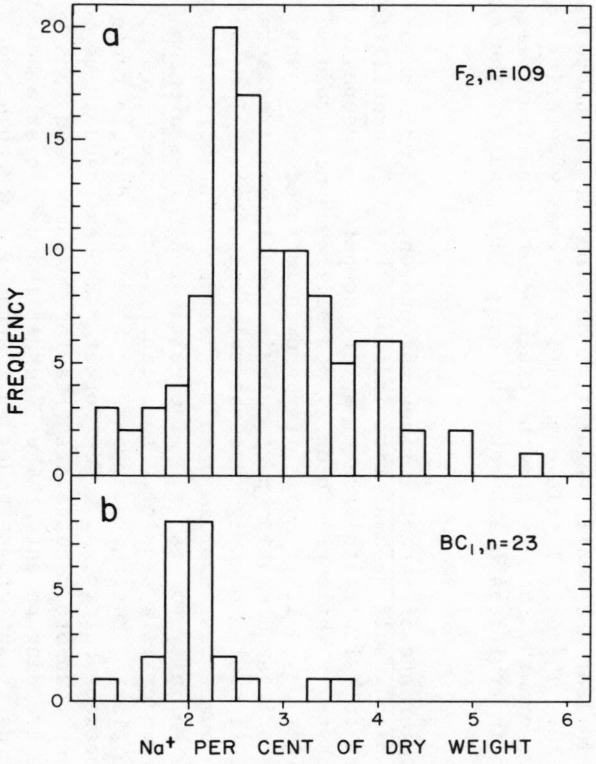

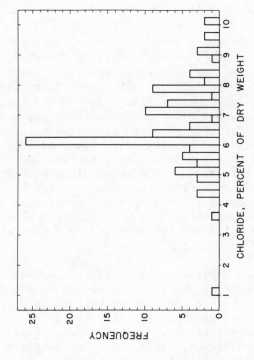

Figure 3. Foliar sodium content frequency distributions in F_2 (A) and BC_1 (B) populations of L. esculentum x S. pennellii.

Figure 4. Foliar chloride content frequency distribution in F_2 population of L. esculentum x S. pennellii.

DISCUSSION

   Wide crosses with tomato. Inquiry into the cytogenetics and
phylogeny of tomatoes was opened by Correll (7) who provided a most
valuable avenue for breeding superior lines of tomatoes when he
discovered S. pennellii. Subsequent hybridization of this species
with L. esculentum by Rick (8) and the collection, genetics and
breeding of other wild taxa of tomato by Rick and his associates
(9-12) and Tal (13) laid the ground work for the extensive research
now proceeding in several directions on the use of genetic materials from wild stocks to improve the performance of tomatoes in
adverse environments. These results were summarized recently by
Rick (14). More recently, Dehan and Tal (15) extended these studies and reported that S. pennellii was more salt tolerant than L.
esculentum. In contrast to L. esculentum plants (16), the growth of
the wild species was not impaired by salinity levels as high as 200
m/molar.

   Perhaps a wider range of studies on salinity effects on tomato
have been carried out in the field using commercial varieties than
by any other procedure. As summarized in a review by Pasternak, et
al (17), tomatoes are especially sensitive to salinity at the young
seedling stage which imposes an irreversible damage to plant growth
rate.   In addition, salinity reduces yield mainly by its direct
effect on fruit weight since salinity reduces the size of constituent cells.   The tomato is to some extent sink limited, so that
both fruit weight and fruit number are little affected by large
variations in leaf area while vegetative development is independent
of fruit load. There is little competition between vegetative and
reproductive growth, and a growth check occurs in fruit loaded
plants and in fruitless plants at the same time and to the same
extent.

   Salt tolerance of parental lines and progeny. Our data on
plant growth and salt accumulation indicate that S. pennellii is,
with respect to salinity tolerance, a very competent mesophyte but
not a halophyte.   While growth of S. pennellii was not reduced by
saline irrigation neither was its growth improved nor was an
increase in salt accumulation particularly tied to salt resistance.
S. penellii is also notably drought tolerant (14).   The well-
developed drought response may also be a strong contributing factor
toward its salinity tolerance.   The cultivated L. esculentum is
itself only moderately sensitive to salinity with respect to other
common crop plants, but there appears to be a considerable range of
tolerance between cultivars.   New Yorker has been found to be more
salt tolerant than other cultivars (Staples, unpubl.) and thus well
suited as a breeding parent.   The F_1 exhibited a favorable hetero-
sis as regards plant growth under salt stress.   Although none of
the advanced selections were as vigorous under salt as the F_1,

there is at least one which is equal to S. pennnellii and which also produced tomato-type fruit.

Resistance mechanisms. The modes of resistance are not yet defined, but it is clear that plants from this gene pool depend in large measure upon protection of the tops from intolerable salt concentrations. This could well be explained by a mechanism involving transfer cells which selectively load the lower stem phloem with undesirable ions from the xylem stream; as has been demonstrated in other mesophytic plants (18-20). While the absolute levels of Na^+ or Cl in the leaves of a particular line may have led to its classification as a salt includer or excluder relative to others in this group, none was able to prevent a substantial increase in either Na^+ or Cl^-. Nor did any line exhibit a salt uptake response proportional to salt application as occurs with halophytic species (4). We feel that salt resistance in both L. esculentum and S. pennellii involves a necessary partial regulation of salt ion accumulation coupled with tolerance of tissues to moderate increases in salt levels. Parameters of water relations and osmotic adjustment under adverse conditions are now being investigated in these species and the advanced breeding lines.

Taxonomic considerations. S. pennellii also offers resistance to sucking insects (21), drought (14), and cold temperatures as well as to salt. There are relatively small genetic barriers to breeding in the early generations but these are not of a magnitude which limit the usefulness of S. pennellii as a source of resistance to several stresses that now impact tomato production. S. pennellii has been found to be closer phylogenetically to L. esculentum than their present classifications would imply (9, 13). Evidence from chromosome pairing, hybrid compatability, hybrid fertility, plant morphology and inheritance of key taxonomic indicator traits suggest that these two species should properly be reclassified within the same genus. If considered as an interspecific cross rather than an intergeneric cross, the L. esculentum x S. pennellii approach to tomato breeding does not seem so radical. As the results to date have shown, the cross promises improved stress-resistant varieties in the near future.

References

1. M. Tal, Australian J. Agric. Research, 22:631-637 (1971).
2. M. Tal and U. Gavish, Ibid, 24:353-361 (1973).
3. Epstein, E. in: "Plant Adaptation to Mineral Stress in Problem Soils", (M. J. Wright, ed.), pp. 73-82, Cornell University Agricultural Experiment Station, Cornell University, N. Y. (1976)

4. D. W. Rush and E. Epstein, Plant Physiol. 57: 162-166 (1976).
5. E. Epstein and J. D. Norlyn, Science, 197:249-251 (1977).
6. D. B. Kelly, J. D. Norlyn and E. Epstein, in: "Proceedings of the Internatinal Arid Lands Conference in Plant Resources", (J. R. Goodin, D. K. Northington, eds.), Lubbock, Texas Tech Univ. Press, pp. 326-340 (1979).
7. D. S. Correll, Madrono, 14:232-236 (1958).
8. C. M. Rick, Proc. Natl. Acad. Sci. (U.S.), 46:78-82 (1960).
9. G. S. Kush and C. M. Rick, Genetica, 167-183 (1963).
10. C. M. Rick and L. Butler, Adv. Genetics, 8:267-382 (1956).
11. C. M. Rick and J. Robinson, Am. J. Botany, 38:639-652 (1951).
12. C. M. Rick and P. G. Smith, American Naturalist, 87:359-373 (1952).
13. M. Tal, Evolution, 21:316-333 (1967).
14. C. M. Rick in: "Proceedings of the 1st International Symposium on Tropical Tomato", (R. Cowell, ed.), Asian Vegetable Research and Development Center, Publication 78-59, pp. 214-224 (1979).
15. K. Dehan and M. Tal, Irrigation Science, 1:71-76 (1978).
16. L. Bernstein, USDA Agr. Inf. Bul. 283, (1964).
17. D. Pasternak, M. Twersky and Y. De Malach, in: "Stress Physiology in Crop Plants", (H. W. Mussell and R. C. Staples, eds.), pp. 127-142. Wiley Interscience, N. Y. (1979).
18. D. Kramer, A. Lauchli, A. R. Yeo, J. Gullasch, Ann. Bot., 41:1031-1040 (1977).
19. A. R. Yeo, D. Kramer, A. Lauchli, J. Gullash, J. Expt. Bot., 28: 17-29 (1977).
20. B. Jacoby, Ann. Bot., 43:741-744 (1979).
21. A. G. Gentile an A. K. Stoner, Economic Entomology, 61:1152-1154 (1968).

PRESENT AND FUTURE APPLICATIONS

CHAIRMAN'S REMARKS

T.E. Brand

Depto. de Biologia Marina
Centro de Investigaciones Biologicas
  de Baja California
Lá Paz, B.C.S., Mexico

Present   and future   applications of   biosaline   research are
diverse and far reaching.   The biosaline   concept, born out of the
need for a   broad integrated approach to solving   complex problems,
has helped focus attention on the necessity of developing, managing
and conscientious harvesting of underdeveloped   and/or non-utilized
resources.   Much of   this work   is being   carried   out within   the
world's tropical   and subtropical   zones where   arid   and semi-arid
land is common   and fresh irrigation water scarce.   The paucity of
water and alkaline soils characteristic of the region cause farmers
to use nonconventional methods of crop cultivation.

One of the most important agricultural pollutants generated in
arid and   semi-arid   acres is   the saline   waste   from fertilizers,
animal wastes and tile drainage.   The chemical   properties of this
water make discharge   into fresh and saline waters   undesirable and
creates   the need   for   some   sort   of   filtering   system.   Oswald
discusses a microalgal system   he has   developed which is   the most
efficient and economical means of controlling the pollutants.

The   future   of enzyme   technology is   discussed   by Zaborsky,
particularly   their   use as   biocatalysts.   It is   emphasized that
enzymes from halophytic microorganisms likely   possess characteristics not found in conventional sources of enzymes, traits which may
be of industrial use.

Serious side effects   caused by   drugs used in   cancer therapy
have underscored   the need   to develop new   anti-cancer cytotoxins.
Natural products isolated from marine organisms may prove useful in
solving this and   other medicinal problems as reported   by Fenical.

339

The importance of mangroves as sources of habitat for crabs, prawns and other food resources is well documented, yet conservation and management programs are practically nonexistant. Teas, in an exhaustive review of the literature and his own data, concludes that mangroves are amenable to management which should lead to increased wood and food resource production. Sordo, et al. have taken an innovative look at the potential for generating freshwater through desalinization by mangroves.

Neushul suggests that macroalgal mariculture has a bright future for both human consumption and industry. Farming of marine algae is highly productive, relatively inexpensive and feasible in coastal waters throughout the world.

As the world's protein shortage becomes more acute in the near future, especially in Third World countries, the importance of developing new sources of food will increase. Innovative technology will be required to turn deserts into productive farmland, to meet demands for potable water, for recycling waste water, and other human needs. Biosaline research will play an integral role in solving these and other problems of the not too distant future.

BIOCATALYTIC CONVERSIONS

Oskar R. Zaborsky

National Science Foundation
Washington, D. C.   20550   USA

INTRODUCTION

Enzymes are macromolecular biological catalysts responsible for a myriad of chemical transformations occurring in nature. For centuries, enzymes have been employed in bioconversion processes to produce desired products, but the exact mode of even the most studied ones has yet to be elucidated. The phrase "biocatalytic conversion" refers to the use of the biological systems, especially enzymes, for converting various forms of energy or for transforming one chemical substance into another.

Figure 1 is a diagrammatic representation of the energy forms for which biocatalysis can play a significant role. As we all know, solar energy is converted to chemical energy through photosynthesis. This is a process of tremendous magnitude and involves the most abundant enzyme on this planet, ribulose 1,5-bisphosphate carboxylase. Once carbon is fixed, it is then converted to a host of other products by enzymes. What is not so well appreciated is that solar energy can also be converted to electrical energy through biochemical systems, and that the existing chemicals can be converted to electricity via biochemical fuel cells. At the moment, biocatalytic systems are the only ones available to us for producing chemical energy (i.e., the host of natural products that we use daily) and for some of the conversion processes used to produce fuels and chemicals (such as ethanol and methane).

Enzymes can be isolated from all living sources--animals, plants, and microorganisms. However, for industrial purposes, microbes offer the best source because they can be grown and handled conveniently, and the concentration of the desired enzymes can be increased by genetic or physiological means. Enzymes are either extracellular or intracellular. Extracellular enzymes are

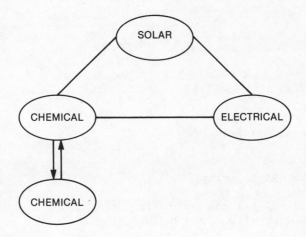

Figure 1.   Bioconversion Pathways of Energy Forms.

excreted by cells into the surrounding environment and can be easily isolated.   Intracellular enzymes are associated with cell-ular membranes or organelles and are much more difficult to isolate.   Yet, it is the intracellular enzymes that are much more prevalent in nature, and only a comparatively few enzymes occur outside living cells.   To date, most enzymes of commercial signifi-cance have been extracellular ones obtained from microbial sources through fermentation (1).

Enzymes exist in a variety of forms which are useful for industrial processes.   Of course, enzymes exist in viable, intact cells of all living matter and, in particular, enzymes in microbial cells offer a very versatile reactor system through fermentation. This is a time-tested approach which has been used to produce a host of products such as antibiotics, organic acids, solvents, vitamins, amino acids and even the enzymes themselves.   A more recent approach is the use of tissue cultures employing viable cell lines.   Alternatively, particular enzymes in intact but nonviable cells can be employed for chemical processing.   In this mode, cells containing the enzymes of interest are first "killed" by chemical or physical treatments and then used as biochemical reactors.   The nonviable cells can be further transformed into immobilized counterparts, a topic which is described later.

Certain enzymes exist in or are attached to cellular bodies; two common examples are chloroplasts and mitochondria.   Although no commercial process is presently based on the use of organelles, these enzyme-containing bodies would probably be employed in an immobilized form. Finally, enzymes can exist in a cell-free environment. The advantages of this form are that the biocatalysts can be used in a concentrated and more controlled manner, devoid of any possibly interfering organic cellular constituents. The use of cell-free enzymes affords higher yields and purity of the product caused by simpler isolation procedures. Reactions are also easier to control than in fermentations and there is essentially no waste disposal problem. Cell-free enzymes have been used commercially in both solution and immobilized forms.

Pros and Cons of Biocatalytic Processes

The advantages of biocatalytic processes are listed in Table 1. In particular, enzymes are superior to conventional catalysts in their high specificity and activity (2, 3). However, other less appreciated advantages are that biocatalysts can conduct unique transformations unattainable by any other means, require low conventional energy input during process conversions and are ubiquitous and renewable. Hence, their use need not be restricted by geographical or political boundaries as is the case with precious metals and minerals. Bioconversion processes afford reactor flexibility by judicious choice of the form of the enzyme; i.e., whether the biocatalyst is contained in a cell, exists in solution or is immobilized. A mixed blessing is the necessity for conducting enzymatic transformation in an aqueous medium.

The disadvantages of biocatalytic processes are given in Table 1. Notable problems are the competing metabolic processes associated with cells (microbial and otherwise) and the necessary rigid control of living entities. Other problems include the inactivation of enzymes (whether in a cell or in an isolated form), coenzyme requirements for many of the synthetic reactions and operational limitations exhibited by any soluble catalyst (whether an enzyme or any other soluble catalyst). These operational difficulties include mechanical loss of the catalyst, limited reuse, inadvertant product contamination and limited reactor options. The problem of operational limitations has received considerable attention during the last 15 years and can be circumvented by immobilization (4).

Immobilized Enzymes

The term "immobilization" refers to the physical confinement or localization of enzyme molecules during a continuous catalytic process. This is largely an operational definition intended to encompass a variety of methods that have been used to achieve this

TABLE 1. Characteristics of Enzymes

Advantages: Disadvantages:

Activity   Unavailability
Specificity   Contamination
Asymmetric Synthesis   Instability
Unique Transformations   Coenzyme Requirements
Renewable and Ubiquitous   Operational Limitations
                                                 (Soluble Catalyst)
                                               - Mechanical Loss
                                               - Limited Reuse
                                               - Product Contamination
                                               - Limited Reactor Options

objective. Classically, the term "immobilization" has been used to describe the process of transforming a water-soluble enzyme into a water-insoluble conjugate--the immobilized enzyme. However, not all methods of immobilization involve the preparation of water-insoluble conjugates. Several methods simply consist of restricting the movement of enzyme molecules to a microspace, but the entrapped molecules still have considerable translational and rotational freedom and often retain their inherent solution characteristics.

Three major reasons exist for immobilizing enzymes. Immobilized enzymes offer a considerable operational advantage over freely mobile ones. They also can exhibit enhanced chemical or physical properties as well as serve as model systems for natural in vivo properties as well as serve as model systems for natural in vivo membrane bound enzymes. General advantages of immobilized enzymes are continuous use of the catalyst with no product contamination, precise product control, flexible reactor and process engineering designs, and enhanced properties. In particular, enhanced thermal stability is often, but not always, observed.

The most commonly employed immobilized enzyme systems are presented in Figure 2. For convenience of classification, chemical methods of immobilization involve the formation of at least one covalent bond or partially covalent bond between groups of an enzyme and a functionalized water-insoluble support or between two or more enzyme molecules. Thus, the immobilization of an enzyme by coupling it to a preformed polymer or by treating it with a bifunctional reagent are chemical methods. Physical methods include procedures that involve localizing the enzyme in any manner

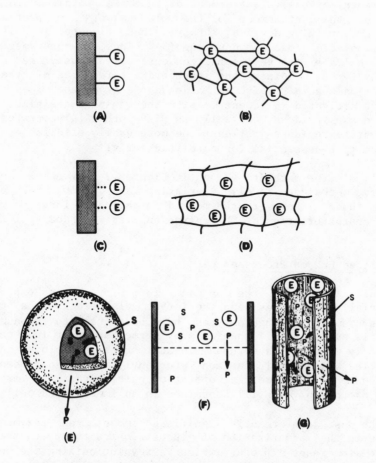

Figure 2:   Schematic representation of immobilized enzyme systems.
            Letters E, S and P represent enzyme, substrate and
            product molecules, respectively.

            A.   Covalently bonded enzyme polymer conjugate;
            B.   Covalently bonded intermolecularly crosslinked
                 enzyme conjugate;
            C.   Absorbed enzyme-polymer conjugate;
            D.   Polymer lattice-entrapped enzyme conjugate;
            E.   Microencapsulated enzyme;
            F.   Ultrafiltration cell-contained enzyme;
            G.   Hollow fiber-contained enzyme.

not dependent on covalent bond formation. Thus, the immobilization is dependent on the operation of physical forces (e.g., electrostatic interactions), entrapment of enzymes within microcompartments or the containment of the biocatalysts in prefabricated membrane-containing devices. Enzymes can also be immobilized by "non-amino acid" covalent coupling (5). Thus, glycoenzymes such as glucose oxidase can be immobilized onto polyaminostyrene beads or films by first oxidizing the carbohydrate residues of the enzyme with periodic acid to give an "aldehydic" enzyme and then contacting the activated glycoenzyme with the amino-containing polymer. With glucose oxidase, as well as with other glycoproteins, the carbohydrate residues (at least the more easily oxidizable ones) do not seem to be essential for catalytic activity.

All of the methods shown in Figure 2 are used, and many reported techniques involve variations and combinations of these basic methods. However, the method of choice is largely dictated by the characteristics of the enzyme and its intended application.

Immobilized Enzyme Processes

Until recently, the dominant form of enzymes employed for commercial purposes has been the intact, viable cell in a fermentation system. Although enzymes have been used to produce fine chemicals and some bulk chemicals by fermentation or in a cell-free form, only recently have these biocatalysts been employed in a more controlled and reusable fashion. Currently about ten processes are reported to be in commercial operation using either an immobilized enzyme or the enzyme still contained in an immobilized cell (3).

The most commercially significant immobilized enzyme process to date is the isomerization of glucose to fructose with the enzyme glucose isomerase, a process which is revolutionizing the sweetener industry (6). Although the isomerization of glucose to fructose can be achieved with alkaline catalysts at high pH, fructose produced by this route was not a commercial success because of the formation of byproducts which imparted a dark color and objectionable flavors. The chemical method employed involved nonspecific and difficult to control reactions, and the formation of byproducts could not be reduced to acceptable levels. The answer to the long sought goal of isomerizing glucose to the sweeter tasting fructose was realized through enzyme technology.

The significant discovery in this development was the enzyme xylose isomerase which converts not only xylose to xylulose but also D-glucose to D-fructose. After the discovery of this enzymic conversion in 1957, industrial development was pursued and the first commercial high fructose corn syrup was introduced ten years later. The product contained 15% fructose and was made by a batch process using a soluble glucose isomerase. In 1968, the Clinton Corn Processing Company introduced a 42% fructose solution made with an immobilized enzyme. A continous system using the immobilized enzyme was put into operation in 1972 which produces a high fructose corn syrup containing 42% fructose, 50% glucose and 8% other saccharides on a dry weight basis. Since that time, a host of enzyme producers and corn wet milling companies have provided significant improvements to the overall process. This year, over 4 billion pounds of high fructose corn syrup will be produced in the U. S.

Coupled to this dramatic development of producing fructose from glucose are the enzymatic steps of converting starch to glucose. Here, two other enzymes are involved. The first, an α-amylase, catalyzes the liquefaction or partial breakdown of starch. The second, glucoamylase, catalyzes the hydrolysis of oligosaccharides (dextrins) to glucose. It is important to note that nearly all commercial glucose (dextrose) produced in the U. S. is made by the enzymatic hydrolysis of starch. The all-enzyme process provides yields of 95-96% glucose compared to 80-85% for the acid-based process.

The switch of catalysts in the hydrolysis of starch to glucose from all acid to all enzyme is a noteworthy event in process engineering. It represents the substitution of the simplest and least expensive catalyst, the hydronium ion, with much more exotic and supposedly fragile biocatalysts. Of course, the substitution is favored because enzymes are more selective and produce higher yields of the desired products.

Reseach Needs

In order to translate the many industrial opportunities offered by enzymes into realities, certain generic problems need to be overcome. Although some of them have already received attention during the last few years, others have largely been ignored.

If enzymes are to become a more common feature in conversion processes, more effective large scale isolation and purification procedures are needed. Presently used enzymes are largely derived from sources of conventional origin (i.e., from mesophilic organisms growing at temperatures from 10-40°C) and from microbes whose characteristics are well established but which have been exploited mostly as research tools. However, enzymes also exist in organisms that survive in unusual habitats, and these may have some decided advantages over those isolated from normal mesophilic organisms. More about this in a later section.

A prime need is also for a more thorough understanding of the inactivation processes of enzymes and the development of effective means for stabilizing biocatalysts. In particular, the in vivo inactivation of enzymes in cells is not well understood nor appreciated. Also, chemical methods for stabilizing enzymes are still more of an art than a science. As with any catalyst, enhancing the stability under actual process conditions is an important goal.

Enzymes involved in synthetic reactions require intermediacy of coenzymes which ultimately must be regenerated if the envisioned process is to become a commercial reality. Although substantial progress has been made, more effective methods for regenerating ATP (adenosine triphosphate) and NAD (nicotinamide adenine dinucleotide) need to be developed. The current costs for these coenzymes are prohibitive for commercial processes, and convenient and inexpensive methods need to be developed.

The operational characteristics of enzymes need to be improved and expanded. Although the development of novel enzyme reactors, especially ones employing immobilized enzymes, has received considerable attention during the past decade, additional research is needed for perfecting large-scale reactors and ones which simultaneously provide for an easier separation of the products from the reactants. More effective and less expensive supports for immobilized enzymes are also necessary.

So far, the use of enzymes in nonaqueous solvent systems has been explored only to a limited extent. Yet, enzymes do show activity in organic solvents or solvents of low water content and even in the solid state. The compatibility of enzymes and microorganisms with organic solvents needs to be examined. Again, enzymes could be modified chemically to expand their operational range in organic solvents.

Additionally, enzymes and conventional catalysts are used most often separately and with no appreciation of attaining the best of both possible worlds. More research is needed to assess the effective interfacing of enzymic and conventional chemical systems (both catalytic and conventional synthetic approaches), so that the limitations, if any, are understood and that traditional barriers are eliminated.

Enzymes from Saline and Hot Environments

As mentioned previously, enzymes currently used in industrial applications are derived largely from mesophilic microbes. Some attention has been given to enzymes isolated from thermophilic bacteria, but enzymes obtainable from other sources, such as halophilic microbes, have not been investigated extensively nor exploited (7). This section presents a brief account of halophilic and thermophilic enzymes and some general considerations for their industrial utilization; however, it is by no means an extensive review of the available literature.

The activity of an enzyme is intimately associated with the tertiary structure of the protein which, in turn, is influenced by such factors as temperature and salts. As we know, proteins can be selectively precipitated by increasing the salt concentration of a solution and they can retain their native conformation during this process. In fact, enzymes can exhibit activity even in the crystalline state. However, high concentrations of NaCl, urea and other compounds are also known to denature proteins.

With regard to temperature, most, but not all, proteins become increasingly unstable above 40-50%°C and begin to denature with a concomitant loss of catalytic activity. However, proteins of thermophilic bacteria are considerably more heat stable than their homologues from mesophilic bacteria. Nearly all proteins of thermophilic bacteria remain in the native state after heat treatment which virtually denatures all the protein of a related mesophile. Thus, we can expect that enzymes produced by thermophilic microorganisms are capable of operating at higher temperatures and, on this basis, would be of greater industrial value in certain situations. The publications by Brock (8) and Zuber (9) provide excellent discussion of thermophilic microorganisms and their associated enzymes. Thermophilic proteins and enzymes have also been discussed elsewhere (10-12).

The reason for the enhanced stability of enzymes obtained from thermophilic microbes in comparison to enzymes from mesophilic organisms is still uncertain and open for debate. Frequently mentioned reasons for enhanced stability at elevated temperatures are an enhanced binding of divalent ions between side chains in the protein, enhanced hydrophobic interactions or increased hydrogen bonding. However, comparative examinations of enzymes from thermophilic and mesophilic microorganisms have, to date, revealed only very minor differences in their physico-chemical properties, and the cause of the enhanced stability must be due to subtle changes.

An explantion for the enhanced stability of enzymes from thermophilic bacteria was given by Perutz in a review on electrostatic effects in proteins (13). His view is that electrostatic effects dominate many aspects of protein behavior and that thermophilic enzymes owe their extra stability mostly to additional salt bridges. Evidence for his view comes from a comparative analysis of the X-ray structure of several proteins. For example, a comparison of the electron transfer protein, ferredoxin, from a mesophilic bacterium (Clostridium pasteurianum) and a thermophilic one (C. thermosaccharolyticum) revealed that the only possible sources of greater heat stability in the thermophilic protein were extra salt bridges on the protein surface, especially bridges linking residues near the amino terminus to others near the carboxyl terminus. Another example is glyceraldehyde phosphate dehydrogenase in which the greater heat stability is due to salt bridges between the four subunits which the mesophilic enzyme does not possess. Serine and glutamine are replaced by arginine and glutamic acid where they form salt bridges between symmetry-related subunits in a cavity shielded from solvent. Another buried ion pair is formed between an arginine residue in one subunit and an aspartic acid residue in a symmetry-related one.

Perutz's interpretation is also consistent on energetic considerations. For example, the extra stabilization energy of thermophilic ferredoxin amounts to no more than 2 kcal/mole and to about 5-10 kcal/mole for glyceraldehyde phosphate dehydrogenase. It has been estimated that an external salt bridge can contribute up to 1 kcal/mol and that an internal one contributes 3 kcal/mole of stabilization energy. Incidentally, a ten-fold increase in the rate at 60°C needs only about 1.6 kcal of extra energy of stabilization.

Although this explanation for the enhanced stability of thermophilic enzymes is appealing and has experimental support, it would not be surprising to discover that other factors also play major roles in other proteins or that the enhanced stability is not only caused by the formation of additional electrostatic ion pairs but by a multiplicity of associated causes. Clearly, much more research is needed in this area.

Microorganisms are not only able to grow at high temperatures but also in salty environments. In fact, some require a high concentration of salts for active reproduction and growth. The halophilic bacteria are extremely well adapted to highly salty environments and can tolerate salt concentrations up to a saturation level (30% NaCl). With regard to the enzymes of halophilic bacteria, nearly all of them exhibit maximal activity between 1-4 M salt (14). At lower salt concentrations, the enzymes lose their activity. Interestingly, optimal activity is not always correlated with stability at a given salt concentration. For example, isocitrate dehydrogenase of Halobacterium cutirubrum and malic dehydrogenase of H. salinasium require 1 M NaCl or KCl for maximal activity, but optimal stability is only achieved in 4 M salt solutions. Also, it is worthy to note that the intracellular millieu of the halobacteria contain mainly KCl and not NaCl. Hence it is not surprising that many enzymes in the halobacteria show

TABLE 2:   Thermophilic Enzymes:

Advantages:

- Enhanced Stability at Elevated Temperatures.
- Easier Isolation and Purification.
- Less Microbial Contamination (Reactor System).
- Less Cooling Requirements (Reactor System) for
  Exothermic Reactions.
- Easier and Less Energy - Intensive Separation of
  Volatile Products.

Disadvantages:

- Low or No Activity at Room Temperature.
- Reactor Incompatibility--Multi-Enzyme Systems.

TABLE 3:   Halophilic Enzymes

Advantages:

- Enhanced Stability in Higher Ionic Media.
- Easier Isolation and Purification.
- Less Microbial Contamination (Reactor System).
- Easier Product Separation (Decreased Solubility in
  Ionic Medium).

Disadvantages:

- Low or No Activity in Normal Ionic Media.
- Reactor Incompatibility--Multi-Enzyme Systems.
- Corrosion of Equipment.

---

higher activity in KCl than in NaCl (especially at concentrations above 2 M and with enzymes involved in protein sysnthesis). The catalytic activity of enzymes of halobacteria is also affected by the anion of the salt.

As with the enhanced stability of thermophilic enzymes at higher temperatures, the enhanced stability of halophilic enzymes toward extreme high salt concentrations is not settled at this time. Examination of the amino acid composition between halophilic and nonhalophilic enzymes revealed an excess of acidic amino acid residues and a deficiency of apolar amino acid residues. As with thermophilic enzymes, the enhanced stability of halophilic enzymes may also, in part, be due to the formation of interior or exterior salt-bridges.

The process engineering advantages and disadvantages of thermophilic and halophilic enzymes are listed in Tables 2 and 3, respectively. For thermophilic enzymes, the advantages are enhanced stability at elevated temperatures (with a concomitant increase in the half-life of the catalyst) and, at times, easier isolation and purification of the desired enzyme (due to its greater thermal and possibly solvent stability). In process engineering terms, use of a thermophilic enzyme in a reactor would mean less microbial contamination of the reactor (especially for prolonged operations), less cooling water requirements for exothermic reactions (especially severe in hot climates) and the possible coupling of the higher temperature operation with easier and less-energy intensive separation of volatile products such as ethanol. The disadvantages of thermophilic enzymes are that little or no activity is exhibited by the enzymes at room temperature, a fact which may preclude their use in some analytical applications. Also, on occasion, one may have some reactor incompatibility when using a thermophilic enzyme along with mesophilic enzymes in a multi-enzymic reactor system.

With halophilic enzymes, somewhat similar advantages and disadvantages are evident. However, with halophilic enzymes, the operational advantage of higher temperatures and simultaneous product separation is not realized. An added disadvantage of halophilic enzymes is the possibility of enhanced corrosion of metal surfaces and fouling of the equipment due to high salt concentrations. However, operation at room temperature is possible.

Conclusions

Several concluding comments are in order:

1. Enzyme technology, although still young and developing, is very beneficial for processing carbohydrates. Large-scale processes for the conversion of starch to glucose and fructose are established.

2. Immobilization provides an effective methodology for circumventing the operational problems of soluble enzymes. With regard to fructose production, immobilized glucose isomerase is the critical step for making the process a commercial reality.

3. Bioconversion technology holds promise for the production and conversion of renewable resources into energy, fuels and chemicals in concert with the production of food, fiber and animal feeds.

4. Critical research needs associated with the use of biocatalysts are the stabilization of enzymes, the regeneration of coenzymes, the interfacing of enzymatic and conventional organic reactions, and the improvement of operational characteristics of enzymes.

5. In this era of catalysis, it is of utmost importance to examine enzymes--the most active and selective catalysts known for their unique characteristics and for new and novel applications. In the world of catalysis, enzymes can no longer be ignored and enzymes from organisms living in extreme environments represent an untapped potential.

References

1. K. J. Skinner, Chem. Eng. News 53:22 (1975).
2. O. R. Zaborsky, in: "Advanced Materials in Catalysis," (R. L. Garten and J. J. Burton, eds.) pp 267-291, Academic Press, New York (1977).
3. O. R. Zaborsky, in: "Future Sources of Organic Raw Materials," Toronto (L. E. St-Pierre, ed.) pp. 513-531, Pergamon Press, Elmsford, N. Y. (1980).
4. O. R. Zaborsky, "Immobilized Enzymes," CRC Press, Inc., Boca Raton, FL (1973).
5. O. R. Zaborsky and J. Ogletree, Biochem. Biophys. Res. Commun. 61:210 (1974).
6. W. H. Mermelstein, Food Technology, June, 20 (1975).
7. R. W. Coughlin and O. R. Zaborsky, in: "The Biosaline Concept: An Approach to the Utilization of Underexploited Resources," (A. Hollaender, J. C. Aller, E. Epstein, A. San Pietro and O. R. Zaborsky, eds.) pp. 333-359, Plenum Press, New York, (1979).

8. T. D. Brock, "Thermophilic Microorganisms and Life at High Temperatures," Springer-Verlag, New York (1978).

9. H. Zuber, (ed.) "Enzymes and Proteins from Thermophilic Microorganisms," Birkhauser Verlag, Basel (1976).

10. R. Singleton, Jr., and R. E. Amelunxen, Bact. Rev. 37:320 (1973).

11. A. R. Doig, Jr., in: "Enzyme Engineering," (E. K. Pye and L. B. Wingard, Jr., eds.), Vol. 2:17, Plenum Press, New York (1974).

12. T. Oshima, Ibid., Vol 4:41 (G. B. Brown, G. Manecke and L. B. Wingard, Jr., eds.) Plenum Press, New York (1978).

13. M. F. Perutz, Science 201:1187 (1978).

14. J. K. Lanyi, in: "Strategies of Microbial Life in Extreme Environments, (M. Shilo, ed.) pp. 93-107, Verlag Chemie, New York (1979).

THE EXPANDING ROLE OF MARINE ORGANISMS IN ANTICANCER CHEMOTHERAPY

William Fenical

Institute of Marine Resources
Scripps Institution of Oceanography
University of California
La Jolla, CA   92093

SUMMARY

Extracts of marine plants and animals possess unusually high
levels of cytotoxicity in the National Cancer Institutes preliminary bioassays. The great species diversity in the sea, and
the unusual composition of marine secondary metabolites, promise to
contribute to the isolation of new and potent cytotoxic agents.
Since selective inhibition of cancer cells is an important goal of
cancer chemotheraphy, investigations of the marine environment may
be expected to expand significantly in the near future.

INTRODUCTION

The combination of diseases known as cancer has evolved over
the last several decades as a major American and international
health problem.   The disease includes many forms of irregular or
neoplastic cell regeneration, including the leukemias and a wide
range of solid tumors, and in 1977 for example, cancer accounted
for over 350,000 deaths in the United States alone (1).

In 1971, Congress enacted the "National Cancer Act" which
entrusted the National Cancer Institute (NCI, one of the National
Institutes of Health) with the development of a nationwide strategy
to combat cancer.   One of the many research and clinical programs
which evolved at NCI is the Natural Products Branch, which became

355

responsible for the extraction and purification of natural materials useful in the chemotherapeutic treatment of the disease. Based upon early medical research associated with World War II, the war gases, nitrogen mustard for example, were shown to exhibit remarkable inhibition of neoplastic cell replication. Based upon these findings, the derivation of new "antineoplastic" drugs from nature was conceived as an important endeavor of the NCI. As more was learned about the nature of malignancy, it became clear that cancer cells often separate from the primary neoplastic source and are transported via the lymphatic and circulatory systems to additional locations, subsequently establishing secondary tumors. Hence, the efficacy of surgery in controlling the disease was questioned, and the importance of systemic chemotherapy was recognized.

The results of a decade of drug development for cancer chemotherapy have been extensive, and these results are available in summary form (2). Many thousands of synthetic compounds and natural products have been scrutinized, and literally hundreds of "active" compounds have been discovered. From this group, over thirty potent agents such as the nitrogen mustard, chlorambucil, the Vinca alkaloids (vincristine, vinblastine) and the antimetabolites, 5-flourouracil and 6-mercaptopurine, have evolved as clinically utilized drugs.

The importance of chemotherapy in combination with prudent surgery and x-ray radiation treatment has been recognized. Currently, approximately 30% of all cancer patients can be expected to subsequently live normal lives, and several of the dreaded cancers, such as Hodgkin's Disease, are completely curable. Although their efficacy has been recognized in many cancer disorders most cancer drugs have limited utility. Not all tumors respond to these antineoplastic drugs, and more importantly, the potent and often general cytotoxicities associated with these substances results in highly unwanted side-effects such as loss of hair, nausea and weight loss. It is frequent that patient death while in cancer chemotheraphy is attributed to the treatment and not the disease.

Problems in New Cancer Drug Discovery

As cancer evolves as a major health threat, it is clear that more potent, less deleterious and more cancer cell-specific chemotherapeutic agents will be needed. It may also be true that cancer cells, like pathogenic bacteria, are capable of developing resistance toward drugs, thus diminishing the utility of existing chemotherapeutics. Therefore, an active NCI program exists Nationwide to explore natural resources for metabolic products useful in cancer cell inhibition.

Unlike the clear-cut methods used to discover a new antibiotic
the discovery of antineoplastic agents is hindered by many problems. Primarily, since cancer is composed of over 100 discrete
diseases involving many tissues, no single preliminary bioassay
system can be considered universally applicable. The NCI has
established several initial screens which are cost efficient and
allow many extracts to be evaluated. These preliminary assays (2)
consist of an in vivo p-388 lymphocytic leukemia assay (mouse) and
several in vitro tissue culture cancer cell lines such as KB (a
human carcenoma of the nasopharynx), PS (p-388 lymphocytic
leukemia) and LE (L1210 lymphoid leukemia). While the latter cell
lines are useful in detecting cytotoxicity, the in vivo p-388
leukemia assay is more highly regarded, as it yields valuable
information in a whole-animal situation, thus affording a measure
of the potential therapeutic value of a new compound. It is
certainly true, however, that none of these preliminary screens
provide a useful extrapolation to the control of hard tumors.

Evaluating plant and animal extracts for antineoplastic activity is a complex process also greatly hindered by the demanding
experimental requirements of the preliminary bioassays. Cancer
cell lines do not transport well and they are difficult to establish outside the environment of a professional tissue culture laboratory. Hence, small samples collected world-wide must be transported and assayed in a central locale, and based upon preliminary
activity leads, collecting sites must be revisited and large
collections made at great expense.

Exploration of the Marine Environment

The exploitation of terrestrial sources for cancer drug development would seem to be less costly than an analogous investigation
of the marine environment. While this may be true, the marine
environment would appear to be a more promising location. An early
report (3) emphasized the high levels of antineoplastic activity in
certain marine invertebrates. Subsequently a survey of over 2,000
marine species confirmed activity in over 10% of those specimens
assayed (4). By comparison, terrestrial sources are known to yield
activity leads in only 3% of the species investigated.

It should also be recognized that the marine ecosystem is
populated with more primitive plants and animals, which are more
likely to have evolved chemical defensive measures. More than one
million marine invertebrates, over 25,000 species of fishes and
over 20,000 species of marine plants are recognized. Recent
chemical studies have yielded exciting information concerning the
secondary metabolites from these organisms (5). While marine
species produce terpenes, acetogenins and other common types of
metatabolic products, new ring systems, halogen substitution and

TABLE 1

CYTOTOXICITIES OF MARINE DERIVED CELL-DIVISION INHIBITORS[+]

Compound	KB (ED$_{50}$)*	PS (ED$_{50}$)*	p-388 (T/C)** in vivo	LE (ED$_{50}$)*	Urchin (ED$_{50}$)*
1 (crassin acetate)	2				4
2 (asperdiol)	24	6			
3 (sinularin)	0.3	0.3			
4 (dihydrosinularin)	16	1.1			
5 (sinulariolide)	20	7.0			
6 (a dihydrospermidine)	1.0	0.37			
7 (pseudopterolide)					4.0
8 (curcuquinone)					4.4
9 (fistularin-3)	4.1	4.3		1.3	
10 (fistularin-3-acetate)		14			
11 (fistularin-1)	26	14		1.3	
12 (fistularin-1-acetate)	26	14		1.3	
13 (spongouridine)					
14 (1-β-D-arabinosyl-cytosine)					
15 (geranylhydroquinone)					16.0
16 (prenylhydroquinone)			138		1.0
17 (aplidiasphingosine)	8.3	1.9			
18 (polyandrocarpidines)		4.8			
19 (debromoaplysiatoxin)			130-180		
20 (aplysistatin)		2.7			
21 (dolotriol)		13			
22 (dolatriol-6-acetate)		10			
23 (bromoobtusenediol)	4.5	10			
24 (elatol)					1.1
25 (zonarol)					1.0
26 (stypoldione)					1.1
27 (spatol)					1.2
28 (bifucarenone)					4.0

[+]Blank spaces indicate the compound was not tested or that data was not available.

ED$_{50}$ values are µg/ml.

**T/C refers to life extension in % of mice versus an untreated control.

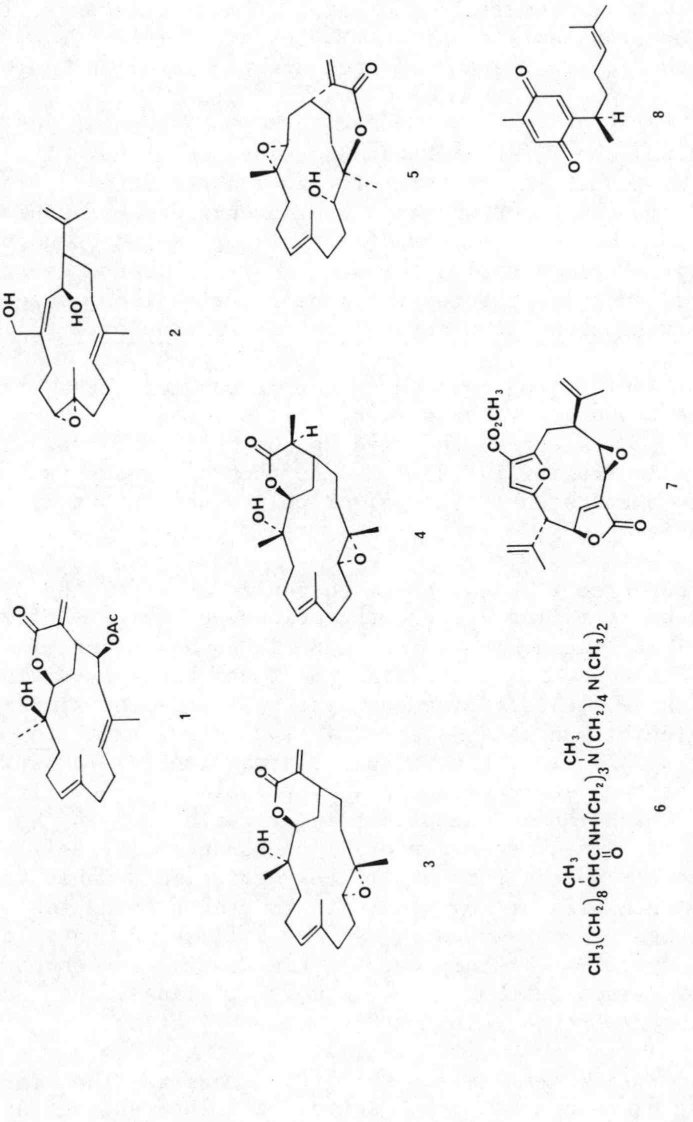

FIGURE 1. Cytotoxic Metabolites Isolated from Marine Corals.

novel new functional groups are more common in marine-derived compounds. Given the high incidence of positive cytotoxicity, and the high probability of isolating novel new structure types, the marine environment should be considered a major area of investigation.

The literature contains many reports of activity from marine organisms, but only a few report the isolation of a purified active component. These results, including recent findings from my research, are summarized below. The cytotoxicities reported or directly measured for these compounds are tabulated in Table 1.

A. Cytotoxins from Marine Corals: The gorgonian corals (sea fans and whips) from the Caribbean Sea were among the first marine invertebrates to be investigated for anticancer drugs. The first cytotoxic compound reported was the cembrenolide, crassin acetate, 1, which occurs in several gorgonians of the genus Pseudoplexaura (6, Fig. 1). More recently, the same research group reported the isolation of another cytotoxic cembrene derivative, asperdiol 2, from the gorgonians Eunicea asperula and E. tourneforti (7).

Octacorals of the related order Alcyonacea (the true soft corals) also produce cytotoxic cembrenolides, and the three related compounds, 3 - 5, have been reported as components of the Australian coral Sinularia flexibilis (8). The soft coral Sinularia brongersmai has also been recently recognized to produce the toxic spermidine derivative 6 (9).

In my research with Caribbean gorgonian corals, the cytotoxic lactone, pseudopterolide 7, and the quinone 8 have resulted (10). Our cytotoxicity measurements are made in an analogous fashion to those at NCI; however, we utilize the synchronously-dividing sea urchin egg as a flexible bioassay system (11). The simplicity of this cell inhibition assay allows it to be utilized in situ to guide field collecting, and recent pharmacological investigations indicate that the sea urchin egg is particularly sensitive toward inhibitors which impede tubulin polymerization (12). We do not feel this cell line is representative of human cancer cells, but it does apparently select for tubulin polymerization inhibitors, and since this mechanism of cytotoxicity is recognized for existing antitumor drugs, the sea urchin egg may well evolve into a highly useful test system. To complement this assay, our compounds are subsequently tested in four human cancer cell lines, the nature of which will be described in a forthcoming publication.

B. Cytotoxins from Marine Sponges: Despite the fact that sponges are a recognized rich source of interesting secondary metabolites, few of these compounds have been reported as possessing cytotoxic activity. The recently described fistularins,

9 -12, from the Caribbean sponge Aplysina fistularis forma fulva,
are an exception (13; Fig. 2). Although the fistularins appear
to be the only recent additions to the list of sponge cytotoxins,
it is of importance to point out the early discovery of the potent
tumor inhibitor spongouridine, 13, isolated from the sponge Crypto-
thetia crypta (14). Subsequent syntheses of analogs of 13 ensued,
and the potent cytotoxin, 1-β-D-arabinosyl cytosine 14, was
produced. This compound, first recognized for its antiviral activity, is a potent antitumor agent and its use has been widespread in
anticancer chemotheraphy (15).

     C. Cytotoxins from Marine Tunicates: Although the tunicates
(Urochordata) have not received extensive chemical investigation,
four compounds have been reported to exhibit at least one form of
antineoplastic activity. As part of an expedition along Pacific
Mexico, I found an undescribed tunicate of the genus Aplidium to
produce large amounts (≈6% dry wt.) of geranylhydroquinone, 15
(16; Fig. 3). The compound was isolated from this natural source
by virtue of its antibacterial properties; however, generanyl-
hydroquinone had earlier been synthesized and scrutinized in
several unique anticancer screens. Geranylhydroquinone is one of
several compounds which could be considered cancer chemo-protective
agents, since when administered to test animals, several common
tumors could not be induced by standard methods (17).

     An Aplidium species from California has also been recognized
to produce an antileukemia compound, identified as prenylhydroquinone, 16. In the in vivo p-388 lymphocytic leukemia assay,
compound 16 showed a T/C value of 138, thus indicating a 38% life
extension in the leukemic mouse (18). Prenylhydroquinone also
showed moderate potency in the urchin egg cytotoxicity assay.

     Tunicates of the genus Aplidium also produce linear, non-
aromatic terpenoids of the sphingosine type. The cytotoxin, aplid-
iasphinogosine, 17, was recently isolated from an undescribed
Aplidium sp. (19). From the point of view of unique structure,
however, certainly the most exciting tunicate-derived cytotoxin is
the polyandrocarpidine mixture, 18, isolated recently from the Gulf
of California tunicate Polyandrocarpa sp. (20). The cyclopropene
functional group in this unusual metabolite is rare, being found in
only one other marine metabolite.

     D.   Cytotoxins Produced by Marine Algae: A considerable
problem encountered in marine natural products investigations lies
in the transfer of secondary metabolites via the food web. Several
cytotoxic metabolites of probable or confirmed algal origin have
been isolated from herbivorous molluscs of the Sub-Class Opistho-
branchia. The cytotoxic phenol, debromoaplysiatoxin 19, for
example, was isolated first from the Hawaiian sea hare Stylocheilus

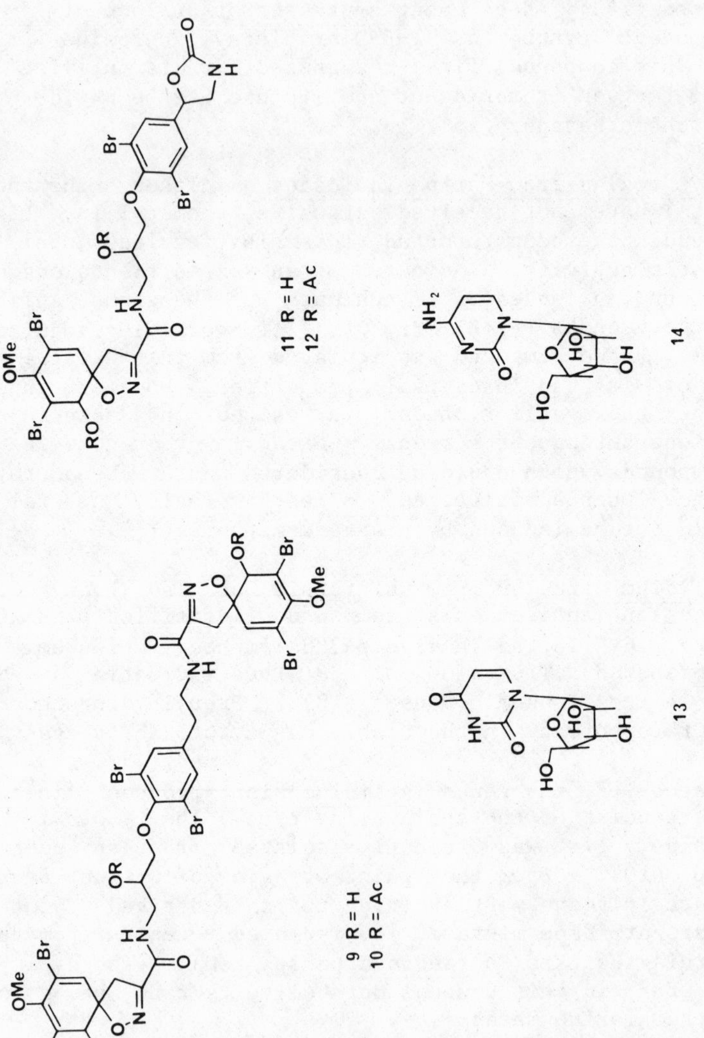

FIGURE 2 Cytotoxic Metabolites Isolated from Marine Sponges.

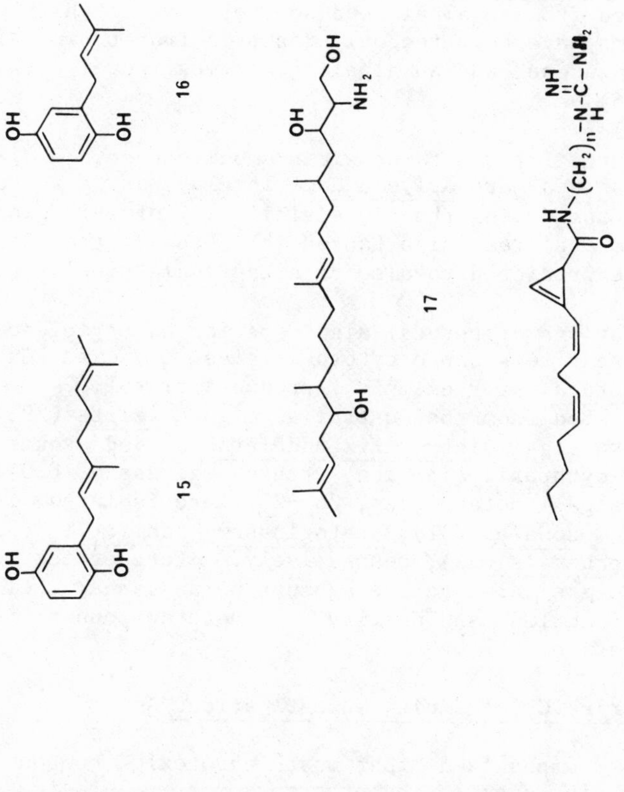

FIGURE 3. Cytotoxic Metabolites Isolated from Marine Tunicates.

longicauda (Fig. 4). In subsequent studies of the sea hare's
diet, compound 19 was shown to be produced by various species of
blue-green algae of the genus Lyngbya (21), and apparently concentrated by the herbivore. The same concept would seem to apply to
the cytotoxic agents 20 -23. The lactone aplysistatin, 20, was
reported as a component of the sea hare Aplysia angasi (22), but 20
was recently isolated from the red alga Laurencia cf. palisada
(23), and considering the now well-known feeding behavior of sea
hares, little doubt exists that most sea hare-derived compounds are
algal metabolites.

An analogous situation exists for the cytotoxic diterpene,
dolatriol 21, and its corresponding acetate, 22, isolated from the
sea hare Dolabella auricularia (24). Dolatriol itself was not subsequently isolated from an algal food source; however, highly analogous ring systems have been recently reported from brown algae of
the family Dictyotaceae, and an algal source must surely exist for
dolatriol (25, 26).

An algal source for the cytotoxic bromoditerpene, 23 isolated
from the Caribbean sea hare Aplysia dactylomela, has also remained
obscure (27). Considering the close similarity of this diol with
obtusadiol, from the red alga Laurencia obtusa, the producing
organism would be predicted to also be a Laurencia species (28).

My investigations of several algal species have yielded metabolites which possess reasonable cytotoxicities. The red Caribbean
alga Laurencia obtusa, for example, produces elatol, 24, a cytotoxin first described from the Australian alga L. elata (29). The
Californian brown alga Dictyoperis undulata yielded zonarol, 25,
which is highly cytotoxic in the urchin egg assay (30). The
remaining three algal metabolites, 26 -28, were isolated from the
algae Stypopodium zonale (31), Spatoglossum schmittii (32) and
Bifurcaria galapagensis (33), respectively. Further testing has
shown that 25 -28 are also active against human cancer cells, and
the mechanism of action and utility of these compounds is under
current investigation.

The Future of Marine Cytotoxicity Investigations

It should be emphasized that most cytotoxic compounds will
never be useful in cancer chemotherapy due to their general
toxicity and low selectivity against cancer cells. The compounds
reported here are probably of this group, and it is therefore not
likely that they will evolve as useful chemotherapeutic agents.
Modest beginnings must be made, however, and it is reasonable that
metabolites with fundamental cytotoxicity be isolated and further
evaluated. The high levels of cytotoxicity, and the stuctural
diversity of marine-derived compounds strongly suggests the marine

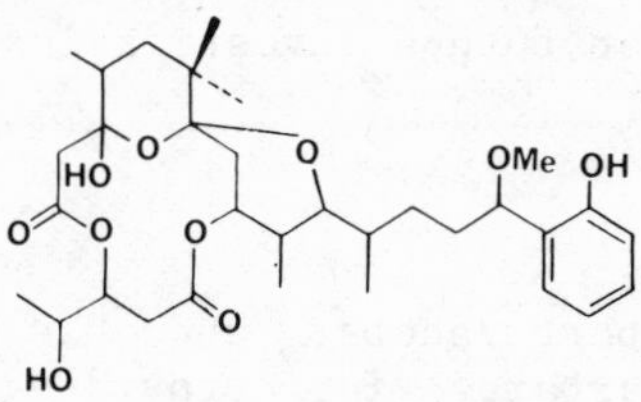

FIGURE 4. Cytotoxic Metabolites Produced by Marine Algae.

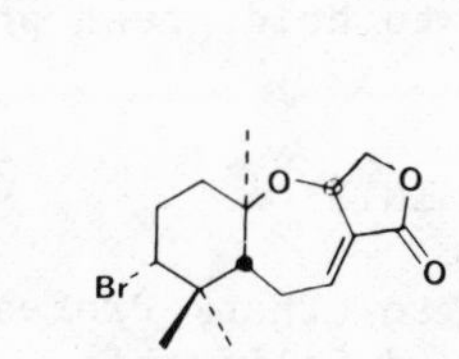

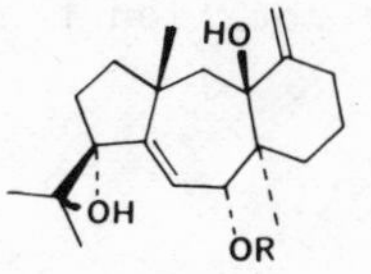

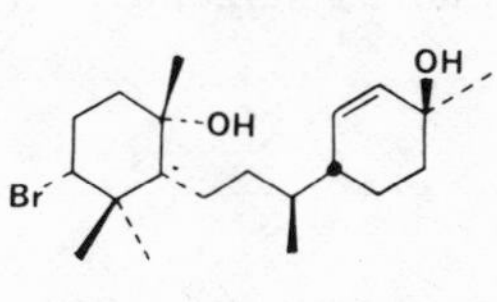

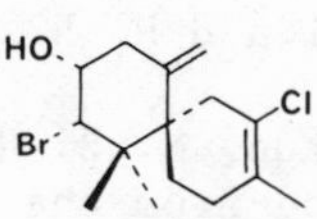

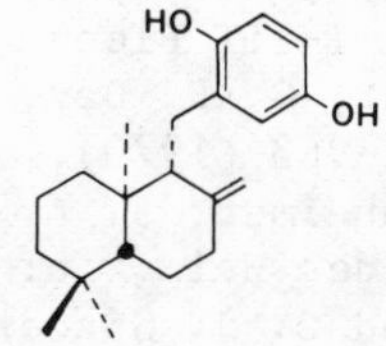

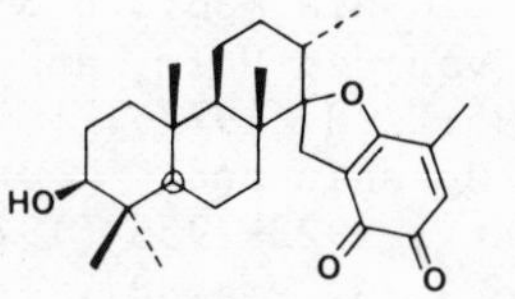

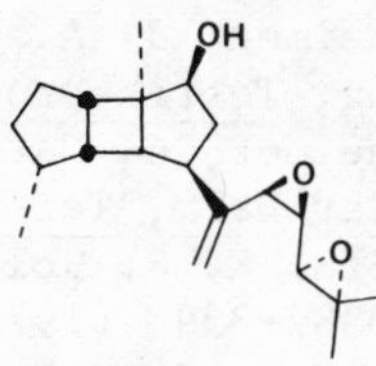

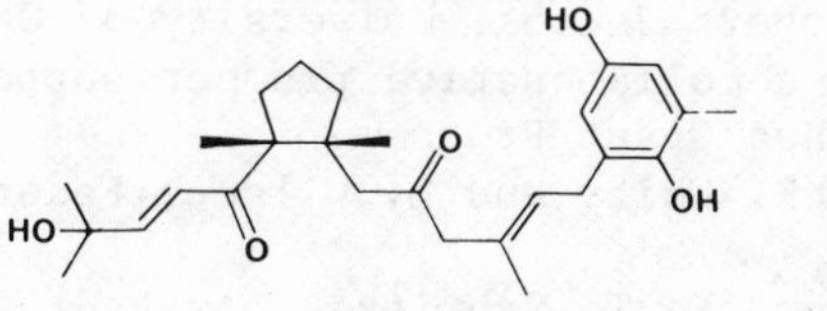

environment to hold great promise for future investigations of this
type.

ACKNOWLEDGEMENTS

I wish to thank Professor Robert Jacobs, and his students,
University of California, Santa Barbara, for providing fruitful
pharmacological collaboration. This work, in part, is a result of
research sponsored by NOAA, Office of Sea Grant, Department of
Commerce, under grant # 04-7-158-44121. The U. S. Government is
authorized to produce and distribute reprints for governmental
purposes, notwithstanding any copyright notation that may appear
hereon.

References

1. R. E. LaFond, "Cancer, The Outlaw Cell," American Chemical
    Society, Washington, D. C. (1978).
2. G. R. Pettit, "Biosynthetic Products for Cancer Chemotherapy,"
    Volumes I-III Plenum Press, New York (1977-1979).
3. G.R. Pettit, G. F. Day, J. L. Hartwell and H. B. Wood, Nature,
    227:962-963 (1970).
4. A. J. Weinheimer, J. A. Matson, T. Karns, M. B. Hossain,
    D. van der Helm, in: "Drugs and Food from the Sea," P. N.
    Kaul and C. J. Sindermann, eds., University of Oklahoma,
    Norman, OK (1978).
5. P. J. Scheuer, "Marine Natural Products; Chemical and
    Biological Perspectives," Volumes I-IV, Academic Press, New
    York (1977-1980).
6. A. H. Weinheimer and J. A. Matson, Lloydia, 38:378-382 (1975).
7. A. H. Weinheimer, J. A. Matson, D. van der Helm, and
    M. Poling, Tetrahedron Lett. 1295-1298 (1977).
8. A. J. Weinheimer, A. Matson, M. B. Hossain and
    D. van der Helm, Tetrahedron Lett., 2923-2936 (1977).
9. F. J. Schmitz, K. H. Hollenbeak and R. S. Prasad, Tetrahedron
    Lett., 3387-3390 (1979).
10. F. J. McEnroe, and W. Fenical, Tetrahedron, 34:1661-1664
    (1978).
11. The urchin egg bioassays are performed under the direction of
    Professor Robert Jacobs, University of California, Santa
    Barbara, in a collaborative project supported by the
    California Sea Grant Program.
12. R. S. Jacobs, S. White and L. Wilson, Federation Proc, 40:26
    (1981).
13. Y. Gopichand and F. T. Schmitz, Tetrahedron Lett., 3921-
    3924 (1979).
14. W. Bergmann and D. C. Burke, J. Org. Chem., 20:1501 (1955).

15. S. S. Cohen, in: "Progress in Nucleic Acid Research and Molecular Biology," J. N. Davidson and W. E. Cohen, eds., Academic Press, New York, Vol 5, pg. 1 (1966).
16. W. Fenical, in: "Proceedings Food and Drugs from the Sea Conference", Marine Technology Society, Washington, D. C. 388-394, (1976).
17. G. Rudali and L. Menetrier, Therapie, 22:895 (1967).
18. B. M. Howard, K. Clarkson and R. Bernstein, Tetrahedron Lett. 4449-4452 (1979).
19. G. T. Carter and K. L. Rinehart, Jr., J. Amer. Chem Soc., 100:7441-7442 (1978).
20. M. T. Cheng and K. L. Rinehart, Jr., J. Amer. Chem. Soc., 100:7409-7410 (1978).
21. J. S. Mynderse, R. E. Moore, M. Kashinagi and T.R. Norton, Science, 196:538-540 (1976).
22. G. R. Pettit, C. L. Herald, M. S. Allen, R. B. Von Dreele, L. D. Vanell, J. P. Y. Kas and W. Blake, J. Amer. Chem. Soc., 99:262-263 (1977).
23. V. J. Paul and W. Fenical, Tetrahedron Lett., 2728-2790 (1980).
24. G. R. Pettit, R. H. Ode, C. L. Herald, R. B. Von Dreele and C. Michel, J. Amer. Chem. Soc., 98:4677-4678 (1976).
25. H. H. Sun, and W. Fenical, Phytochem., 18:340-341 (1979).
26. M. Ochi, M. Watanabe, I. Miura, M. Taniguchi and T. Tokoroyama, Chemistry Lett., 1229-1232 (1980).
27. F. J. Schmitz, K. H. Hollenbeak, D. C. Carter, M. B. Hossain and D. van der Helm, J. Org. Chem., 44:2445-2447 (1979).
28. B. M. Howard and W. Fenical, Tetrahedron Lett., 2453-2456, (1978).
29. J. J. Sims, G. H. Y. Lin and R. M. Wing, Tetrahedron Lett. 3487-3490 (1974).
30. W. Fenical, J. J. Sims, D. Sqautrito, R. M. Wing and P. Radlick, J. Org. Chem, 38:2383-2386 (1973).
31. W. H. Gerwick, W. Fenical, N. Fritsch and J. Clardy, Tetrahedron Lett., 145-148 (1979).
32. W. H. Gerwick, W. Fenical, D. Van Engen and J. Clardy, J. Amer. Chem. Soc., 102:7991 (1980).
33. H. H. Sun, N. M. Ferrara, O. J. McConnell and W. Fenical, Tetrahedron Lett., 3123-3126 (1980).

SALINE SILVICULTURE

Howard J. Teas

Biology Department
University of Miami
Coral Gables, Florida   33124

SUMMARY

Mangroves are the trees and shrubs that grow in saline waters along the edges of the world's seas and estuaries. They are an important renewable resource that requires little fossil fuel for its production. Under suitable growing conditions mangroves are as efficient as the average upland temperate and tropical forests.

Mangroves are effective competitors of upland plants in the ordinary marine saline environment, but do not survive salinities of more than 3 times seawater. Engineering methods have proven effective for establishing tidal circulation and normal salinities in some natural hypersaline areas.

Investigations suggest that it should be possible to increase mangrove wood yield and to expand the regions in which they grow by a combination of engineering applications, silvicultural management and plant breeding.

INTRODUCTION

Mangroves are the trees and shrubs of tropical and subtropical shores and estuaries. Although mangroves are the main feature of low energy shorelines between 20° North and South of the equator, single species are found as far south as 37° (1) in New Zealand and 35° on Kyushu Island in Japan (2). Silviculture with saline water almost exclusively involves mangroves.

Mangroves are a diverse group of woody land plants that have

369

in common that they have developed salinity tolerance; they did not evolve in the sea. There is some variation in the species considered to be mangroves by different authors, but the group includes at least 10 plant families, 14 genera and 50 species, with the greatest species diversity occurring in the Indo-Pacific region (3, 4).

Mangroves serve for wood and charcoal as well as a variety of more specialized uses such as tannin, livestock fodder, food and medicinals. Mangroves are important sources of fuel along many tropical shorelines. In addition, mangroves are also a source of secondary production in the form of invertebrates and vertebrates that derive energy from a mangrove litter detritus cycle.

MANGROVE SILVICULTURE

Silviculture of mangroves involves the planting, management and harvesting of mangrove forests. Mangrove silviculture was practiced in the Andaman Islands in the 19th century (5) and in Malaya since 1904 (6). Noakes (7) in 1955 estimated that more than 80% of the mangroves of Malaya were under sustained yield management in forest preserves. Mangroves of the family Rhizophoraceae are most often managed. Species of Rhizophora are the most highly desired mangroves for wood and charcoal productions. Figure 1 shows a Rhizophora forest and Figure 2 shows mangrove charcoal.

It is clear that mangroves can be planted and grown in appropriate sites for silviculture. Watson (6) reported the use of hand planting in mangrove forest regeneration, and Lang (8) cited French literature for the earlier planting of 38,000 ha of mangroves in Vietnam. Recently at least 5,000 ha of mangrove have been replanted in Vietnam (P.H. Ho, pers. comm.). In addition to silvicultural practice, mangroves have also been planted to control erosion (9), to stabilize canal banks (3), to reclaim land from the sea (3), and for restoration or mitigation as required by government agencies (Fig. 3; 10).

MANGROVE PRODUCTIVITY

Primary Net Production: Well developed Rhizophora mangrove forests may produce more dry matter per ha than do the average tropical or temperate forests (Table 1). However, the local conditions are of critical importance since hypersaline scrub Rhizophora mangrove forests may have rates that are 3% or less of those of a well developed coastal forest.

Secondary Production: Mangrove forests serve as habitat for many species and are important sources of reduced organic matter in

Figure 1. Well developed Rhizophora Forest in Miami, Florida area.

Figure 2. Mangrove charcoal in market at Esmeraldas, Ecuador.

Figure 3.   A site replanted with Rhizophora mangle for restoration.
St. Croix, U.S. Virgin Islands.

TABLE 1.   Total net primary production for several plant
communities. Data from Golley and Lieth (11) and from Teas (12,
13). Mangrove data calculated from litter X 3.

Community	Dry wt (mt/ha/yr)
Temperate forests	
Deciduous (average)	12
Coniferous (average)	28
Tropical forests (average)	25
range	13-123
Mangroves	
Rhizophora, coastal, Florida	32
Rhizophora, sparse scrub, Florida	1

the form of leaves, twigs, bark, fruits, flowers, branches, trunks, etc. that provide the energy for a detrital cycle. It is this cycle that is responsible for the production of several species of animals important to man, including crabs, prawns, shellfish, and finfish (14, 15).

SALT TOLERANCE IN MANGROVES

Mangrove species vary in their salt preferences and tolerances. Macnae noted that Sonneratia caseolaris·grows only if the salinity is less than 10 o/oo (parts per thousand); Bruguiera parviflora grows optimally at about 20 o/oo; and Rhizophora mucronata and Sonneratia alba prefer waters near normal seawater salinity, i.e. ca. 35 o/oo. Macnae (3) also reported that Avicennia marina and Lumnitzera tolerate salinities greater than 90 o/oo.

Although it tolerates salinities above those of ordinary seawater, Rhizophora mangle grows more luxuriently in the middle reaches of estuaries, where the salinity is 10-20 o/oo, than in full seawater (16). Higher salt in the soil solution appears to increase the ratio of respiration to photosynthesis (17) and to reduce transpiration (18). Watson's (6) recommendation that mangrove forests be ditched to improve productivity is probably based on a reduction of interstitial salinity from the improved tidal circulation.

The two major mechanisms for salt tolerance in mangroves are salt exclusion and salt excretion. In salt exclusion, roots function as reverse osmosis systems that separate freshwater or almost freshwater from seawater. In mangroves the energy for freshwater separation is supplied by the negative xylem pressure generated by evaporation of water from leaf stomates, i.e. from transpiration. Rhizophora mangle, R. mucronata, Ceriops tagal, Bruguiera gymnorrhiza, B. parviflora and Kandelia candel are salt-excluding species (19).

Salt-secreting species have root membranes that exlude ions but are not so effective in filtering seawater as those of salt-excluding species. However, they have glands that secrete salts by active transport mechanisms. The following genera contain one or more species that are salt-secreting: Avicennia, Aegialitis, Agiceras, Acanthus and Laguncularia (19, 20).

Mangroves appear to have evolved the strategy of salt tolerance as a method of competition. In a tropical lowland moist forest there may be more than 200 woody species in 1 ha (21) whereas in a nearby mangrove forest there are frequently no more than 2-5 species of woody (mangroves) plants per ha. Indeed there

are extensive mangrove forests that are monocultures, or near
monocultures.

SALINITY RANGES IN NATURE

Natural hypersaline areas occur at many places in arid regions
of the world.   Guilcher (22) has described hypersaline vegetation-
free zones in the spring tide area between the mangroves and
continental or upland vegetation in Madagascar.   Fosberg (23)
correlated the occurrence of hypersaline vegetation-free zones that
are found between the mangroves and uplant vegetation with tidal
range, rainfall and length of dry season.   He noted the occurrence
of hypersaline zones near Queensland, Australia; Guayaquil,
Ecuador; in Salvador; in Honduras; along the Gulf of California;
along the Persian gulf coast of Saudi Arabia; and along the Guajira
Peninsula in the Caribbean.   All of these areas are characterized
by low rainfall and long dry seasons.

On the western coast of peninsular Florida near Punta Gorda,
vegetation-free salt barrens are found between the mangroves and
the upland vegetation.   At Punta Gorda there are 6 relatively dry
months per year, i.e. months that average less than 55 mm/mo rainfall (24).   Interstitial salinities in such salt barrens have been
found to be in the range of 88-142 o/oo (Teas, unpub.).   A salt
barren south of Punta Gorda is shown in Figure 4.   Such salt
barrens appear to be relatively stable; they are identifiable on
maps drawn before 1900.

Although salt barrens are too saline to grow any plants,
except blue-green algae in the relatively fresh surface water after
rainfall, they may be made productive for mangroves by engineering
means.   Figure 5 shows the site of a former salt barren that was
dug out and ditched so upland runoff and tidal waters could enter.
The lagoon so formed is less than seawater salinity and is suitable
for growth of mangroves.

In Ecuador a salt barren south of Guayaquil that had an interstitial salinity of 168 o/oo (Teas, unpub.) has been ditched and
leached of excess salt for prawn mariculture.

PROSPECTS FOR BREEDING MANGAOVES

There has been little work published on the genetics or
reproduction of mangroves.   Sidhu (25) reported that _Rhizophora_
species were diploid (36 chromosomes) and that meiotic divisions
were regular.   The finding by Teas and Handler (26) of more than 50
Rhizophora _mangle_ trees in south Florida and Puerto Rico that

Figure 4. Aerial view of a salt barren zone (white sand) near Punta Gorda, Florida. The mangroves show to the right, toward the bay and the upland vegetation can be seen to the left.

Figure 5. A lagoon excavated at the site of a former salt barren, near Punta Gorda, Florida.

produced approximately 3 normal green to 1 albino (white or yellow, i.e. non-green) propagule indicates that, although R. mangle is adapted for wind pollination (27), it is self-compatible. Further, it appears that self-pollination must be frequent. Normal and albino propagules from one of these heterozygous albino trees is shown in Figure 6. A leaf pattern mutation that appeared on one branch of a R. mangle tree is shown in Figure 7.

Personal communications from several researchers have established the occurrence of albino and normal propagules in other species of Rhizophora and in Ceriops. In addition, the author (unpub.) has found occasional albino propagules of Bruguiera gymnorrhiza floating and washed ashore near Suva, Fiji.

Albino propagules of Avicennia germinans seed have been observed occasionally in the shoreline wrack of south Florida, and one tree has been found that produces approximately 3 normal to 1 albino seed (Fig. 8).

Inherited characters are probably involved in other features of the plants than seeds, propagules and leaves. McMillan (28) tested the cold tolerance of Avicennia marina and A. germinans from a variety of locations. As can be seen in Table 2, those collected from the poleward portions of their ranges were much more resistant to cold than similar seedlings from the more tropical latitutes. He concluded that chilling resistance is based on inherited properties.

Thus, the main groups of mangroves, i.e. Rhizophoraceae and Avicenniaceae, appear to be normal diploids and show evidence of genetic diversity. The relatively short generation time (for a tree crop) or 3 1/2-4 years for Rhizophora mangle, Avicennia germinans and Laguncularia racemosa in Florida (Teas and Jurgens, unpub.) suggest that collection, breeding and selection may be fruitful approaches to the production of mangroves that are more rapidly growing, more cold tolerant, more salinity tolerant, etc. In addition, mutation breeding, cell hybridization and single cell selection methods may be useful in developing superior mangroves.

WAYS TO INCREASE WORLD MANGROVE PRODUCTION

Management of Existing Mangroves: Mangrove forests produce optimally when managed and protected rather than left to the vagaries of harvest by fishermen, wood cutters and filling for land development. The silvicultural technology for mangroves was well developed more than 50 years ago (6). Unmanaged mangrove forests may be underexploited or overexploited, even to the point of disappearance. Clearly, management under government regulation

Figure 6. Normal (green) and albino (yellow) Rhizophora mangle propagules from a tree in Puerto Rico.

Figure 7. Mutant leaf pattern in Rhizophora mangle.

Figure 8. Normal and albino Avicennia germinans seeds.

TABLE 2. Effect of twelve successive 2–4°C nightly cold shocks on seedlings of Avicennia marina and A. germinans collected from sites at different latitudes (From 28).

Species and collection locality	Latitude of collection	Percentage of Plants	
		Severe leaf and stem damage	No visible damage
Avicennia marina			
Guam	13°30'N	100	0
Darwin, Aust.	12°23'S	100	0
Swansea, Aust.	33°05'S	0	100
Auckland, N. Z.	36°55'S	0	100
Avicennia germinans			
Belize, C. A.	17°29'N	100	0
Veracruz, Mex.	19°11'N	100	0
Brownsville, Tex.	26°94'N	0	100
Corpus Christi, Tex.	27°50'N	0	100

Figure 9. Rhizophora mangle forest on the island of Oahu, Hawaii.

Figure 10.   Dead mangroves in a hypersaline basin in South Florida.

would increase mangrove forest productivity compared to non-regulation.

Introduction of mangroves: Mangroves can be introduced successfully into areas that have no mangroves. They were brought into Hawaii early in this century and Rhizophora mangle is now growing on all five of the main islands (Fig. 9).

There are other islands that appear to have suitable habitat but lack mangroves, for example, eastern Polynesia (29).

Reintroduction of mangroves: Mangroves might be reintroduced into areas where they have been overexpoited by man. For example, literature (cited in 3) indicates that there may have been Rhizophora growing along the Red Sea 4,000 or more years ago in areas where today only Avicennia are found. He also noted that the western coast of India has few mangroves because they were destroyed hundreds of years ago by the inhabitants.

Changing the Local Environment: Mangroves can be grown successfully in areas where hypersalinity has been ameliorated by restoration of tidal circulation. Candidates for such treatment are the hypersaline ponds in mangrove islands along arid shores (20), salt barrens, as noted above, and anthropogenic hypersaline basins such as the one shown in Figure 10, which was caused by cutting off tidal circulation in the process of road construction.

Mangroves have become established along formerly excessively high energy shores after breakwaters of jetties have been built. Engineering means of reducing wave and current energy would be useful in increasing mangrove areas.

REFERENCES

1. V. J. Chapman and J. W. Ronaldon, "The Mangrove and Saltgrass Flats of the Auckland Isthmus," New Zealand Dept. Scient. Industr. Research, Bull. 125, 79 pp. (1958).
2. C. G. G. J. van Steenis, Koninkl. Ned. Akad. Wetensch. Ser. C. 65:164-169 (1962).
3. W. Macnae, Advances in Marine Biology 6:73-270 (1968).
4. V. J. Chapman, Tropical Ecology 11:1-19 (1970).
5. J. Banerji, World Forest. Congress 3:425-430 (1958).
6. J. D. Watson, Malayan Forest Records 6:1-275 (1928).
7. D. S. O. Noakes, Ibid, 18:23-30 (1955).
8. A. Lang, "The Effects of Herbicides in South Vietnam," Part A, National Academy of Sciences, Wash., D.C., 372 pp. (1974).
9. V. Mac Caughey, Hawaiian Forestry Agr. 14:361-366 (1917).

10. H. J. Teas, Environ. Conservation 4:51-58 (1977).
11. F. B. Golley and H. Leith, in: "Tropical Ecology," F.B. Golley
    and R. Misra, eds., University of Georgia, Athens, Georgia,
    pp. 1-26 (1972).
12. H. J. Teas, in :"Biscayne Bay; Past, Present and Future,"
    University of Miami Sea Grant Spec. Report 5, University of
    Miami, pp. 103-113 (1976).
13. H. J. Teas, Mangroves of Biscayne Bay, Mimeo, Dade County,
    107 pp. (1974).
14. W. Macnae, in: "Reports IOFC, International Indian Ocean
    Fisheries Survey and Development Programme," No. 74/34,
    35 pp. (1974).
15. W. E. Odum and E. J. Heald, Bulletin Marine Science 22:671-738
    (1972).
16. J. H. Davis, Jr., in: "The Ecology and Geological Role of
    Mangroves in Florida," Carnegie Inst. Wash. Publ. 32:305-412
    (1940).
17. D. B. Hicks and L. A. Burns, in: "Proc. 1st. Int. Sympos. on
    Biol. and Management Mangroves," Univ. Florida Press,
    pp. 238-255)1975).
18. H. H. M. Bowman, Proc. Amer. Philosophical Soc. 56:589-672
    (1917).
19. P. F. Scholander, Physiologia Plantarum 21:258-268 (1968).
20. H. J. Teas, in: "The Biosaline Concept," A. Hollaender, ed.,
    Plenum Publ. Co., pp. 117-161 (1979).
21. K. A. Longman and J. Jenik, "Tropical Forest and Its
    Environment," Longmans, New York, 196 pp. (1974).
22. A. Guilcher, "The Sea," Vol. 3, Interscience Publishers, pp.
    620-654 (1963).
23. F. R. Fosberg, U.S. Geological Survey Professional Papers, No.
    365, pp. D-216-218 (1961).
24. F. L. Wernstedt, "World Climatic Data," Climatic Data Press,
    Lemont, Pennsylvania, 523 pp. (1972).
25. S. S. Sidhu, Caryologia 21:353-357 (1968).
26. H. J. Teas and S. H. Handler, in: "Proc. of the International
    Sympos. on Marine Biogeography and Evolution in the Southern
    Hemisphere," 2:357-361 (1979).
27. P. B. Tomlinson, R. B. Primack and J. S. Bunt, Biotropica 11:
    256-277 (1979).
28. C. McMillan, in: "Proc. 1st. Int. Sympos. on Biol. and
    Management of Mangroves," Univ. Florida Press, pp. 62-68
    (1975).
29. F. R. Fosberg, Ibid, pp. 23-42 (1975).

SALT EXCRETION IN THE MANGROVE Avicennia germinans

C. M. Sordo, G. Padilla and L. A. Romero

Centro de Investigaciones Biologicas
   de Baja California
La Paz, Baja California Sur, Mexico

INTRODUCTION

Mangroves exhibit an extraordinary adaptation capacity to salinity changes in the substrate. They are found in high salinity soils (35 o/oo) or in fresh water, though more frequently in marine environments. Changes in salt concentration in the area surrounding mangroves are related to tide, rivers and climate as pointed out by Davis (1), Macnae and Kalk (2), Diaz-Piferrer (3), Thom (4), Vegas-Velez (5) and Rodriguez (6).

About 60-75% of the tropical coast is covered by mangroves (7) and their general distribution seems to be related to some basic aspects of marine environments such as air temperature, shallow water, marine currents, soil composition, etc.

Along the Mexican coast, there are four species of mangroves; Rizophora mangle, Lagunacularia racemosa, Avicennia germinans and Conocarpus erectus (8). A. germinans and L. racemosa excrete salt through their leaves (9). In contrast, R. mangle does not excrete salt through any of its organs. Histological studies of the leaves of these mangrove species show that the two excretory species possess salt glands (Fig. 1a). The salt gland cells of L. racemosa form a circle around a cavity in which salt granules are sometimes observed. In cross sections, the salt glands are spaced along the leaf and resemble an epidermis invagination (Fig. 1b). On the other hand, the salt glands in A. germinans appear smaller, are closer one to another and apparently more associated with the epidermis cells. In contrast to the round-shaped epidermis cells of A. germinans, its salt gland cells are elongated (10 ; Fig 1c).

As noted above, the abundant salt excretion by A. germinans

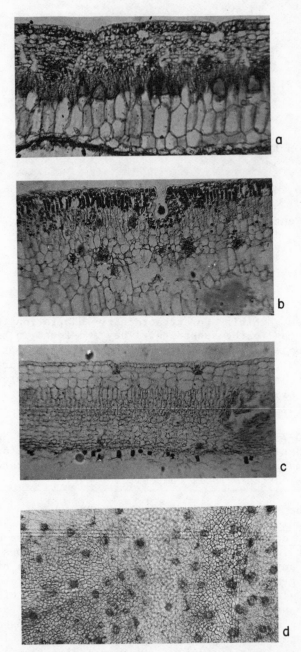

Figure 1.   a) Cross section of Rizophora mangle leaf.   b) and c)
Cross section of Laguncularia racemosa and Avicennia
germinans leaves respectively.   d) Epidermis smear of
Avicennia germinans leaf obtained by enzymatic
activity.

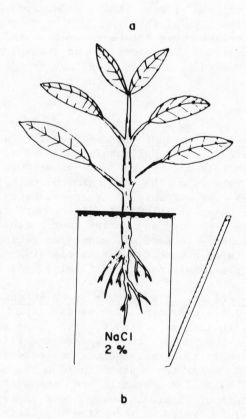

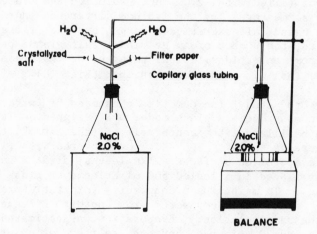

Figure 2.   Experimental designs for measuring water absorption and
             salt excretion: a) in a mangrove plant;   b) in a glass
             model.

leaves motivated the present study. Our aim was to estimate the desalination capacity of this particular mangrove species and to determine the relationship between excretion and transpiration, as well as the possible mechanisms of ion transport in these plants. In a comparative study, a glass model was designed.

Materials and Methods

a) Salt excretion in field mangroves: The area of study was the Bay of La Paz, Baja California Sur. The selected branches, having approximately 150 leaves, were washed with a brush and 50 ml of distilled water and the washings collected in 100 ml flasks. Salt excretion was determined by argentometry (11) and/or conductivity (12) using a YSI Model 31 conductivity meter. Samples were obtained every third day during four months. The results were extrapolated to one hectare of mangroves taking into account the tree density, using a number and size classification for leaves and for trees, as previously described (13-14).

b) Salt excretion in mangroves grown in the laboratory: Four months old plants obtained from seeds germinated in individual containers, with sand as support, were used as follows: First, they were watered with fresh water and finally with saline solution. The salt thus excreted was collected and dissolved in 30 ml of distilled water for quantification by the methods already mentioned. The values reported are the average of 48 plants; Second, the roots of 12 plants were placed in centrifuge tubes (1.4 cm ID x 10 cm) to which a graduated capillary tube was coupled. The system was filled with 2.0% NaCl solution and sealed. Water absorption was measured by the volume difference noted in the capillary tube and salt excretion by argentometric titration (Fig. 2a); and Third, the number of A. germinans salt glands per cm^2 was calculated from an epidermis leaf smear using Macerase (Onozuka, Tokyo, Japan) (15) with a Neubawer chamber (Fig. 1d).

Finally, to study the osmotic phenomena of water and salt absorption in mangrove, a glass model was designed. It consisted of a 30 cm capillary tube center piece (the "stem") to which 4 "branches" were attached. The ends were covered with filter paper to serve as evaporation surfaces (like leaves) of 5 cm in diameter. The model was placed in a sealed 500 ml Erlenmeyer flask with 2.0% NaCl solution. To maintain a constant hydrostatic pressure, the top of the "stem" was connected to another Erlenmeyer flask containing the same solution. Evaporation was estimated by weight difference (every 30 minutes) and the salt excretion through the filter papers every 2 hrs. (Fig. 2b).

Osmotic phenomena with the glass model were studied using a

1.0 ml graduated pipette with a dialysis bag attached to its end.
The dialysis bag contained a 3.2% NaCl solution to simulate the
internal salt concentration in A. germinans mangrove (16). The
model was then introduced into a 1 1 beaker with different salt
solutions (0.01% and 1.0% NaCl) as shown in Figure 3. Water
diffusion was estimated from the difference in water column height
and conductivity changes in the bulk were measured every 10 minutes
at 22°C.

RESULTS

Salt excretion in the leaves of field mangroves is about 0.67
mg NaCl/leaf/day, whereas in those grown in the laboratory it is
about 0.58 mg NaCl/leaf/day. Previous studies indicate that the
mangrove density in the area of study is about 84,950 trees/hectare
(14). Thus, the calculated salt excretion per hectare is about 821
Kg NaCl/day.

Comparing the water absorption cycle in the glass model and
the mangroves grown in the laboratory (Fig 4a), it is inferred that
temperature increases water absorption most likely by enhancing
evaporation. The salt excretory capacity in both systems exhibit a
similar tendency. Nevertheless, the salt excreted by the model is
less than that in the mangrove plant, if the gland area (1 cm^2 =
1,600 salt glands) is considered (Fig. 4b).

The effect of salt concentration on salt and water absorption
is shown in Figure 5. While salt absorption increases with
increasing salt concentration, the volume of water absorbed by the
plant diminishes. In contrast, in the glass model, both water and
salt absorption decrease as a function of salt concentration (Fig.
5). This fact suggests different mechanisms for water and salt
absorption in mangrove plants that probably involve active transport rather than simple diffusion, as occurs in the glass model.

By extrapolation of the data shown in Figure 5, we conclude
that salt excretion in mangroves grown in the laboratory (0.75 mg
NaCl/leaf/day) is similar to that observed in mangroves irrigated
only with sea water and growing along the coast (0.67 mg NaCl/leaf-
/day). Thus, the amount of sea water absorbed should be about 0.46
ml/leaf/day. Further, the desalination capacity of a one hectare
A. germinans mangrove field is 1.4 tons/day (Table 1).

DISCUSSION

From the studies carried out during the last five years on the
mangroves found in Baja California, we have elucidated the natural

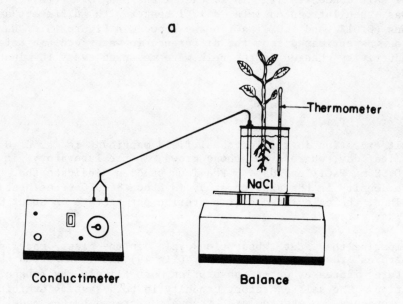

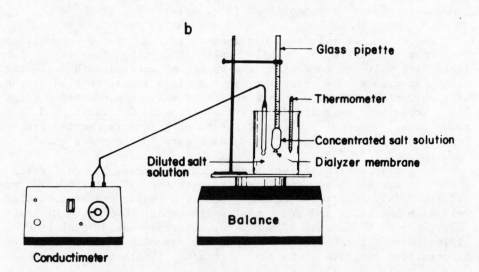

Figure 3.   Experimental design for measuring water and salt absorption, a) in a mangrove plant;   b) in a glass model.

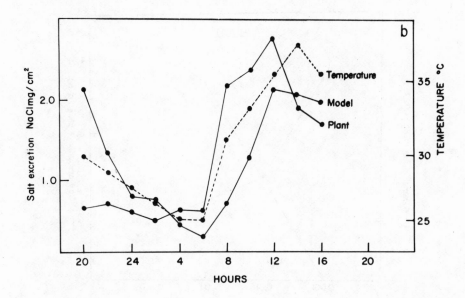

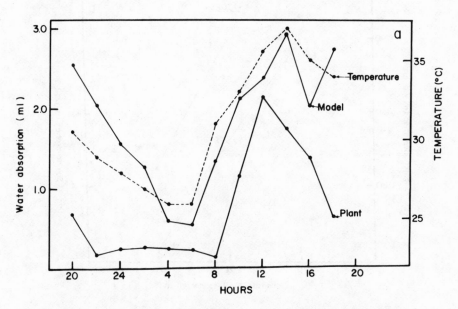

Figure 4. Mangrove plants and glass model comparison of:
a) water absorption, and b) salt excretion.

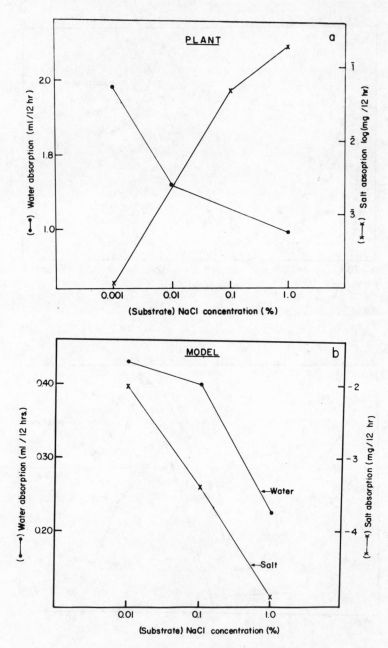

Figure 5.   Effects of salt concentration on salt and water absorption in: a)   the   mangrove   plants;   b)   the   glass
model.

TABLE 1:   Desalination capacity of <u>Avicennia</u> <u>germinans</u> mangrove per hectare*

Salt excretion	821 Kg NaCl/day
Salt absorption	922 Kg NaCl/day
Water absorption	562,000 liters/day
Sea water desalination capacity	1.4 tons/day

*Mangrove density calculated according to (14).

conditions under which this kind of halophyte plant grows. The lack of a secondary flux of fresh water gives rise to especially high saline soils, thus creating an unusual environment for the mangrove in this region. In spite of this, some mangrove species, such as <u>R</u>. <u>mangle</u>, <u>L</u>. <u>racemosa</u> and <u>A</u>. <u>germinans</u>, show a large capacity to excrete salt through their leaves; <u>A</u>. <u>germinans</u> is more efficient than <u>L</u>. <u>racemosa</u> in this respect (9). The capacity of <u>A</u>. <u>germinans</u> to excrete salt is apparent even in its early stages of development, either under the natural environment or in the laboratory. Salt excretion in these mangroves seems to be independent of the chemical composition of the substrate. They appear to possess an extraordinary ability to remove excess salts from the water absorbed and are able to grow under extreme saline conditions.

The temperature effect on salt excretion is directly reflected in an increase in transpiration. At higher temperatures, there is a greater transpiration and, therefore, a larger water absorption. This finding is an agreement with previous studies done by Bowman in <u>R</u>. <u>mangle</u> (17). As shown in Fig 5a, <u>A</u>. <u>germinans</u> mangroves grown under hypersaline conditions are able to absorb a large amount of salt which is excreted through the leaves (18). When compared to the glass model described here, the capacity of <u>A</u>. <u>germinans</u> to excrete salt appears to be the result of a complex mechanism which involves both passive and active transport.

Further studies are now in progress to determine if the desalination capacity of <u>A</u>. <u>germinans</u> can be useful as a biological device for obtaining from the sea water of a quality appropriate for agriculture.

Acknowledgements

We thank Angel Acosta, Felipe Ascencio and Dr. Jose Luis Ochoa for helpful assistance in this research and preparation of the manuscript.

References

1. J. H. Davis, Pap. Tortugas Lab. 517(32):303-412 (1940).
2. W. Macnae, Adv. Mar. Biol. 6:73-270 (1968a).
3. Díaz-Piferrer, Las algas superiores y fanerogamas marinas, in: "Ecología Marina," pp. 273-307, (monografia 14), Fundación La Salle de Ciencias Naturales, Caracas, Venezuela (1967).
4. B. G. Thom, I. Ecol. 55:301-343, (1967).
5. M. Vegas-Velez, "Introducción a la Ecología de Bentos marinos," (Serie de biologia, monografia No. 9), O.E.A. Washington, D.C. (1971).
6. G. Rodríguez, Las Comunidades bentonicas, in: "Ecología Marina" Fundacion La Salle de Ciencias Naturales, Caracas, Venezuela, pp. 563-600, (monografia No. 14) (1972).
7. J. T. McGill, Map of coastal landforms of the world, Geogr. Rev. 48:402-405 (1958).
8. L. F. Mendez, "Los manglares de la laguna de Sontecomapan, Los Tuxtlas," p. 98, Ver. Estudio floristico ecologico, Tesis prof. Fac. Ciencias, Univ. Nal. Auton., Mexico (1976).
9. C. M. Sordo, T. Castellanos, G. Padilla, L. A. Romero, Informe General de Labores, pp. 81-96, Centro Investigaciones Biologicas, La Paz, B.C.S., Mexico (1978).
10. C. M. Sordo, T. Castellanos, G. Padilla, L. A. Romero, A. Solis, Ibid., 65-79 (1978).
11. R. A. Vogel, "Química Analitica Cuantitativa," (S. A. Kapeluz, ed.) 2a. edicion (1960).
12. M. C. Rand, A. E. Greenberg, M. J. Taras, "Standard methods for the examination of water and wastewaters," 14th edition (1975).
13. G. Cintron, Biotropica 10(2):110-120 (1976).
14. C. M. Sordo, T. Castellanos, G. Padilla, L. A. Romero, Informe General de Labores, pp. 71-81, Centro de Investigaciones Biologicas, La Paz, B.C.S., Mexico (1979).
15. E. C. Cocking, Int. Rev. Cytol. 28:89-124 (1970).
16. C. M. Sordo, G. Padilla, L. A. Romero, Informe General de Labores, Centro de Investigaciones Biologicas, La Paz, B.C.S., Mexico (In Press).
17. H. H. M. Bowman, Proc. Amor. Pheos. Soc. 56:589-672 (1977).
18. C. McMillan, Salt tolerance of mangroves and submerged aquatic plants, "Ecology of halophytes," pp. 379-390, Academic Press, Inc. (1974).

MACROALGAL MARICULTURE

B. W. W. Harger

M. Neushul

Marine Science Institute
University of California
Santa Barbara, CA 93106 USA

Neushul Mariculture Inc.
275 Orange Avenue
Goleta, California 93117 USA

SUMMARY

Macroalgal mariculture in coastal waters is highly productive. Successful marine farming of marine macrophytes is carried out in Japan, Korea, the Philippines and China. These countries now produce crops that exceed the yields from the wild populations that have traditionally been the sources of algin and carrageenan. The Chinese kelp harvest alone is presently 280,000 dry tons per year, an average of 15 dry tons being produced per hectare of farm. The history of macroalgal farming is very recent, since truly scientific mariculture has been practiced for only the last two decades. The future potential of scientifically based macroalgal mariculture is considered, using as specific examples potential farm sites in California and Mexico.

INTRODUCTION

Marine macroalgae are important sources for the phyco-colloids known as alginates, carrageenans and agars. Until recently, the entire world supply of these valuable chemicals came from the wild, rather than from a cultivated crop. Unfortunately, this is still the case for agar, which no one has yet been able to farm. However, carrageenans and algins are now derived, in part, from farmed crops.

The purpose of this paper is to consider macroalgal domestication and farming as it is now practiced and to illustrate the steps that can be taken to provide basic information about a potential crop plant and sites where it might be farmed. Our recent work

393

with the agarophyte, <u>Gelidium</u>, the carrageenophyte, <u>Eucheuma</u>, and
the alginophyte, <u>Macrocyctis</u>, has led us to develop new tools and
techniques that have proved to be useful. Successful planting in
the sea depends on a clear understanding of the resource needs of
the plant and whether or not resource supplementation will be
required at a specific farm site.

EVALUATING A POTENTIAL MARINE FARM SITE

The land-based farmer, in evaluating a plot of land as a site
for growing a crop, draws upon generations of prior experience in
judging the soil, the climatic conditions and other factors. The
would-be mariculturalist must proceed without the advantage of such
prior experience. The mariculturalist like the agriculturalist
must be aware of meteorologically induced seasonal changes, but
must measure and be able to predict oceanographic changes as well.
In a sense, the mariculturalist must measure two climates that are
obviously very different. For example, in place of wet or dry
seasons, the marine farmer has high and low nutrient "seasons" to
contend with and may well have to develop methods for "nutrient
irrigation." He must be aware of the many factors that can damage
a crop at a particular farm site. A host of deleterious biological
factors such as epiphytism, animal encrustation, grazing and
disease can damage a crop. The effects of physical factors such as
low or high water motion, light, micronutrient availability and
sedimentaiton can also be important.

The measurement of the physical factors in the aquatic
environment that influence plants has been reviewed by Wheeler and
Neushul (1). In evaluating a farm site, basic measurements should
be made. For example, the degree of water motion should be
measured so that a coefficient of turbulence can be calculated,
since turbulence influences both mass transport to and from the
plants and exerts drag forces on them. Turbulence results in the
transport of dissolved gasses and nutrients through the thin boundary layer adjacent to the plant surface. The thickness of this
layer is thus very important. For example, kelps can be put under
severe stress when held in quiet water, at velocities of less than
6 cm per second (2). At the other extreme kelp plants are damaged
or dislodged at high water velocities during storms (3).

The level of illumination reaching the crop is of obvious
importance. Deep-growing macroalgae are able to survive at light
levels reduced to 1% of that reaching the sea surface, and some
encrusting forms can survive at 0.01% of that amount of light that
reaches the sea surface over a year. However, the mariculturalist
will obviously be seeking higher light levels, and may well be concerned about solarization that might occur at or near the sea
surface.

A systematic study of potential farm sites for the red alga, Porphyra, in Japanese waters was undertaken by Matsumoto (4), who wondered why Korean farms flourished in non-estuarine areas while in the Seto Inland Sea of Japan one could farm only in the estuarine regions. Matsumoto first studied Porphyra in the laboratory and in the sea and determined optimum temperatures, nutrient levels, water motion levels and salinity for the plant. He then showed that conditions in the Seto Inland Sea, away from estuaries, were unsuitable because of high temperatures, low nutrients and low water motion. In estuarine habitats temperatures were lower, nutrient levels were higher and favorable water currents developed. Current speeds of 30 cm per second were required for maximum plant growth when nutrient levels were low, while only 5 cm per second was required when nutrient levels were high. Based on his laboratory and field measurements Matsumoto concluded that high water motion habitats, other than those near estuaries, would be suitable for Porphyra cultivation. He tested this hypothesis by experimental plantings in the Strait of Onomichi Suido, showing that high water motion substituted for having only low nutrient levels.

In evaluating potential farm sites, Matsumoto focused his attention on the four major environmental factors that influence macroalgae: (1) water temperature; (2) dissolved nutrients; (3) illumination; and (4) water motion. He showed how variations in these four factors influence the color, luster, toughness, flavor and other features of the Porphyra. He attempted to express the relative importance of these factors as a ratio of water temperature: illumination: dissolved nutrients: water current, and found the ratios to be 30:10:35:25 or 35:10:35:20. Matsumoto's careful study exemplifies a logical approach to marine farm site evaluation.

EVALUATING A POTENTIAL MACROALGAL CROP PLANT

The mariculturalist who seeks to domesticate a new marine crop plant must first study the natural conditions under which it grows and reproduces. Particular attention must be given to the resource requirements of the plant and the availability of these resources in the sea. A basic requirement is the identification of the plant as a winter or summer annual, or a perennial. It is also important to know the season when recruitment occurs, since this provides essential clues as to how the life history might be manipulated and controlled, and allows one to determine the best times to plant and harvest.

Once information about the potential crop plant and its environment is assembled, a manipulative phase should be initiated. Phycologists have traditionally approached the cultivation of algae by first isolating reproductive cells, usually spores, so as to

obtain unialgal cultures. These dish cultures are studied to
identify all of the phases of the life history. While this is an
essential first step in the domestication of an alga, the cultiva-
tor must go far beyond this and must learn how to produce large
amounts of seed stock and to control the timing of dormancy, growth
and reproduction. Plants like Porphyra and Macrocystis have a
microscopic, filamentous life history phase that can be stored and
manipulated. The light levels required for growth and reproduction
in these filamentous stages have been studied for Californian kelps
by Luning and Neushul (5). The agarophytes and carrageenophytes,
in contrast, have a tri-phasic life history where haploid and
diploid phases are morphologically equivalent. Polne, et al (6)
have been able to store very young tetrasporophytes and gameto-
phytes as germlings, whose dormancy can be broken by appropriate
culture conditions in bubble flasks or tubes. Thus it seems that
these young sporeling stages can serve as "storage" phases for the
mariculturalist. Hopefully this approach will be applicable to
agarophytes as well.

    After selecting a farm site and developing nursery techniques
for a potential crop plant the mariculturalist must develop in-
the-sea test farms. While the commercial farms of China and Japan
can serve as models, some special adaptations are needed to handle
and measure individual plants which would, of course, not be done
on a commercial farm. Also, the experimental farming structure
should be much more adjustable than a commercial farm would need to
be. Such farms need not be complex. The farm structures that we
have used successfully for test planting in California are comparatively simple, being made of rope and plastic pipe. The genetic
potential of a wild plant can also be considered now that hybridization methods have been developed for red algae (7,8) and brown
algae (9,10).

MACROALGAL FARMING TOOLS AND TECHNQUES

    As noted above, in evaluating a farm site, the mariculturalist
must make careful measurements of light, turbulence, temperature,
sedimentation, nutrients and other factors. The tools for doing
this are not very sophisticated. Sediment traps, a secchi disk, a
current meter and appropriate thermometers are needed. The
measurement of nitrates requires a spectrophotometer and some
specialized glassware but the procedure is not overly difficult.
These measurements have been made in California at Goleta for the
last year (Figure 1). One can define the stormy winter period and
the calm summer period in terms of daily wave height and current
measurements. Water temperatures provide a good indication of a
third period when upwelling and nutrient enrichment occurs.

    Tools for measuring a potential crop plant are not complex.

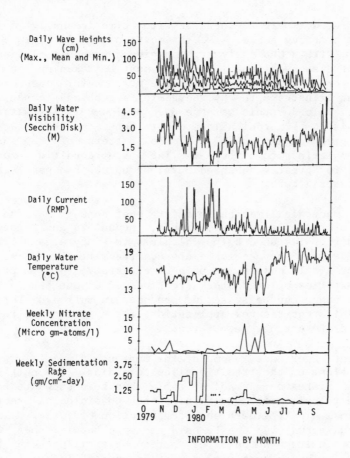

INFORMATION BY MONTH

****missing data

Figure 1.   Measurements over a 12 month period made in Goleta Bay,
California, at an   experimental marine   farm site, show   daily wave
height, water   visibility (as   Secchi Disk   depth), current (as
current   meter revolutions   per minute), water   temperature, weekly
nitrate concentration   and sedimentation rate (in grams   per square
centimeter per day).

     Seasonal   growth   conditions   shown   here fall   into: (1) A
stormy, high sediment, high current, high wave period (November –
January); (2) A calmer, low temperature, medium sediment, medium
current, low wave period (March –May);   (3) A calm, high temperature, low current, low wave period (June –September).   It is particularly noteworthy that the three low temperature periods in April,
May and June coincide with the high nitrogen peaks, this being due
to the   upwelling of cold, nutrient-rich water.   Thus, this   farm
site has a   nutrient rich season and two nutrient-poor seasons.
This is why nutrient irrigation has been tried.

For example, wet weight increase with time is easily measured by hand with a spring scale. Measurements of the nitrogen content of the plant, with a CHN analyzer, can give a valuable clue as to the amounts of dissolved nutrients that must be taken up in order to produce a crop. Perhaps some of the most useful measurements are those that illustrate the hydrodynamic "scrubbing" capabilities of the plant. To do this we have used a low-velocity water tunnel (11). This complex tool allows one to determine the water speed at which a given alga "trips" the flow pattern of the surrounding water from laminar to turbulent. The latter condition provides the potential for greatly enhanced nutrient uptake and gas exchange and hence greater growth.

The macroalgal mariculturalist must depend on the standard array of microscopes, culture dishes, incubators and related equipment traditionally used by phycologists to study algal life histories. However, as mentioned above, the mariculturalist must go beyond the simple completion of a life history, to the point where large quantities of seed stock are produced on one hand, and where a dormant stage can be selected, produced in bulk and stored on the other. This requires new applications for the standard tools and a somewhat different new set of tools.

In our experience, some of the most useful cultivation tools are extensions of the traditional culture dish. The cultivation of plants in dishes on a gradient table has been shown to be a very effective technique for determining the physiological requirements for optimal algal growth and reproduction (12, 13). The bubble flask of Hasegawa (14) can be enlarged into bubble tubes or cylinders, wherein large numbers of spores or germlings can be raised (Figure 2). The culture dish can be "expanded" into a larger sized tank culture, and this can be installed in a greenhouse, for either large scale bubble culture, or hydrodynamically controlled cultivation (16, 17).

As mentioned earlier, planting in the sea can best be tried experimentally with simple racks and lines, much like those used for operational farms in the Orient. We have used several designs where individual plants are planted on collars that can be easily removed and reclipped in place to facilitate their measurement and manipulation (Figure 3).

TWO SPECIFIC EXAMPLES OF MACROALGAL FARM POTENTIAL

It is all very well to talk in general terms of asking how fast a plant grows, how it reproduces, what its resource needs are and how the mariculuturalist would design his farming stategy to meet thse needs. The fact remains that a successful farmer must be able to accurately predict what the yield of his specific crop will

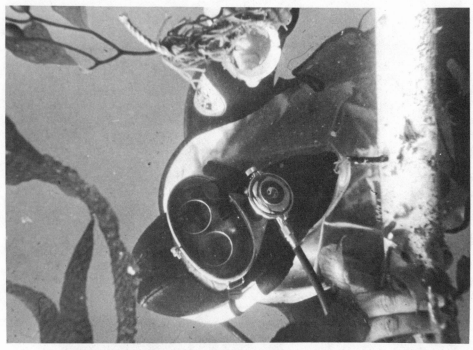

Figure 2 (Above) -Eucheuma germlings are raised to a few mm in size and then put into tubes with air supplied from below to provide agitation. They are raised here from germlings to outplantable size.

Figure 3 (Right) -Experimental outplantings of kelps and other macroalgae are carried out using rope pieces attached to clips that snap around plastic-pipe farming substrates (shown in fore-ground).

be, after he has successfully tapped the natural patterns of nutrient cycling and energy flow at a specific site.

Some experimental farming work and site evaluation that we have done over the past year in Goleta Bay allows us to make some predictions about yield at this site. We have also worked on a carrageenophyte (Eucheuma uncinatum) which was collected from Bahia de Los Angeles in Baja California, which shows considerable promise as a future crop plant. Here also some predictions of potential yield are possible. While the farm designs and farming strategies for these two areas would most certainly differ, the exercise of developing these farming plans illustrates the steps that the mariculturalist might take to develop actual farms.

1. The Potential of Macroalgal Farming in Goleta Bay, California: Goleta Bay is in Santa Barbara County, on the mainland and the northern end of the southern California bight, facing the Santa Barbara Channel. The waters here are periodically enriched by upwelling and influenced by storms during the winter (see Figure 1). Large kelp forests occur along the coast and illustrate its suitability as a site for macroalgal mariculture.

Our exprimental farming has been carried out at two sites in Goleta Bay, one in the calmer waters within the bay and one at a point on the western end of the bay, where water motion levels are higher. Nitrogen levels at both sites is generally lower than 2 microgram atoms per liter for most of the year, so nutrient irrigation has been tried.

Giant kelps (Macrocytis angustifolia) and agar weeds (Gelidium nudifrons and Gelidium robustum) have been farmed. Gelidium, an important source of agar, has been the main focus of attention. The best growth rates for Gelidium in a low nutrient period (July -October; see Table I) has been 1.26% per day, with fertilizer supplementation provided by porous tubes of a slow-release, resin-coated fertilizer (osmocote), made by Sierra Chemical Co. Control plants not provided with fertilizer grew at only 0.37% per day.

Given the farm site information for the past year, and the test growth rates of 1.26% per day for nutrient irrigated Gelidium plants, it is possible to select an optimum planting time and to identify those periods when nutrient irrigation might be needed. If high growth rates are possible, via more effective fertilization and/or the selection of faster growing strains, it is conceivable that productive agarweed farms could be established in southern California. Since the current price for dry Gelidium is about $2 a pound and is likely to increase with a predictable supply, such farms might be profitable.

TABLE I. The growth (as % increase in wet weight per day) of
three test farmed and nutrient irrigated macrophytes in Goleta Bay,
California (Aug - Oct '80).

	Sample Size	Ave.	Stand. Error	Degrees of Freedom	t-test	Probability Value
Gelidium nudifrons						
Fertilized*	16	1.262	0.270	29	2.4570	0.0214
Unfertilized	15	0.373	0.238			
Gelidium robustum						
Fertilized*	14	0.357	0.107	22	4.7402	< 0.0010
Unfertilized	10	-1.540	0.453			
Macrocystis angustifolia						
Fertilized*	5	3.800	0.192	10	2.6935	0.0233
Unfertilized	7	2.928	0.235			

*Slow released resin-coated fertilizer was placed in, and diffused
from, containers floated next to the plants.

2. Macro-algal Farming in Bahia de Los Angeles, Baja California: Bahia de Los Angeles is off Baja California, near the boundary between the northern and southern portions of the Gulf of California, a region characterized by large islands and rapid tidal currents. The Gulf of California has some of the most extreme environmental variations of any of the world's seas, with a seasonal change in temperature that ranges from 9° C at the mouth of the gulf to 22° C at the northern-most end. A striking change between winter and summer floras has been documented by Norris (18).

The gulf is protected from the influence of the Pacific Ocean by the mountains of the Baja California peninsula, hence the climate is not oceanic but continental. In winter, northwesterly winds produce upwelling on the Mexican mainland coast. In summer, southwesterly winds produce upwelling on the peninsular Gulf coast. This simple pattern is significantly modified near the midriff islands where there is considerable mixing due to tidal currents and upwelling and low temperatures occur at all seasons. Solodov (19) has diagrammatically indicated where these regions occur relative to Bahia de Los Angeles. The lowest tides are from February to April and the highest from July to September. One might assume that maximum tidally induced upwelling would occur at these times. Van Andel and Shor (20) discuss the oceanography of the Gulf of California in general. It would be necessary to make careful measurements in order to see if there is frequent upwelling and nutrient enrichment in or near Bahia de Los Angeles, as seems likely.

The lush beds of large Sargassum plants, and the large naturally occuring plants of Eucheuma uncinatum that grow in Bahia de Los Angeles, suggest that conditions there are frequently very favorable for these plants. Polne et al (6) have studied the growth and reproduction of Eucheuma uncinatum collected from this site and have measured growth rates approaching 10% increase in wet weight per day under greenhouse conditions with nutrient levels over 10 μg atoms/1. They have also been able to produce sporelings at will from both gametophytic and cystocarpic plants, and to hold these at low light levels until needed. Then dormancy can be broken and through the use of bubble tubes (Figure 2) a seed stock crop can be produced. Thus crop production would seem to be possible.

An alternative farming strategy for Eucheuma in Bahia de Los Angeles would be that which has been successfully developed in the Philippines (21, 22). Doty (pers. comm.) suggests that one might introduce E. spinosum and E. cottonii from Zamboanga, to provide seed stock for vegetative propagation. Test farm substrates would be constructed of nylon monolines installed at likely farm sites in Los Angeles Bay. A similar test planting of native E. uncinatum would also be possible.

Regardless of whether native or non-native species are farmed, it seems likely that the fast growing carrageenophyte, Eucheuma, could be grown in or near Bahia de Los Angeles. A major unknown factor is the degree of natural upwelling and nutrient enrichment that occurs at this potential farm site. If year round tidally-induced upwelling does occur then very productive macroalgal farms could be developed. The current price for Eucheuma depends on the availability and price of the Philippine crop so the economic incentives are less obvious than those for agarophytes.

CONCLUSIONS

We cannot advocate immediate initiation of farming work in either of the above mentioned locations because the costs of farming are not yet well known and only limited site characterizations and species characterizations have been done. Also studies of optimum planting density are needed. Studies of natural populations suggest that dense plantings are possibly favorable (23). However, the approach described here, where seasonal site characterization and crop characterization are followed by test plantings, might be used as a guide to avoid overlooking some of the major pitfalls that a marine farmer must ultimately face. Obviously, a basic array of equipment is required, and measurements of water motion, nutrient levels and other factors discussed above are essential. Growth rates of 1.26% per day for nutrient irrigated Gelidium and nearly 10% per day for fertizilzed greenhouse grown Eucheuma are promising. A systematic, quantitative approach, rather than a trial and error one, is certainly advantageous and will speed the experimental process that must inevitably provide the basis for successful macroalgal mariculture.

REFERENCES

1. W. N. Wheeler and M. Neushul, in: "Encyclopedia of Plant Physiology: Interactions of Plants with the Physical Environment," E.L.Lange, P. Nobel, B. Osmond and H. Ziegler, eds., (In Press).
2. W. N. Wheeler, Mar. Biol. 56:103-110 (1980).
3. R. J. Rosenthal, W. D. Clarke and P. K. Dayton, Fish Bull. 72:670-684 (1974).
4. F. Matsumoto, in: "Mem. Dept. Fish. Animal and Vet. Sci.," Hiroshima Univ. J. Fac. Fish. Animal Husb. 2, 249-327 (1959) (In Japanese with English).
5. K. Luning and M. Neushul, Mar. Biol. 45:297-309 (1978).
6. M. Polne, M. Neushul and A. Gibor, Pro. Int. Seaweed Symposium 10 (In Press).

7.   A. R. Polanshek and J. A. West, in: "Handbook of Phycological
     Methods, Developmental and Cytological Methods," E. Gantt,
     ed., Cambridge Univ. Press.: Cambridge, England (1980).
8.   J. P. van der Meer, Phycologia 18:47-54 (1979).
9.   Y. Sanbonsuga and M. Neushul, J. Phycol. 14:214-224 (1978).
10. Y. Sanbonsuga and M. Neushul, in: Handbook of Phycological
     Methods, Developmental and Cytological Methods, E. Gantt,
     ed., Cambridge University Press., Cambridge, England (1980).
11. A. C. Charters and S.M. Anderson, J. Hydronautics 14:3-4
     (1980).
12. P. Edwards, in: "Contributions in Phycology," B. Parker and M.
     Brown, eds., Univ. of Texas Press, Houston (1971).
13. P. Edwards and C. van Baalen, Bot. Mar. 13:42-43 (1970).
14. Y. Hasegawa, J. Fish. Res. Board Can. 33:1002-1006 (1975).
15. B. E. Lapointe and J. H. Ryther, Bot. Mar. 22:529-537 (1979).
16. A. C. Charters and M. Neushul, Aqu. Botany 6:67-78 (1979).
17. M. Neushul, Proc. Int. Seaweed Symp. 10 (In Press).
18. J. Norris, Ph.D. Dissertation, University of California, Santa
     Barbara (1975).
19. V. A. Solokov, Calif. Cooperative Oceanic Fisheries
     Investigations Report, 17:92-96 (1974).
20. T. H. van Andel and G. G. Shor Jr., eds., "Marine Geology of
     the Gulf of California," Am. Ass. Petrol. Geol. Memoir
     3:1-408 (1964).
21. M. S. Doty, Micronesica 9:59-73 (1973).
22. M. A. Ricohermoso and L. E. Deveau, (Paper distributed at the
     IXth International Seaweed Symposium, unpublished).
23. D. R. Schiel and J. H. Choat, Nature 285:324-326 (1980).

POLLUTANT AND WASTE REMOVAL

William J. Oswald

Sanitary Enginering and Public Health
University of California
Berkeley, CA 94720 USA

SUMMARY

The saline wastes from irrigation return flows and tile drainage, together with miscellaneous animal and avian wastes, constitute major agricultural pollutants in arid and semi-arid zones. High residual organics, nitrates, and sodium ratios make these waters undersirable for discharge into both saline and fresh water receiving bodies. In the San Joaquin Valley of California, saline marshes have been considered as a method of controlling pollutants, reducing volume and propagating wildlife. Three marsh types are considered: terminal, fill and draw, and flow through. Terminal systems were rejected because they become increasingly saline and ultimately become unproductive salt beds. Fill and draw systems were selected for special study and in this presentation resultant quality changes and environmental impacts are projected and compared with a flow through microalgal system. The microalgal system appears to meet more of the desirable objctives of the system using less energy and at a lower cost.

INTRODUCTION

The introduction of high-quality irrigation water into the western and southern San Joaquin Valley has necessitated long-range planning to remove large amounts of naturally occurring soluble salts accumulated in the soils and to prevent accummulation of salts originating from the irrigation water. The conditions of high soil salinity result mainly from a large excess of evaporation over precipitation in the Valley. As shown in Figure 1, the western San Joaquin is largely in the rain shadow of the South

405

Fig. 1
Map of California showing San Joaquin Valley
location and location of Key points near the
Bakersfield – Sacramento Axis.   Referenced
in Fig. 2.

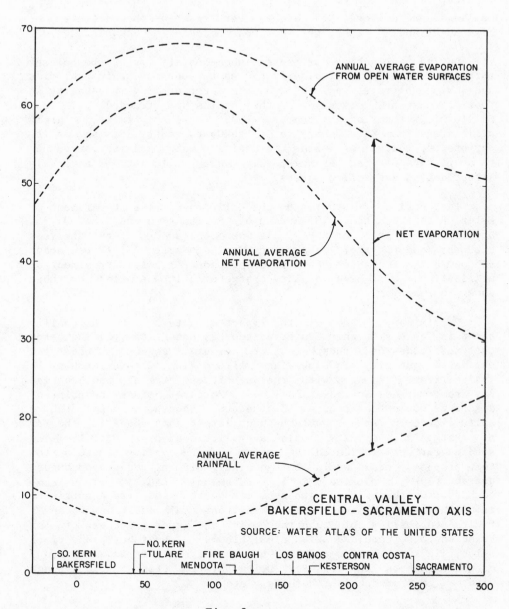

Fig. 2.
Mean annual rainfall and Evaporation Profiles
between Bakersfield and Sacramento,
California.

Coast Range. There, as shown in Figure 2, evaporation exceeds precipitation by up to 60 inches per year in the vicinity of Coalinga and Hanford, 55 inches per year in the vicinity of Firebaugh, and 42 inches per year near Modesto and Oakdale.

During irrigation of crops, good-quality water percolates through the saline soil dissolving salts such as sodium chloride and sodium nitrate, and, consequently deteriorating in quality so that it must be collected in the drains and removed. In the California Water Plan, by the year 2000 or slightly later, a drain will extend from Kern County to Contra Costa County, having as its purpose the removal of subsurface drainage water and its conveyance to a portion of San Francisco Bay which will not be adversely affected by these saline waters.

In a series of studies by the California State Department of Water Resources, the U.S Bureau of Reclamation and the U. S. Department of Interior (EPA), it was concluded that by the year 2000 the quantity of subsurface drainage water will be about 600,000 acre feet per year which, without special treatment or utilization, will have quality characteristics similar to those shown in Table. 1.

Preliminary studies of the impact of waters with the quality shown in Table 1 on the San Francisco Bay and Delta (1) indicated that pesticides and phosphates were of small concern since their estimated quantity in subsurface tile drainage water would be insignificant compared with the quantities entering San Franciso Bay from surface sources (2, 3,). Moreover, since the point of discharge would be selected so that the concentrations of those salts listed in Table 1 would be much larger than those in the Bay at that point, salinity could have little adverse effect. The studies further indicated the major adverse quality factors from the proposed drain would be the inorganic nitrogen compounds, nitrate (NO_3^-), nitrite (NO_2^-), and ammonium (NH_4^+). These compounds, together with phosphates, carbonates, and trace nutrients from other sources, under special environmental conditions having a finite probability of occurrence in the upper reaches of San Francisco Bay, could nourish algal blooms in enormous proportions which, in turn, could cause a host of adverse environmental and economic impacts in the Bay Delta System (fish kills, beach windrows, H_2S production, botulism, etc.).

The conclusion that inorganic nitrogen compounds could trigger algal blooms during quiescent periods in the Bay led to the proposed standard of a limit of 2 mg of N per liter for wastewaters entering San Francisco Bay and initiation of studies of micro-biological techniques such as algal stripping (4-10) and bacterial denitrification (11) for the removal of nitrogen from drainage waters destined to enter the Bay.

TABLE 1. Estimated Chemical Substances in Untreated Subsurface
Tile Drainage Waters at Antioch 2020*

Substance	Milligram per liter
Total Dissolved Solids	3,000
Salts	
Sulfate	700
Chloride	900
Calcium	100
Magnesium	50
Bicarbonate	100
Potassium	10
Boron	3
Nutrients	
Total Nitrogen	20
Total Phosphate	0.15
Pesticides	Less than 0.001

*Yearly average with drain in full operation (no treatment or
holding reservoirs (1)

As studied, algal stripping could not reach the desired
concentration of 2 mg N per liter during much of the year and,
while bacterial denitrification did attain the required levels for
nitrate, it required a complex and expensive upflow filter system
and a sustained source of organic carbon such as methanol to
operate. Due to inflation and crude oil price increases, methanol
has since quadrupled in cost. Each process had an estimated cost
in the range of 30 1969 dollars per acre foot which, when
multiplied by 600,000 acre feet per year, would involve the commitment of at least 18 million 1970 dollars per year for treatment
alone. Although there was some evidence that algal stripping could
lead to the production of algal material worth as much as the cost
of its production, and that improvements in cultivation techniques,
together with skillful pond and storage management, could result in
effluents having N concentrations consistently below 2 mg total N
per liter, studies were continued in search of a more economical
method of nuitrient removal.

An algal-bacterial symbiotic process, a grass bacterial
symbiotic process, and a combined algal-grass system were studied.

It was concluded from the studies that "algal-grass" system would be most dependable, effective, and economical and that cost of such treatment would be on the order of 12 1972 dollars per acre foot (12).

The success of grass plot studies in apparent nutrient removal, coupled with the work of Cederquist (13) and others (5, 6, 14), indicates that such areas can provide habitats for wild fowl, reduce wild fowl grazing pressure on cultivated lands, improve the recreational potential of marshes, and provide opportunities for improved flow regulation. These and other potential benefits suggested that marshes should be explored in detail as a potential system for treatment and partial disposal of San Joaquin Valley drainage water.

A detailed study of the feasibility of marshes was then made by Ives et al (15) who concluded that brackish marshes are of substantial value to water fowl and other wildlife and that sufficient information is available to predict the vegetation that will occur and to design the marsh. On the other hand, Ives, et al were in doubt about management and felt that pilot marsh units would be required to gain management experience and provide information on quality changes and problems which may arise such as botulism and fish kills. As will be indicated later, the contents of this paper also indicate that pilot plant studies should precede any final decision to utilize marshes as a major treatment element in the system.

In the Ives study three types of marshes were defined:

Type I: Without flow-through circulation, water added to make up for evapotranspiration and percolation losses;

Type II: A continuous flow-through of water in flooded units with continual inflow and outflow adjusted so that a salt balance is achieved;

Type III: Utilizes either agricultural crop or native marsh plants, flooded and drained in spring for irrigation purposes, continuously flooded into spring for salinity and vegetation control.

In evaluating the utility of each type of marsh, Ives, et al concluded that Type I is not really applicable for highly saline drainage water as salinity would build up to a point that it would be excessive for plants which produce food for wildlife. This, of course, must be the mode of operation in certain segments until the drain is completed when it will be possible to operate them as Type

II or III marshes. Ives, et al, did feel that Type I marshes could be operated in the San Joaquin Basin and Tulare Basin East Side areas where water quality will be sufficiently good to allow for a great deal of management flexibility. Type II marshes were recommnded for the kern area, for the Tulare Lakes areas, and for the San Luis area until the quality of the subsurface drainage improves. Type III marshes were recommended for the Delta Mendota service area west of the San Joaquin River and for the area north of the Delta Mendota service area west of the San Joaquin River.

One ultimate program would be as shown in Figure 3 with seven marsh areas: South Kern, North Kern, Tulare Lake, Mendota, Los Banos Grass Lands, Kesterson Reservoir, and a Contra Costa Marsh. Kesterson could serve to provide cooling water for up to 3 power plants and also act as a salt sink. In this scenario all marshes would be the seasonally flooded Type III systems to maximize wildlife habitat.

OBJECTIVES

Long-range objectives of the IDP program, not necessarily in order of importance, are to protect agricultural productivity, protect water resources, promote beneficial uses of drainage water, avoid adverse environmental impact, avoid seepage and lateral movement of saline waters, minimize energy use, promote integrated uses of drainage water, make maximum use of existing facilities, minimize land severance, and avoid encroachment on rare and endangered species (16).

The avoidance of adverse environmental impact is a major objective of advanced study and planning of the marshes and evaporation ponds. Of particular concern are the projected changes in water quality at the discharge point resulting from the use of drainage waters for marsh management and power plant cooling. This is the basis for undertaking the work described herein.

One technique employed in projecting ultimate quality is to theoretically isolate each of the component facilities and to conduct a mass balance for each constituent within each facility. This, of course, has required the prediction of water temperatures and photosynthetic microbiological and physicochemical reactions as influenced by cyclic and transient phenomena for each component facility as a function of time. However, terminal discharge quantities and qualities for the above two combinations of facilities, as later modified, were of major interest.

Specific monthly quantities of drainage flow at specific monthly concentrations of TDS, nitrogen, phosphorus, and BOD were

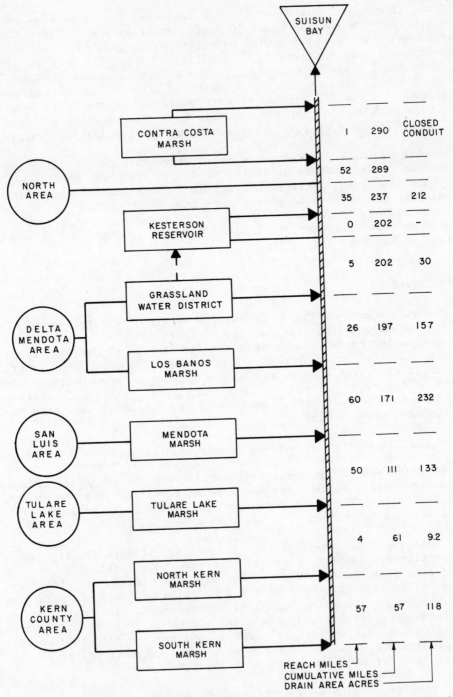

Figure 3. Assumed Drain and Marsh Pattern After Extension of the Drain to Kern County.

assumed and provided by the Planning Group of the Department of Water Resources. These are presented in Tables 2A and 2E inclusive.

Water quality parameters of major interest are a) biochemical oxygen demand, b) total organic carbon, c) nitrate nitrogen, d) ortho phosphate, and e) algal biomass. Water quality parameters of secondary interest are f) ammonia nitrogen, g) organic nitrogen, h) inorganic phosphorus, i) organic detritus, j) chlorophyll a, k) silica, l) total dissolved solids, and m) total suspended solids (16).

Marsh Criteria

Certain criteria were used by IDP to select and size marsh and evaporation pond sites. Their criteria follow (16):

1. Site Selection

   a. Lands either not farmed or of low productivity
   b. Tight soils (low permeability) as determined from soil maps
   c. Located adjacent to or near Valley Drain alignment and elevation relative to drain that minimizes pumping

2. Size

   a. Evaporation of 4 ft per year
   b. Gross area 120 percent of net area

3. Design criteria for the marsh and evaporation facilities

   a. Ponds divided into 100-acre cells
   b. Transfer structures sized to drain or fill a cell in four days. This will minimize feather edges and botulism problem.

   c. Deep percolation from cells minimized by natural or artificial barriers. Any excess seepage will be intercepted before it can affect good water quality.

4. Operational criteria for the marsh and evaporation facilities

   a. Manage facilities such that salinity and volume impacts of discharge on receiving waters of the western Delta will be minimized.
   b. Move water sequentially through evaporation ponds so the

TABLE 2A

Assumed Input Flows and Quality Factors
Kern County Area – Year 2000

Month	Q/Mo AF	EC µmhos/cm	TDS mg/l	TDS lbs/AF	Inorganic N mg/l	Inorganic N lbs/AF	Inorganic P mg/l	Inorganic P lbs/AF	BOD$_5$ mg/l	BOD$_5$ lbs/AF	Cumulative Flow Sept Base AF
Jan	1,020	8,800	7,040	19,148	10	27.2	0.5	1.36	2.0	5.45	6,052
Feb	2,380	8,200	6,560	17,843	10	27.2	0.5	1.36	2.0	5.45	8,432
Mar	3,060	8,700	6,960	18,931	10	27.2	0.5	1.36	2.0	5.45	11,492
Apr	4,080	8,200	6,560	17,843	10	27.2	0.5	1.36	2.0	5.45	15,572
May	4,080	8,500	6,800	18,496	10	27.2	0.5	1.36	2.0	5.45	19,652
Jun	4,080	7,400	5,920	16,102	10	27.2	0.5	1.36	2.0	5.45	23,732
Jul	4,080	7,200	5,760	15,667	10	27.2	0.5	1.36	2.0	5.45	27,812
Aug	3,740	7,000	5,600	15,232	10	27.2	0.5	1.36	2.0	5.45	31,552 / 34,000
Sep	3,060	7,400	5,920	16,102	10	27.2	0.5	1.36	2.0	5.45	612
Oct	2,040	7,600	6,080	16,537	10	27.2	0.5	1.36	2.0	5.45	2,652
Nov	1,360	8,500	6,800	18,496	10	27.2	0.5	1.36	2.0	5.45	4,012
Dec	1,020	8,500	6,800	18,496	10	27.2	0.5	1.36	2.0	5.45	5,032

TABLE 2B

Assumed Input Flows and Quality Factors
Tulare Lake Service Area - Year 2000

Month	Q/Mo AF	EC μmhos/cm	TDS mg/l	TDS lbs/AF	Inorganic N mg/l	Inorganic N lbs AF	Inorganic P mg/l	Inorganic P lbs/AF	BOD$_5$ mg/l	BOD$_5$ lbs/AF	Cumulative Flow Sept Base AF
Jan	2,370	11,600	9,280	25,241	10	27.2	0.5	1.36	2.0	5.45	14,062
Feb	5,530	10,800	8,640	23,500	10	27.2	0.5	1.26	2.0	5.45	19,592
Mar	7,110	11,400	9,120	24,806	10	27.2	0.5	1.36	2.0	5.45	26,702
Apr	9,480	10,800	8,640	23,500	10	27.2	0.5	1.36	2.0	5.45	36,182
May	9,480	11,100	8,880	24,153	10	27.2	0.5	1.36	2.0	5.45	45,662
Jun	9,480	9,800	7,840	21,324	10	27.2	0.5	1.36	2.0	5.45	55,142
Jul	9,480	9,500	7,600	20,672	10	27.2	0.5	1.36	2.0	5.45	64,622
Aug	8,690	9,100	7,280	19,801	10	27.2	0.5	1.36	2.0	5.45	73,312
Sep	7,110	9,800	7,840	21,324	10	27.2	0.5	1.36	2.0	5.45	79,000 / 1,422
Oct	4,740	10,000	8,000	21,760	10	27.2	0.5	1.36	2.0	5.45	6,162
Nov	3,160	11,100	8,880	24,153	10	27.2	0.5	1.36	2.0	5.45	9,322
Dec	2,370	11,100	8,800	24,153	10	27.2	0.5	1.36	2.0	5.45	11,692

TABLE 2C

Assumed Input Flows and Quality Factors
San Luis Area Year 2000 Mendota Marsh

Month	Q/Mo AF	EC μmhos/AF	TDS mg/l	TDS lbs/AF	Inorganic N mg/l	Inorganic N lbs AF	Inorganic P mg/l	Inorganic P lbs/AF	BOD5 mg/l	BOD5 lbs/AF	Cumulative Flow Sept Base AF
Jan	1,920	7,200	5,760	15,662	30	81.6	0.04	0.11	2.0	5.44	11,200
Feb	4,480	6,700	5,360	14,579	30	81.6	0.04	0.11	2.0	5.44	15,680
Mar	5,760	7,100	5,680	15,449	30	81.6	0.04	0.11	2.0	5.44	21,440
Apr	7,680	6,700	5,360	14,579	30	81.6	0.04	0.11	2.0	5.44	29,120
May	7,680	6,900	5,520	15,014	30	81.6	0.04	0.11	2.0	5.44	36,800
Jun	7,680	6,000	4,800	13,056	30	81.6	0.04	0.11	2.0	5.44	44,480
Jul	7,680	5,900	4,720	12,838	30	81.6	0.04	0.11	2.0	5.44	52,160
Aug	7,040	5,700	4,560	12,403	30	81.6	0.04	0.11	2.0	5.44	59,200 / 64,000
Sep	5,760	6,000	4,800	13,056	30	81.6	0.04	0.11	2.0	5.44	960
Oct	3,840	6,200	4,960	12,491	30	81.6	0.04	0.11	2.0	5.44	4,800
Nov	2,560	6,900	5,520	15,014	30	81.6	0.04	0.11	2.0	5.44	7,360
Dec	1,920	6,900	5,520	15,014	30	81.6	0.04	0.11	2.0	5.44	9,280

TABLE 2D

Assumed Input Flows and Quality Factors
Delta Mendota Area Year 2000* - Los Banos Marsh

Monthe	Q/Mo AF	EC µnhos /cm	TDS mg/L	TDS lbs/AF	Inorganic N mg/l	Inorganic N lbs/AF	Inorganic P mg/l	Inorganic P lbs/AF	BOD5 mg/l	BOD5 lbs/AF	Cumulative Flow Sept Base AF
Jan	1,680	5,500	4,400	11,968	25	68	0.04	0.11	2.0	5.44	9,968
Feb	3,920	5,200	4,160	11,315	25	68	0.04	0.11	2.0	5.44	13,888
Mar	5,040	5,500	4,400	11,968	25	68	0.04	0.11	2.0	5.44	18,928
Apr	6,720	5,200	4,160	11,315	25	68	0.04	0.11	2.0	5.44	25,648
May	6,720	5,300	4,240	11,532	25	68	0.04	0.11	2.0	5.44	32,368
Jun	6,720	4,700	3,760	10,227	25	68	0.04	0.11	2.0	5.44	39,088
Jul	6,720	4,500	3,600	9,792	25	68	0.04	0.11	2.0	5.44	45,808
Aug	6,160	4,400	3,520	9,574	25	68	0.04	0.11	2.0	5.44	51,968 56,000
Sep	5,040	4,400	3,520	9,574	25	68	0.04	0.11	2.0	5.44	1,008
Oct	3,360	4,800	3,840	10,445	25	68	0.04	0.11	2.0	5.44	4,368
Nov	2,240	5,300	4,240	11,532	25	68	0.04	0.11	2.0	5.44	6,608
Dec	1,680	5,300	4,240	11,532	25	68	0.04	0.11	2.0	5.44	8,288

* Not including tile drainage generated through 1980 which goes to Grass Lands.

TABLE 2E

Assumed Input Flows and Quality Factors
Northern Area — Crows Landing to Delta Year 2000

Month	Q/Mo AF	EC µmhos /cm	TDS mg/1	TDS lbs/AF	Inorganic N mg/1	Inorganic N lbs AF	Inorganic P mg/1	Inorganic P lbs/AF	BOD$_5$ mg/1	BOD$_5$ lbs/AF	Cumulative Flow Sept Base AF
Jan	900	3,900	3,120	8,486	10	27.2	0.15	0.41	0.5	1.36	5,250
Feb	2,100	3,600	2,880	7,833	10	27.2	0.15	0.41	0.5	1.36	7,350
Mar	2,700	3,800	3,040	8,268	10	27.2	0.15	0.41	0.5	1.36	10,050
Apr	3,600	3,600	2,880	7,833	10	27.2	0.15	0.41	0.5	1.36	13,660
May	3,600	2,960	8,051	8,051	10	27.2	0.15	0.41	0.5	1.36	17,250
Jun	3,600	3,300	2,640	7,180	10	27.2	0.15	0.41	0.5	1.36	20,850
Jul	3,600	3,200	2,560	6,963	10	27.2	0.15	0.41	0.5	1.36	24,450
Aug	3,300	3,000	2,400	6,528	10	27.2	0.15	0.41	0.5	1.36	27,750
Sep	2,700	3,300	2,640	7,180	10	27.2	0.15	0.41	0.5	1.36	30,000 450
Oct	1,800	3,300	2,640	7,180	10	27.2	0.15	0.41	0.5	1.36	2,250
Nov	1,200	3,700	2,960	8,051	10	27.2	0.15	0.41	0.5	1.36	3,450
Dec	900	3,700	2,960	8,051	10	27.2	0.15	0.41	0.5	1.36	4,350

most saline water is maintained only in a limited area of the total evaporation site and salt storage likewise occurs only in a limited area. This will allow marsh habitat to become established in a large portion of the evaporation ponds and facilitate conversion of evaporation ponds to marsh habitat (the increase in soil salinity will be minimized.)

c. Marsh operation (water depth, time of flooding and drain-ing-flushing procedures, levee maintenance practices, and irrigation) will maximize waterfowl babitat.

To carry out the indicated marsh program would require a system of the type shown in Figure 4 in spite of the fact that it greatly complicates the apparent simplicity of the marsh concept.

Studies of factors such as nitrification and dentrification, nitrogen fixation, contribution to nutrients of decomposing marsh polants, nutrients from birds and other wildlife, and concentration of nutrients and other chemical components due to an excess of evaporation over precipitation characteristic of the western San Joaquin Valley in Figure 4 type systems are summarized in Table 3.

From Table 3 the indication is that although a great reduction in the volume of water would be accomplished in Type III marshes the reductions are accompanied by increased concentration of all conservative substances to an extent that is undesirable at the drain outlet. A summary type projection of effluent quality under the type III marsh regime is presented in Table 4.

For a detailed development of the complex analysis leading to Table 4 refer to the original report from which portions of this paper are drawn (17).

One alternative to marshes for fixed nitrogen removal is microalgal culture. Instead of converting fixed nitrogen gas, as occurs in marshes, microalgae take up nitrate and incorporate it into their cell protein thereby rendering it harvestable and reclaimable (18).

In the scheme to be described here each drainage area would be associated with a microalgal growth system, optimized for rapid algal growth with the primary purpose of removal of nitrate from saline drainage water and the secondary purpose production of methane which could be used as an energy source in the central valley. A byproduct would be nitrogen rich digester residue which could be used to replace or supplement the chemical ferilizers now used at ever increasing expense.

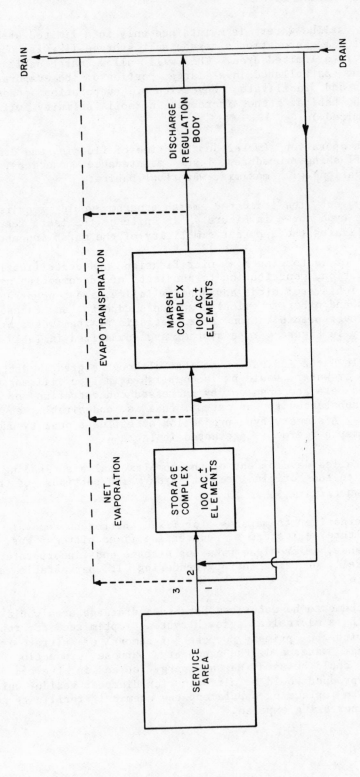

Fig. 4.
Assumed Flow diagrams for service area, storage complex,
marsh complex, discharge regulation body, and drain.

TABLE 3: Estimation of Marsh and Reservoir Losses with the Southern Extension and Resultant Flow and TDS to the Drain.

Marsh	Inf. Q. Ac-Ft/Yr	Influent Reservoir Active Area [1] Ac	Influent Reservoir Enet Ft/Yr	Influent Reservoir Loss Ac-Ft/Yr	Marsh Active Area Ft/Yr	Marsh Loss Factor Ft/Yr	Marsh Loss Ac-Ft/Yr
South Kern	22,000	300	4.29	1,287	2,700	4.17	11,259
North Kern	12,000	200	5.08	1,016	1,400	4.38	6,132
Tulare	79,000	1,000	5.08	5,080	9,500	4.38	41,610
Mendota	64,000	800	4.62	3,696	7,900	4.12	32,548
Los Banos	56,000	700	4.04	2,828	7,400	3.69	27,306
North Area	30,000	500	2.83	1,415	4,400	3.02	13,288

Marsh	Effluent Reservoir Active Area Ac	Effluent Reservoir Enet Ft/Yr	Effluent Reservoir Loss Ac-Ft	Total Loss Ac-Ft/Yr	Total Discharge Ac-Ft/Yr	Service Area to Influent Storage TDS mg/L	Service Effluent Storage to Drain TDS mg/L	Q to Drains Ac-Ft/Day
South Kern	400	4.29	1,716	14,262	7,738	6,400+510	18,195+1,449	21.2
North Kern	200	5.08	1,016	8,164	3,836	6,400+510	20,020+1,580	10.5
Tulare	1,300	5.08	6,604	53,294	25,706	8,406+660	25,833+2,015	70.4
Mendota	1,100	4.62	5,082	41,326	22,674	5,231+417	14,714+1,177	62.1
Los Banos	1,100	4.04	4,444	34,578	21,422	4,006+337	10,472+879	58.7
North Area	700	2.83	1,981	16,684	13,316	2,806+222	6,321+0,499	36.5

[1] Active influent and effluent reservoir areas are assumed to be 1/2 of the actual area to allow for periods of dryness. However, they are rounded off to the next highest 100 acres, in recognition of the fact that losses will occur until the reservoir is actually empty.

[2] Based on $C_2 = \dfrac{V_1}{V_2}$

TABLE 4: Summary of Projected Concentrations of Substances in Drainage Water at Suisan Bay under the Indicated Conditions

Estimated Projected Mean Values in Milligrams per Liter

Parameter	With Drain at Suisan without Southern Extension — No Power Plant	With Drain at Suisan with Southern Extension			
		No Power Plant	1 Power Plant[1]	2 Power Plants[1]	3 Power Plants[1]
5-day biochemical O$_2$ demand[2]	20 ± 7	20 ± 7	20 ± 7	20 ± 7	20 ± 7
Total Org. Carbon	20 ± 7	25 ± 9	20 ± 15	20 ± 15	20 ± 15
Orthophosphate	1.0 ± 0.6	1.25 ± 0.7	1.25 ± 6	1.5 ± 7	1.5 ± .7
Algal Biomass	17 ± 5	17 ± 5	17 ± 5	17 ± 5	17 ± 5
Ammonia Nitrogen	1.3 ± 1.0	1.7 ± 1.25	1.7 ± 1.25	1.3 ± 1.0	1.0 ± 0.8
Organic Nitrogen	4.0 ± 1.5	5 ± 1.09	6 ± 1.9	7 ± 2	8 ± 2.5
Inorganic Phos.	0.2 ± 0.14	0.25 ± 0.18	0.3 ± 0.20	.35 ± 0.23	0.40 ± 0.26
Organic Detrius	39 ± 14	50 ± 18	50 ± 18	50 ± 18	50 ± 18
Chlorophyll a	0.17 ± 0.05	0.17 ± 0.05	0.17 ± 0.05	0.17 ± 0.05	0.17 ± 0.05
Silica (dissolved)	8.0 ± 2.5	10 ± 3.0	11.6 ± 3.5	13.5 ± 4.5	15.5 ± 4.5
Total Suspended Solids	65 ± 18	80 ± 23	80 ± 23	80 ± 23	80 ± 23
Total Dissolved Solids	12,000 ± 1,000	15,500 ± 1,200	18,000 ± 1,400	21,000 ± 1,700	25,000 ± 2,000
Nitrate Nitrogen	17 ± 12	22 ± 15	22 ± 15	20 ± 15	20 ± 15

[1] Assuming that water passing through power plants is completely treated to remove particulates and particulates are disposed separately.

[2] BOD of residually warmed water is exerted more rapidly but is inhibited by salt conc.

The details of microalgal growth systems have been described elsewhere (19) but are briefly reviewed here for completness. Required are shallow (15-15 cm) channelized ponds with a hard lining and with provision for continuous mixing by linear flow with velocities in the range of 10 to 15 cm per second. Paddle wheels have been found to be ideal for such mixing because their gentle action also tends to enhance bioflocculation of the algae and their subsequent sedimentation and concentration in settling chambers. Culture depth is controlled by overflow weirs and inputs are adjusted so that the residence time in the growth pond is 2-5 days depending on the solar energy flux and temperature.

By varying depth the concentrations of microalgae to be grown in the ponds can be adjusted to the concentration of nitrate to be removed. The projected nitrate concentration in drainage water from the service areas varies from 10 mg/liter in the Kern County area to 30 mg/liter in the San Luis service area. Thus in the Kern area to reach a desirable lower concentration of 3 mg per liter of nitrate 7 mg per liter of nitrate would need to be incorporated in algal cells and removed, whereas in the San Luis area 27 mg/liter would need to be incorporated and removed. Inasmuch as algae are about 8% nitrogen the indicated concentrations are 7/.08 = 87.5 mg per liter at Kern and 27/.08 = 337 mg/liter at San Luis.

Experimental work indicates that attainable concentrations are inversely propotional to culture depth, the relationship being approximately:

$$d = \frac{6000}{Cc}$$

in which d is the depth in cm and Cc the concentration of algal cells in mg per liter. From equation 1 the calculated culture depth at Kern would be 6000/87.5 = 68 and at San Luis 17.8 cm. We will use 60 cm for Kern and 18 cm for San Luis in the following presentation.

A second requirement is that sufficient time be allowed to permit algal cells to incorporate visible solar energy in their tissues to provide the required concentrations of biomass. The expression for minimum required residence time Θ in days, is:

$$\Theta = \frac{dhCc}{1000F(S-So)}$$

in which d is the depth in cm, h the heat of combustion of the algae (usually 5.5 cal/mg), Cc is the algal concentration in mg/l, F is the solar energy conversion efficiency expressed as a decimal fraction, S is the total solar energy flux in $cal/cm^2/day$ and So is the total solar energy flux at which productivity approaches 0.

This flux is apparently a function of temperature, the relationship being:

$$So = 8T$$

in which T is the temperature in degrees centigrade. Because of variations in light, temperature and the volume of drainage water one must examine the residence time and other factors required each month of the year to determine the area of land to be set aside for algae growth systems.

This is done for Kern and San Luis areas in Tables 5 and 6. It is interesting to note that because the nitrogen concentration in the Kern area is 1/2 of its magnitude at San Luis, Kern algae ponds can be 3.23 times as deep as the San Luis algae ponds and therefore for 1/2 the flow, require only 15 percent of the San Luis area (161 acres compared with 1105 acres).

Projected areas of flow through microalgae ponds are compared with projected marsh sizes in Table 7.

DISCUSSION

The problems involved in design, construction and operation of the marsh-drain system make it apparent from the foregoing analyses that implementation of the regimen prescribed for the marshes in Figure 4 will require a physical plant and control system of unprecedented size and complexity. Designs will have to include not only reservoirs and marshes but also pumping stations and distribution and collection for control systems. Construction and grading for the marshes will require substantial amounts of precise surveying, grading, and compaction. A large system of roadways will be required for inspection, operation, and maintenance of the elements, and mosquito control will be a major public health requirement. The integrated operations and maintenance involved in the filling and drawing of multiple 100-acre ponds (in the case of Tulare and the Mendota Marshes, approximately 100 100-acre marsh elements and up to 40 100-acre storage elements each) will require a high level of mangement of conduits and channels. Because of this complexity, a rigorous engineering feasibility study of the system should be made before a final commitment is made to build such a system. Preliminary engineering designs will be required for the entire system, construction costs determined, the system subjected to hypothetical model operations for O & M costs, and to permit determination of the monitoring and control systems involved. A comprehensive engineering study of the design model and the operations model would take several years and perhaps more than a million dollars for its completion but is an essential

TABLE 5
Kern County Service Area Predicted N Concentration
10 mg/1 Required Algal Concentration 87.5 mg/1

Month	S	T	So	S-So	1000F	(1)	dhCc	Θ days	Q(2) Acft/day	V(3) Acft	A(4) Acres
Jan	183	8.6	69	114	55	6,270	28875	4.6	33	152	76
Feb	284	10.6	85	199	53	10,547	"	2.7	85	229	114
Mar	348	12.5	100	248	51	12,648	"	2.3	99	227	114
Apr	482	15.7	126	356	49	17,444	"	1.65	236	224	112
May	537	20.2	162	375	47	17,625	"	1.63	131	213	106
Jun	531	25.0	200	331	45	14,895	"	1.93	136	262	131
Jul	568	29.6	237	331	45	14,895	"	1.93	131	253	126
Aug	496	28.5	228	268	47	12,596	"	2.29	120	275	137
Sep	400	26.6	213	187	49	9,163	"	3.15	102	321	161
Oct	302	21.2	170	132	51	6,732	"	4.28	66	282	141
Nov	218	14.6	117	101	53	5,353	"	5.39	45	242	121
Dec	180	9.3	74	106	55	5,830	"	4.95	33	163	82

(1) 1000F (S-So)
(2) Q acre ft/day
(3) V = QΘ
(4) A = QΘ/d

TABLE 6
San Luis Service Area Predicted N Concentration
30 mg/1 Required Algal Concentration = 337 mg/1

Month	S	T	So	S-So	1000F	(1)	dhCc	Θ days	Q(2) Acft/day	V (3) Acft	A(4) Acres
Jan	180	7.4	59	121	55	6,655	33368	5.1	61.9	316	533
Feb	281	9.5	76	205	53	10,865	"	3.1	160	496	838
Mar	342	11.4	91	251	51	12,801	"	2.6	186	484	817
Apr	480	14.6	117	362	49	17,738	"	1.9	256	486	820
May	536	18.7	150	386	47	18,142	"	1.8	248	446	753
Jun	530	23.1	185	345	45	15,525	"	2.1	256	537	907
Jul	555	27.8	222	333	45	14,985	"	2.3	248	570	962
Aug	490	26.2	210	280	47	13,160	"	2.5	227	567	957
Sep	395	24.0	192	203	49	9,947	"	3.4	192	653	1,103
Oct	300	19.1	153	147	51	7,497	"	4.4	124	546	922
Nov	215	12.8	102	113	53	5,989	"	5.6	85	476	804
Dec	178	8.3	66	112	55	6,160	"	5.4	62	335	566

(1) 1000F (S-SO)
(2) acre feet per day
(3) V = QΘ
(4) A = V/d   d = 0.592 ft

TABLE 7. Summary of Projected Areas for Flow Through Algal
Ponds and Type III Fill and Drain Marsh Systems

Area	Projected Peak Flow Ac-ft day	Projected N Conc. mg/l	Projected Required Algae Ponding Area Acres	Projected Marsh plus Inlet and Outlet Area Acres
Kern County	135	10	161	6,300
Tulare Lake	316	10	362	14,100
San Luis Area				
Mendota Marsh	256	30	1,103	11,700
Delta Mendota				
Los Banos Marsh	224	25	791	11,000
Northern Area	120	10	130	6,800
Totals	1,031	−	2,547	49,900

element of future planning to arrive at a rational estimate of
costs, benefits, and cost benefit ratios for this system and for
certain alternatives.

One method of improving downstream quality in the systems
studied would be to decrease the size of the marshes and to
increase the size of the drain. Another would be to establish
multiple intensive treatment systems in conjunction with each
drainage area. These would be short-detention-time algal systems
which could quickly remove nutrients without greatly increasing
TDS. Such treatment systems, operated with short retention periods,
would minimize evaporation and contamination through wildfowl
importation of nutrients. Type II wildfowl marshes could be
operated in series with such systems if desired but would increase
costs.

It is quite important to recognze that there is little in the
literature about nitrification and dentriification at high salinities or about nitrogen fixation at high salinites.

Both wildfowl nutrient import and microbial dentrification are also assumed in this report to occur systematically in the marshes and holding reservoirs, and these assumptions have had a powerful influence on marsh plant growth and the projected quality results. Another area of doubt is the utility of marsh plant yield data in conjunction with projected evapotranspiration. Unfortunately, both marsh plant growth and wildfowl nutrient contributions are speculative as to consistency and magnitude. To determine the real impact of these factors, it would be necessary to establish an experiment of sufficient scale and duration to examine these phenomena and to accumulate data with which to make more substantial future projections. At least one 100-acre storage reservoir and one 100-acre marsh should be operated for several years to obtain such information. Effluent storage reservoirs would be difficult to model and would require an even larger system.

The evidence presented herein indicates that the use of marshes managed as indicated in Figure 4 will have profound influences on the physical, chemical, and biological quality of drainage waters, some beneficial but most adverse. There is little question regarding the projected changes in total dissolved solids concentrations and volume, but the changes in other, less conservative materials such as nitrogen and phosphorus would need further examination in large-scale experimental systems.

The "fill and draw" operation of the marshes contributes greatly to the degradation of quality and to the projected size, costliness, and complexity of the system. It follows that, if further long-range projection studies are made, consideration should be given to quality changes which would result from the operation of the elements as Type II marshes (as opposed to the Type III marshes assumed in this study). Reconsideration should also be given to the fast-treatment systems without the marshes and following the marshes in which newer techniques for microalgal growth and separation and more salt-tolerant species of algae are subjected to cost benefit evaluation. Important breakthroughs in algal propagation and separation have been made in the past few years, and these should not be ignored or considered comparable to the first studies in the late 1960's (10).

While it has apparently been established that continuously submerged marshes improve water quality (5, 6, 7, 13, 14), this treatment is either a short-term phenomenon or the operation of marshes as fill and draw systems is entirely different from continuous submergence. There is little question that integration of nutrients in peaty biomass, in well-buried or chemically bound sediments, or conversion to gases will affect their removal, but the intermittent wetting and drying proposed for Type III marshes and reservoirs operated as in Figure 4 will produce an effluent greatly concentrated and enriched with organic, as well as inorganic carbon and nitrogen. If wild fowl, either indigenous or migrant, frequent the system, it will also integrate phosphorus and, without silt burial, chemical binding, or gasification, the phosphorus sediments and decomposing vegetation will eventually release more phosphorus than is in the original input waters. The final water from such systems will almost certainly require treatment for nutrient and detrius removal before it can be used in ion exchange beds, cooling towers or ponds, in power plants, or be discharged into limited receiving bodies. All of these factors certainly must be carefully studied, as to technical, engineering, economic, and environmental feasibility before a final rational decision can be made to construct the proposed system.

In the case of microalgae systems an established technology exists for their growth, their separation and their fermentation to methane. Extensive descriptions of this technology are available elsewhere (20).

Estimates of the amounts of microalgae that could be grown on drainage water are speculative but the amount should be on the order of 50,000 tons per year with a gross energy content of 10^{12} BTU and as much as 5×10^{11} BTU of reclaimable energy in the form of methane. At $5.00 per 10^6 BTU the value would be 2.5×10^6 dollars which would amortize a 25 million capital investment in algae ponds. If such ponds could be built for $10,000 per acre, their cost could be amortized by the energy they produce and only the land and the operation and maintenance of the systems would require taxpayer support.

Since the marshes required are 20 times as extensive an area as the algae systems their capital cost would need to be as low as $500 per acre to be competitive. It is unlikely that 100-acre elements with pumps and piping could be constructed for less than $3,000 per acre. In this case the cost of the marsh system would approach $100,000,000 with little capital return, and considerable O & M for their complex operation. Actual detailed engineering designs of each system would be required before accurate cost comparisons and cost benefit analysis could be made. It is difficult to imagine that a marsh system would have tangible benefits that offset the apparent degradation of drainage water quality. Pilot experimental work should provide needed answers both with respect to marshes and microalgal systems.

Acknowledgements

This work was supported by the California State Water Resources Control Board, Planning Division, Sacramento, California.

I am indebted to Mr. Peter Lee of the Department of Water Resources and Mr. Doug Albin of the California Department of Fish and Game for their support during this work. I also wish to acknowledge Mr. Louis Beck and Dr. Randolph Brown of the Resources Agency for their interest in this work.

References

1. Anon. "Effects of the San Joaquin Master Drain on Water Quality of the San Francisco Bay and Delta," Central Pacific Basins Comprehensive Water Pollution Control Project, U.S. Dept. of the Interior, Federal Water Pollution Control Admin., S.W. Region, San Francisco, CA (1967).

2. W. J. Oswald, D. J. Crosby and C. G. Golueke, "Removal of Pesticide and Algal Growth from San Joaquin Drainage Waters" (A Feasibility Study), Rpt. of San Joaquin Dist., CA Dept. of Water Res. (1964).

3. D. G. Crosby, "The Photodecomposition of Pesticides in Water, Fate of Organic Pesticides in the Aquatic Environment," Advances in Chemistry Series, Amer. Chem. Soc. III (1972).

4. Anon. Collected Papers Regarding Nitrates in Agricultural Wastewaters, Federal Water Quality Admin., U.S. Bureau of Reclamation, Calif. Dept. of Water Resources, Project No. 13030 ELY (1969).

5. Anon. Wastewater Reclamation and Reuse, Pilot Demonstration Program for the Suisun Marsh, California, Progress Rpt., U.S. Dept. of Interior, Bur. of Reclamation (1974).

6. Anon. Ibid., Progress Rpt. (1975).

7. J. F. Arthur and M. Ball, "Entrapment of Suspended Materials in the San Francisco Bay-Delta Esturay, U.S. Dept of Interior, Bur. of Reclamation, Sacramento, CA (1978).

8. M. D. Ball, "Phytoplankton Growth and Chlorophyll Levels in the Sacramento-San Joaquin Delta through San Pablo Bay," Rpt. to the U.S. Dept. of Interior, Bur. of Reclamation, Mid-Pacific Region Water Quality Branch, Sacramento, CA (1977).

9. J. R. Benemann and M. A. Murray, "Review of the Literature on Biomass and Productivity of Aquatic Vascular Plants," Summary of Biomass and Productivity of Emergent Aquatic Plants (1978).

10. R. L. Brown and L. A. Beck, "Field of Evaluation of Anaerobic Dentrification in Simulated Deep Ponds," Bioengineering Aspects of Agricultural Drainage, CA State Dept. of Water Res., Bull. 174-3 (1969).

11. R. L. Brown, Ibid., 174-10 (1971).

12. D. N. Cederquist and M. Rumboltz, Ibid., 174-18 (1976).

13. N. Cederquist, "Wastewater Reclamation and Reuse," Pilot Demonstration Program for the Suisun Marsh, Calif. Prog.

Rpt., U.S. Dept. of Interior, Bur. of Reclamation, Mid-Pacific Reg. Water Quality Branch, Sacramento, CA (1977).

14. F. C. Demgen and B. Lubaugh, "Marsh Enhancement," Pilot Program, Progress Rept. No. 3, Mt. View Sanitary Dist., P. O. Box 2366, Martinez, CA 94553 (1977).

15. J. H. Ives, C. R. Hazel, P. Gaffney and A. W. Nelson. "An Evaluation of the Feasibility of Utilizing Agriocultural Tile Drainage Water for Marsh Management in the San Joaquin Valley, California, Prepared for U.S. Fish and Wildlife Service and U.S. Bur. of Reclamation by Jones & Stokes Assoc., Inc. and Hydroscience, Inc. (1977).

16. L. A. Beck, P. Lee and D. Albin, "Agricultural Drainage and Salt Management in the San Joaquin Valley Interagency Drainage Program, U.S. Dept. of Interior, Bur. of Reclamation, Calif. Dept. of Water Resources, Calif. State Water Resources Control Board, Sacramento, CA (1979).

17. W. J. Oswald, "Projected Changes in Quality of San Joaquin Valley Subsurface Drainage Waters in a Proposed Marsh and Canal Transit System." Planning Div., Calif. State Water Resources Control Board, Sacramento, CA (1978).

18. W. J. Oswald, "Pilot Plant High-Rate Pond Study of Waste Treatment and Algal Production," Rpt. V to World Health Organization, Reg. Ofc. for the Western Pacific, United Nations Ave., 12115 Manila, Philippines (1978).

19. W. J. Oswald, "Engineering Aspects of Microalgae." In: "Handbook of Microbiology," Vol. II, Fungi, Algae, Protozoa, and Viruses, pp. 519-552, CRC Press, Inc., Florida (1979).

20. W. J. Oswald, D. M. Eisenberg, J. R. Benemann, R. P. Goeble and T. Tiburzi, "Methane Fermentation of Microalgae," in: "Proc. 1st Intl. Symposium on Methane Fermentation," pp. 123-135, Univ. of Cardiff, Wales (1979).

CONTRIBUTED PAPERS

AIR LAYERING METHOD FOR VEGETATIVE PROPAGATION OF JOJOBA

(Simmondsia chinesis)

M. L. Alcaraz

Centro de Investigaciones Biologicas
   de Baja California
La Paz, Baja California Sur, Mexico

In Baja California the saline nature of the soil and the scarcity of fresh water for agriculture is a serious problem as the non-saline soil and water used for this purpose are becoming exhausted. Therefore, it is necessary to study the possibility of the utilization of plants with a natural ability for growth in saline soil and drought water. One plant with such characteristics is jojoba (Simmondsia chinensis), a wild shrub originally of the Sonoran Desert which requires only 5-12 inches of rain annually (4).

The physicochemical similarity of the liquid wax of jojoba seeds and whale sperm (3) makes this plant also valuable as a raw material for industry, in addition to its evident ecological importance.

S. chinensis (Link, Schneider) is a dioecious plant. Flowering occurs after two to three years of germination. Throughout this time, its sex remains unknown (2). Since seed production is determined by the plant sex, its forecasting is important. In crop lands the optimal male : female distribution for high productivity has been found to be 1:5 (2). Thus, asexual propagation might constitute the best resource for culture planning. This problem has been approached by different techniques such as culture tissue (2, 8), grafting (5), cuttings (6, 7) and layering (5). In this paper the air layering method is discussed.

Jojoba plants three years old grown from seeds of wild shrubs around La Paz, Baja California Sur Bay were used. Air layering

method was followed by means of an anular ripping of 2 cm in the
stems, 15-20 cm from the apical extreme. A mixture of lanoline and
1.5×10^{-2} or 2×10^{-2} M Indol-3-Butyric Acid (IBA) (Sigma Chemical
Co., St. Louis, MO, USA) was applied. The ripping was covered with
50% soil and 50% vermiculite contained in 15 x 10 cm black
polyethylene sheets with both ends tied. Three layerings per plant
and 23 repetitions of hormonal treatment were carried out.   They
were watered every three days using tap water.   After 113 days the
air layerings were examined and those which developed the longest
roots (2 cm length roots) were cut from the mother plant.   Another
dose of IBA was applied before transplanting to flower pots with
50% soil and 50% vermiculite.

Figure 1 shows the IBA effect in jojoba rooting after 113
days.   An inducing effect by this hormone in root development was
clearly observed.   However, no great differences in rooting development at the two hormone concentations used, could be detected.
The lengths of adventicies roots are about 2-15 cm and these were
shown to be longer and stronger than those in the control plants.

A total of 36 plants were transplanted and 28 were still alive
after 2 months.   These results indicate an efficiency of 77% in
transplanting.

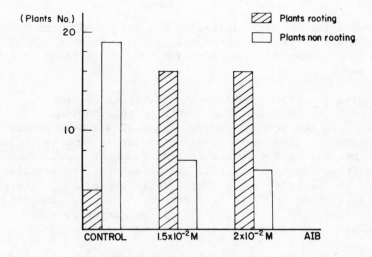

Fig. 1. Root development in Jojoba Plants after 113 days.

In comparison with cutting, this method is less expensive because it does not require special installations such as a greenhouse with constant mist and controlled temperature. Thus, air layering might be considered a good alternative for plant propagation in arid and semi-arid zones.

ACKNOWLEDGEMENTS

The assistance of Dr. Felix Cordoba Alva and Biologist Teodoro Reynoso G. in the preparation of this manuscript is gratefully acknowledged. I also thank Irene Munoz and Albertina Cota for permitting the experiments in the jojoba plot they developed.

REFERENCES

1. E. Birnbau, in: " Proc. Second Int'l. Conf.
    on Jojoba and its uses," Ensenada, Mexico, Ed. Council of
    Science and Technology (CONACYT) pp. 91-94 (1976).
2. M. De la Vega, "La Jojoba, domesticacion de un cultivo poten-
    cial," Centro de Investigaciones del Noroeste INIA, SARH,
    Campo Agricola Experimental de la Costa de Hermosillo,
    Circular CIANO No. 92, (SARH) Mexico (1977).
3. K. Foster, A processing scenario for alternative jojoba
    products, Arizona Jojoba Conference, University of Arizona.
    Tucson, Octubre 15-16 (1979).
4. H. S. Gentry, Economic Botany 12:261-295 (1958).
5. C. J. Hansen and H. T. Hartmann. Propagation of temperate zone
    fruit plants. University of California. CA Agricultural
    Experiment Station. Extension Service. Circular 471. pp
    37-40 (1966).
6. L. Hogan, C. W. Lee, A. D. Palzkill and R. W. Feldman.
    Recent progress in the propagation of jojoba by stem
    cuttings, Proc. Third Int'l. Conf. on Jojoba, Riverside. CA.
    pp 1-4 (1978).
7. A. A. Maisai. Factors affecting the rooting and transplanting
    of jojoba Simmondsia chinensis (Link) Shneid. Master's
    thesis, Univ. of Arizona (1966).
K. J. Tautvdas, in Proc. Third Int'l. Conf. on Jojoba.
    Riverside. CA. pp. 25-28 (1978).

THE TOTAL LIPID CONTENT OF MICROALGAE AND ITS USE AS A POTENTIAL

LIPID SOURCE FOR INDUSTRY

T. Berner[1], A. Dubinsky[2] and S. Aaronson[1]

[1]Biology Dept., Queens College, City University
   of New York, Flushing, N.Y. 11367 USA
[2]Dept. of Life Sciences, Bar-Ilan University
   Ramat-Gan, Israel

Lipids (oils, fats and their derivatives) are an important commodity in world trade. Lipids of higher plants and animals presently serve as sources of food, oil for cooking, surfactants (soaps, etc.), fatty nitrogen compounds, industrial grease, surface coatings and are used in the rubber, plastic, textile, food, cosmetic pharmaceutical and other industries (1). World production of lipids reached 51.5×10^6 metric tons by 1975 (2). The value of the world's imports and exports of vegetable oils (soft and non-soft) came to \$4.1 and \$3.9 $\times 10^9$, respectively, in 1975 (3). The present sources of lipids are mainly plant (ca. 70% of the total) and animal (ca 30%) and these sources are highly vulnerable to scarcity and increased cost because they depend on the earth's depleting supply of petroleum for a large part of the energy for their production and for plant fertilizer. In the recent past when a nation's supplies of lipids for food and industry were threatened, as for example Germany in the First and Second World Wars, they turned to lipid production by fungi grown on cheap available nutrients (2). Algae were also considered but not utilized. With increasing awareness that the earth's energy and raw material resources are limited (Aaronson, this volume), we must find new sources of raw materials such as lipids, etc. that will spare or replace our depleting energy stores and raw materials. We suggest here that microalgal lipids may, in time, replace the use of higher plant and animal lipids to be used solely for human use. The use of lipids is not a novel idea for others suggested it more than twenty-five years ago but the time was not propitious (4).

Microalgae have been associated with and possibly responsible for some types of petroleum-containing rock for millions of years.

The green alga, Botryococcus braunii, is associated with Bog head
coal, lignite and paleozoic oil bearing rocks (5) and coccolith
blooms appear to be the origin of the North Sea oil in the Kimmer-
idge Clay (6). More recently we noted (7, 8) that higher plant
agriculture and animal husbandry would continue to require larger
amounts of fuel and fertilizer making the lipid product increasingly expensive. We suggested that microalgal sources be exploited
for lipids. In this paper we offer further evidence for the use of
microalgae as a source for lipids.

Examination of the scientific literature for total lipid production by algae proved to be somewhat frustrating for various
workers extracted lipids by at least 17 different methods ranging
from the highly nonpolar solvents: ether, petroleum ether, hexane-
chloroform, to the relatively polar solvent mixtures: chloroform-
methanol-water (1:3:1, v/v) and methanol. As these solvents
extract different lipids as well as other molecules from cells, and
as the algae used varied enormously in age, growth conditons, etc.,
the total lipids reported were only tentative amounts. Nevertheless, a number of generalizations may be made from the data of
295 assays on 184 species summarized in Table 1: photosynthetic
bacteria contain more lipid than other bacteria (perhaps they contain more membrane); most microalgae generally, but not always,
contain significantly more lipid than all macroalgae examined; macroalgae were unusually low in lipid and this bears further investigation for some macroalgae contained trace amounts; several microalgal species were unusually rich in lipids (over 20%) and these
may be likely candidates for exploitation for lipids (Table 1).
Microalgae may store lipid as an energy reserve. The lipid content
of microorganisms is subject to environmental as well as internal
variation in vitro and in nature due to nitrogen depletion, gaseous
atmosphere, temperature, light intensity, and senescence (7). Only
nitrogen depletion and senescence seem to increase the lipid content significantly. Unfortunately no single algal species has been
examined in detail for environmental manipulations of lipid
content.

The major sources of lipids containing polyunsaturated fatty
acids for industry are presently plant oils. Several microalgal
species, however, can be grown in bulk and their unsaturated fatty
acids content compares reasonably well with that of plant oils
(Table 2). In many microalgae, more than 25% of the total fatty
acids were polyunsaturated (9). It is pertinent to note here that
microalgae may be involved in the stripping of nutrients from
wastewater in high rate sewage oxidation ponds where large quantities of microalgae may be harvested and usable water for agriculture and/or industrial use may be generated (10, 11). The lipid
content of the microalgae in several high rate sewage ponds
has been determined and significant quantities of lipids useful for

TABLE 1: Range of total lipids in photosynthetic microorganisms.

Microbial Groups	No. of Assays	Total Lipids (% Cell dry wt) Range	Mean	No. of Species with over 20% Lipid
Prokaryota				
Bacteria, non-photosynthetic	17	2-18	8	0
Bacteria, photosynthetic	12	9-23	13	1
Bacteria, blue-green	27	1-29	8	1
Eukaryota				
Bacillariophyceae	52	2-44	17	19
Chlorophyceae	101	0-79	17	24
Chrysophyceae	2	29-35	33	2
Cryptophyceae	5	1-44	19	1
Dinophyceae	11	3-36	17	4
Euglenophyceae	6	10-37	21	1
Phaeophyceae	23	0-9	3	0
Prasinophyceae	3	3-21	9	1
Prymnesiophyceae	14	2-48	21	7
Rhodophyceae	15	0-14	3	0

TABLE 2: A comparison of microalgal fatty acids with major
           commercial oils and fats.

Source	Fatty Acid Composition (% w/w)* 14:0	16:0	16:1	18:0	18:1	18:2	18:3	20:1+
Microalgae**								
Blue-green bacteria		6+	5+	+	4+	4+	4+	0
Cryptophyceae	2+	3+	2+	+	+	2+	3+	4+
Bacillariophyceae	2+	3+	5+	+	+	+	+	4+
Chlorophhyceae	+	3+	2+	+	3+	4+	3+	+
Commercial oils and fats***								
Soybean oil	tr	2+	-	+	3+	6+	+	tr
Cottonseed oil	tr	3+	tr	+	2+	6+	tr	tr
Groundnut oil	tr	2+	-	+	5+	3+	-	+
Palm oil	tr	5+	tr	+	4+	+	-	-
Coconut oil	2+	+	-	+	+	+	-	-
Linseed Oil	-	+	-	2+	2+	2+	5+	-
Rapeseed oil	-	+	tr	tr	2+	2+	+	6+
Animal lard	+	3+	+	2+	4+	+	tr	-
Fish oil (herring)	+	2+	+	+	+	+	tr	5+

*Key: up to 10% = +   30% = 3+   50% = 5+   tr = trace 1%,
         20% = 2+   40% = 4+   60% = 6+   or less
**   Ref. 8.
*** Ref. 2.

industry may be obtained (up to 23%) (7). It may be argued that the lipids of algae grown in domestic wastewater might be enriched in hydrocarbon pesticides, etc. and this is likely if they are found in the wastewater but these same hydrocarbons accumulate in extracted oil of soybeans from fields treated with these pesticides (12).

The growth of microalgal biomass and the harvesting of lipids represents potential for harvesting solar energy products and supplying industrial raw materials and possibly reutilizable wastewater while minimizing the utilization of expensive fossil fuels and fertilizer and doing a necessary societal function - nutrient stripping of domestic wastewater to prevent eutrophication. The problems that are involved in the utilization of microalgal biomass are discussed elsewhere in this volume (Dubinsky and Aaronson).

The writing of this paper was supported by a grant No. PFR-7919669 from the National Science Foundation.

References

1. A. G. Johanson, J. Am. Oil Chem. Soc., 54:848A-852A (1977).
2. C. Rutledge, in: "Economic Microbiology", Vol. 2, (A. H. Rose, ed.) pp. 263-302, Academic Press, London (1978).
3. 1976 Yearbook of International Trade Statistics, United Nations, N.Y. (1977).
4. H. W. Milner, in: "Algal Culture from Laboratory to Pilot Plant", (J. S. Burlew, ed.) pp. 285-302, Carnigie Inst. of Washington, Publ. 600, Wash., D. C. (1953).
5. J. R. Maxwell, A. G. Douglas, G. Eglinton and A. McCormick, Phytochemistry, 7:2157-2171 (1968).
6. R. W. Gallois, Nature, (Lond.), 259:473-475 (1976).
7. Z. Dubinsky, T. Berner and S. Aaronson, Biotech. Bioeng. Symp., 8:51-68 (1979).
8. S. Aaronson, T. Berner and Z. Dubinsky, in: "The Production and Utilization of Microalgae Biomass", (G. Shelef, C. J. Soeder and M. Balaban, eds.), Elsevier/North-Holland Biomedical Press, Amsterdam (1980).
9. B. J. B. Wood, in: "Algal Physiology and Biochemistry", (W. D. P. Stewart, ed.), pp. 236-265, University of California Press, Berkeley (1974).
10. W. J. Oswald and C. G. Golueke, in: "Algae, Man and the Environment", (D. F. Jackson, ed.), pp. 371-390, Syracuse Univ. Press, Syracuse (1968).
11. R. Moraine, G. Shelef, A. Meydan and A. Levin, Biotech. Bioeng, In Press.
12. M. M. Chaudry, A. I. Nelson and E. G. Perkins, J. Amer. Oil Chem. Soc., 55:851-853 (1978).

GROWTH CONDITIONS AND PROTEIN SYNTHESIS BY A MODERATE

HALOPHILIC BACTERIA CIB-1

T. Castellanos*, R. Sanchez*, F. Mendoza[+],
G. Alfaro[+], and L. Diaz de Leon[+]

*Centro de Investigaciones Biologicas
 La Paz, Baja California Sur, Mexico
[+]Dept. de Biologia del Desarrolo
 Inst. de Investigaciones, Biomedicas, UNAM
 Mexico City

INTRODUCTION

One of the most outstanding features of marine and saline bacteria is their capacity to grow in an environment containing relatively high salt concentrations. This implies changes in their genic expression, protein synthesis and structure of their constitutive proteins and enzymes to overcome such an environment hostile to other organisms. An increase in acidic residues and a deficiency of hydrophobic residues in the protein of extremely halophilic bacteria stabilize these structures in high concentrations of salt (1). Many proteins show marked preference for KCl over NaCl (2). This is consistent with large gradients of K^+ and Na^+ across the cell membranes of the halobacteria. Maximal enzyme activities and protection against thermal inactivation are affected by different salt concentrations (3, 4). A recent review (5) describes the development of a cell-free system derived from H. halobium and some of the properties and features of protein synthesis in halophilic bacteria. In the present communication the growth conditions, temperature, salt tolerance and protein synthesis were studied in a bacterium tentatively classified as moderately halophilic: CIB-1, isolated from La Paz Bay. This bacterium was compared with an extremely halophilic bacterium H. halobium, from which total RNA was also isolated and its mRNA activity evaluated in a reticulocyte derived cell-free system. The Halobacterium halobium bacteria R-1 strain, was kindly provided by Dr. Walter Stoeckenius and cultures were grown in 213 halobacterium medium (ATCC Catalog) at 25°C.

RESULTS

Protein biosynthesis by H. halobium cultures. When bacteria
were cultured in M-9 medium a 70% decrease in cell growth was
observed, compared to that obtained in 2216 medium (Fig. 1a). This
decrease was also accompanied by a diminution of total protein
synthesis in the M-9 medium cultures (Fig. 1b). Protein synthesis
was linear throughout the growth cycle in cells cultured in
complete medium. In contrast, the M-9 medium cultures reached a
maximum at the beginning of the stationary phase, and then showed a
significant decrease during the late stationary phase. When the
incorporated ^{3}H-leucine was determined in the different fractions,
it was observed that membrane fraction and secreted protein showed
similar patterns of synthesis as for total proteins. In both
conditions, the soluble fraction had the lowest incorporation of
the isotope.

Growth and protein synthesis by CIB-1. Before the labeling
experiments were performed, the effects of varying NaCl concentrations on cell growth were studied in CIB-1 bacteria. Similar
growth curves were obtained at 25°C in absence and with 1M NaCl
(Fig. 2a), but growth was significantly retarded in 2M NaCl.
Subcultures adapted to 1M NaCl also showed a better growth than no
salt cultures grown in 1M NaCl. The values obtained for total
protein synthesis were higher at 1M NaCl than in the no salt
cultures, but "adapted" cultures showed the highest incorporation
of ^{3}H-leucine (Fig. 2b). The lowest incorporation of label was
obtained in cultures grown in 2M · NaCl. Similar to the results
obtained with H. halobium, the membrane fraction and secreted
proteins had the major incorporation of the isotope. In all
fractions analysed protein synthesis was linear throughout the
growth cycle, except for the "adapted" cultures that reached a
maximum at the beginning of the stationary phase.

Total RNA and mRNA activity in H. halobium. When total RNA
was quantitated from the different growth phases, a linear increase
in the 213 Medium condition was observed. In contrast, the RNA
in the cells grown in M-9 medium reached a plateau at the beginning
of the stationary phase, the amounts of RNA being approximately 30%
of the values found in 213 medium cells. Nevertheless, when total
RNA was expressed as µg RNA/O.D. 560, it was observed to remain
relatively constant throughout the growth cycle in the M-9 medium
cultures, whereas in the 213 medium its value decreased initially
and augmented progressively to reach the M-9 cultures value at the
end of the growth cycle (Fig. 3a). When total RNA from H. halobium
was evaluated for its ability to promote protein sysnthesis in a
reticulocyte derived cell-free system, it showed little activity of
mRNA over the background levels. However, PAGE-electrophoresis of
cell-free products obtained with RNA isolated from cultures of both

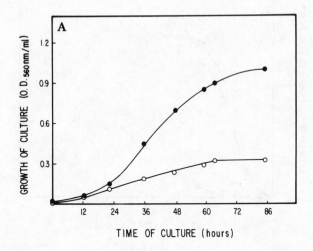

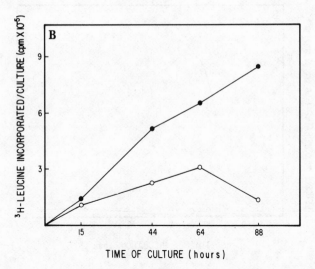

Fig. 1 A) Growth curves of Halobacterium halobium cultured in 213
medium (● - ●) and M-9 medium supplemented with casamino
acids (o - o).

B) Total protein synthesis by H halobium cultures during the
growth cycle. Cells cultured in 213 medium (● - ●) and in
M-9 medium plus casamino acids (o - o).

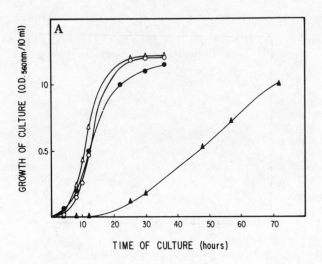

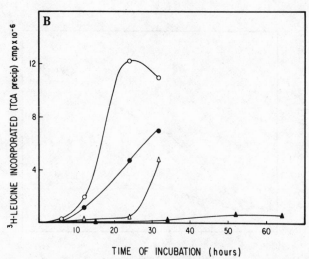

Fig. 2 A) Growth curves of CIB-1 bacteria cultured at different NaCl concentrations: (● - ●) Basal, no NaCl added; (o - o) 1 M NaCl; (Δ - Δ) 1M NaCl "adapted" subcultures; (▲ - ▲) 2 M NaCl.

B) Total protein synthesis by CIB-1 bacteria cultured under different NaCl concentrations: (Δ - Δ) Basal, no NaCl added; (● - ●) 1 M NaCl; (o - o) 1 M NaCl "adapted" subcultures; (▲ - ▲) 2 M NaCl.

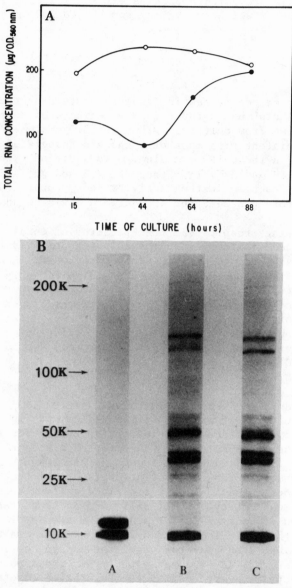

Fig. 3 A) Changes in total RNA/OD$_{560}$ during growth cycle of H. halobium cultured in: (●-●) 213 Medium and (o-o) M-9 Medium plus casamino acids.

B) Slab gel electrophoresis of H. halobium RNA cell-free products labeled with ^{35}S-methionine. (A) No RNA added; (B) 5 μg of RNA from cells grown in 213 Medium; (C) 5 μg of RNA from cells grown in M-9 medium plus casamino.

M-9 or 213 media, revealed approximately 14 radioactive bands with apparent molecular weights ranging from 15K to 200 K daltons (Fig. 3b).

DISCUSSION

The observed decreases in growth, protein synthesis and RNA content of H. halobium cultured in M-9 medium when compared to the values obtained from cultures grown in 213 medium can be interpreted as consistent with metabolic changes associated to depletion of nutrients in the culture medium. In addition, the M-9 medium contains less K^+ and Mg^{+2} ions than the 213 medium, and this could be a crucial factor in the activity and proper functioning of enzymes and protein needed for an adequate division and growth.

A surprising result was the patterns of total protein synthesis obtained in the different cultures of CIB-1 bacterium. Subcultures "adapted" to 1M NaCl showed the best incorporation of ^{3}H-leucine whereas the incorporation was significantly retarded in the basal cultures (without NaCl) (Fig. 2b). These results can be interpreted as: 1) A better uptake of ^{3}H-leucine, and availability of the incorporated isotope for protein synthesis; 2) Larger pools of free leucine in the basal cultures with dilution of the isotope and an apparent retardation in rates of protein synthesis; and 3) A stress induced change due to NaCl with increase in leucine-rich proteins at 1M NaCl.

In regard to the changes in total RNA/O.D. 560 observed with H. halobium cultures grown in complete medium, the initial decrease in RNA/O.D. could be due to a lesser dependence for rRNA or tRNA compared to the M-9 condition. However, the most striking observation was the expressed capacity of H. halobium RNA to promote protein synthesis in an eukaryotic derived cell-free system. It has been reported that halophilic derived cell-free systems function only in concentrations of salt approaching saturation (6). This situation of course is far apart from the optimal conditions employed in the reticulocyte translation system. Although it has been implied that the aparatus for protein synthesis in halobacteria resembles that in eukaryotes more than prokaryotes (5), there are other groups suggesting that the halophiles together with the methanogens comprise a third kingdom, named archaebacteria (7). Thus, the observed mRNA activity in a relative low salt containing translation system, makes it interesting to explore the nature of the translation products compared to those obtained in halophilic cell-free systems.

Acknowledgements

We wish to acknowledge the help of Dr. W. Stoeckenius for reviewing this manuscript.

References

1. J. K. Lanyi, Salt tolerance in microorganisms, in: "The Biosaline Concept." (A. Hollaender, J. C. Aller, E. Epstein, A. San Pietro and O. R. Zaborsky, eds.) pp. 217-233 Associated Universitites, Inc., Washington, D.C. and Plenum Press, N.Y. (1979).
2. J. K. Lanyi, Bacteriol. Rev. 38:272-290 (1974).
3. M. M. Lieberman and J. K. Lanyi, Biochem. 11:211-216 (1973).
4. J. S. Hubbard and A. B. Miller, J. Bacteriol. 99:161-168 (1968).
5. S. T. Bayley and R. A. Morton, CRC Crit. Rev. Microbiol. 6: 151-205 (1978).
6. S. T. Bayley and E. A. Griffiths, Biochem. 7:2249-2256 (1968).
7. C. R. Woese, Proc. Nat. Acad. Sci. USA 74:5088-5090 (1977).

ARSENATE DETOXICATION BY DUNALIELLA TERTIOLECTA

R. V. Cooney, J. M. Herrera-Lasso and A. A. Benson

Scripps Institution of Oceanography
La Jolla, CA 92093 USA

INTRODUCTION

Arsenic is found in the world's oceans at a relatively constant concentration of approximately 2×10^{-8} M (1). Surface waters of much of the open ocean as well as some tropical reef areas contain phosphate concentrations less than or equal to the arsenate concentration (2). Because of the chemical similarity between the two ions, arsenate is absorbed by algae through the phosphate transport system (3). Marine algae have been shown to accumulate arsenic to levels in excess of 100 ppm (4). However, values are commonly found on the order of 20 to 30 ppm in regions of higher phosphate concentration (5). Accumulation of a toxic element such as arsenic would eventually lead to decreased productivity and possibly cell death if allowed to continue. Consequently, marine algae have evolved a mechanism for the detoxification of arsenate which involves methylation and conversion into a variety of organo-arsenic compounds.

Interestingly, the conversion of arsenic into a membrane lipid appears to be a uniquely marine phenomenon. Although some fungi and bacteria are capable of arsenic reduction and methylation (6,7) there are no reports of the formation of organo-arsenicals by freshwater algae. Because of this we sought to examine the effect of salinity on arsenic metabolism in Dunaliella tertiolecta. This organism was chosen for both its known salt tolerance and its unique ability to convert more than 90% of absorbed arsenic into lipid-soluble forms.

Results and Discussion

Dunaliella tertiolecta readily absorbs arsenate which is then converted primarily into one of three different lipid compounds. One of these, a phospholipid (R_f = 0.65), was identified as O-phosphatidyltrimethylarsoniumlactate and is found in most marine algae we have examined (8). The other main arsenic lipid (R_f = 0.8) is of yet unknown structure, but contains no fatty acid esters. The third compount (R_f = 0.97) chromatographs with the neutral lipids and pigments. To observe the effect of salt concentration on Dunaliella tertiolecta, a series of cultures ranging in salinity from 0% to 10% were prepared. Each contained 10 μM phosphate, 20 μM nitrate, 1.0 μM arsenate (10^6 cpm/nanomole) and an equal number of cells. The tubes were illuminated and shaken for 24 hours, at which point the cells were alternately centrifuged and washed and then extracted with ethanol. After counting an aliquot of the ethanol extact in a scintillation counter, the sample was dried under nitrogen and partitioned with $CHCl_3$: CH_3OH : H_2O (2:1:1). Each phase was then counted (Table 1).

The chloroform extract of each sample was chromatographed on a silica gel-60 TLC plate with chloroform (130) : methanol (70) : water (8) : ammonia (0.5) as the developing solvent. An auto-radiogram was made of the plate and each spot observed was counted (Table 2). Figure 1 illustrates the effect of salinity on the amount of arsenic contained in the neutral lipid fraction.

External salt concentration appeared to have little or no effect on the uptake of arsenic. Some slight stimulation of uptake was observed at salinity values below that of seawater. From 3% to 10% salt concentrations, aresenic uptake remained the same. Additionally, the ratio of arsenic distributed between lipid and water-soluble compounds remained constant for all samples (94%), with the exception of those grown in distilled water which were apparently lysed and were unable to metabolize arsenic.

Chromatographic analysis of the lipid fraction indicated that the relative amounts of the two major arseno-lipids were not affected significantly by salinty. Conversely, the [74]As-labeled compound that chromatographed with the neutral lipids and pigments was produced in direct relation to the external salt concentration. This may represent the production of an arsenic analogue of a non-organic containing compound normally produced in response to osmotic stress. Less likely is the possibility that a unique organoarsenic compound is produced with a specific function in osmotic regulation.

Although arsenate is absorbed into the cell because of its similarity to phosphate, once inside the cell it is converted to lipids in which the arsenic more closely resembles nitrogen with regard to function. This has particular significance in those areas of low nutrient availability where utilization of arsenic in

TABLE 1. Influence of salinity on the uptake and formation of
          arsenolipids in D. tertiolecta.

S%	$HAsO_4^=$ uptake (10^{-1} nanomoles)	% of ^{74}As in lipid
0	0.08	–
1	3.6	95
2	3.3	94
3	2.5	94
4	2.5	93
6	2.8	93
8	2.8	93
10	2.8	93

TABEL 2. Distribution of ^{74}As-activity between various
          arsenolipids.

S%	Arsenolipids (%)		
	Phospholipids I $R_f = 0.65$	Lipid II $R_f = 0.80$	Lipid III $R_f = 0.97$
0	–	–	–
	59	40	1.2
2	55	43	2.0
3	56	40	3.5
4	62	34	4.0
6	59	35	5.5
8	58	35	6.6
10	57	36	7.5

certain compounds such as lipids may free more nitrogen for other
metabolic needs which can use only nitrogen.   The salt induced
arsenic compound we have observed may therefore have evolved in
response to the dual conditions of low nutruient availability and
high salt concentration.   This implies that a non-arsenic

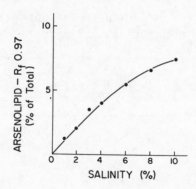

Fig. 1. Effect of salinity on the production of a nonpolar
         arsenolipid.

containing lipid analogue may also be produced in response to
osmotic stresss.   Clearly, additional investigation is warranted
into the structure of this salt-induced compound as well as the
general effects of salinity of lipid production.

        This work was supported by a grant from the Natinal Science
Foundation #OCE 78-25707 and Consejo Nacional de Ciencia y
Technologia, Mexico.

References

1. H. Onishi, Handbook of Geochemistry, (K.. H. Wedepohl, ed.),
       Vol. II-2, Chapter 33.
2. D. L. Johnson and M. E. Q. Pilson, J. of Marine Research
       38:140-149 (1972).
3. J. G. Sanders and H. L. Windom, Estuarine and Coastal Marine
       Science 10:555-568 (1980).
4. G. Lunde, J. Sci. Food Agric. 21:416-418 (1970).
5. J. G. Sanders, Estuarine and Coastal Marine Science 9:95-99
       (1979).
6. F. Challenger and C. Higginbottom, Biochem. J. 29:1757-1778
       (1935).
7. B. C. McBride and R. S. Wolfe, Biochemistry 10:4312-4317
       (1971).
8. R. V. Cooney, et al, Proc. Natl. Acad. Sci. 75:4262-4264 (1978)

PROTEIN BIOSYNTHESIS BY Bouvardia ternifolia

CELL CULTURES ADAPTED TO NaCl

L. Diaz de Leon, L. Arcos, J.L. Diaz de Leon[1], H. Soto

Departamento de Biologica del Desarrollo, Instituto de
Investigaciones Biomedics, UNAM, Mexico City and
[1] Centro de Investigaciones Biologicas, La Paz, B.C.,
Mexico

INTRODUCTION

In a previous report from our group an increase in the intracellular free proline pools was observed and also a decrease in
protein-bound hydroxyproline during NaCl stress in cultures of a
nonhalophytic plant, B. ternifolia (1). Considerable accumulation
of free proline in plants during water deficit or salinity stress
has been reported for several species (2). It has been suggested
that proline acts as a solute for intracellular osmotic adjustment
(3), as well as a storage compound for energy and reduced nitrogen
and carbon to be used during post-stress metabolism (4). It has
been also reported that an increased accumulation of proline
parallels a better resistance to water stress. The principal
source of proline accumulated under stress is de novo synthesis
from glutamate as well as a decrease in proline catabolism (5).
Changes in protein synthesis also have been reported for tissue
culture cells and protoplasts under severe osmotic stress. In the
present communication the protein synthetic capacity of B.
ternifolia cells during the growth cycle was studied in cell
suspension culture by incorporation of ^{3}H-leucine. Possible
changes in total RNA content were also examined. Basal cultures
and cells grown in 100 mM NaCl were compared for their ability to
incorporate ^{3}H-proline into proteins and for the variations in
their intracellular free-proline pools during the growth cycle.

RESULTS

Protein biosynthesis in basal cultures. The incorporation of
^{3}H-leucine was measured as an index for total protein synthesis and

also for the synthesis of constitutive and secreted proteins. Total protein synthesis was maximal at the beginning of the stationary phase. When the incorporated ^{3}H-leucine was determined in the different fractions, it was observed that the percentage of secreted proteins remained constant throughout the whole cycle and incorporation was the lowest of all the fractions. The membrane fraction (10K rpm pellet) showed the highest incorporation of the isotope. The results obtained are illustrated in Figure 1a. Differential and total protein synthesis did not increase considerably after reaching the stationary phase. When the proteins synthesized by the cultures at the stationary phase were analysed by SDS-PAGE electrophoresis, the fluorographs revealed approximately 6-12 radioactive proteins in the different fractions (Fig. 1b). The apparent molecular weights of the radioactive proteins varied from 15K to 210K daltons. Although there were some coincident bands in the soluble and membrane fractions, in general the electrophoretic pattern was quite distinct for each fraction.

Incorporation of ^{3}H-proline in basal cultures and cells grown in 100 mM NaCl. Prior to the evaluation of the protein synthesizing capacity of the cultures, it was decided to compare the growth curves in the two conditions in order to use equivalent stages of the growth cycle in the labeling experiments. Growth was retarded at 100 mM NaCl and reached confluency at about 60% of the levels observed in the basal cultures (Fig. 2a). In spite of this difference, when total protein synthesis was determined throughout the growth cycle, similar profiles were obtained. Total protein synthesis was maximal at mid-logarithmic phase (Fig. 2b), in contrast with the results obtained with ^{3}H-leucine in which maximal protein synthesis occurred at the beginning of the stationary phase. When the incorporated ^{3}H-proline was analysed in the different fractions, the membrane fraction had the highest incorporation of the isotope. When the values obtained in the two cultures were compared, a gradual decrease was observed in the incorporation of ^{3}H-proline in the membrane fraction of 100 mM NaCl cultures contrary to the increased radioactivity in basal cultures. This decrease showed a slight correlation with an increase of ^{3}H-proline incorporated in the 10K soluble fraction. No significant differences were found in the protein patterns when the different fractions were analysed by SDS-PAGE electrophoresis.

Changes in total RNA. When the values obtained for total RNA isolated from basal cultures was compared, a decrease in the amount of total RNA/mg dry weight was observed at mid-logarithmic phase. Nevertheless, it returned to the initial value at the end of the growth cycle. Attempts to evaluate the activity of total RNA to promote protein synthesis in a cell-free system derived from rabbit reticulocytes were not successful.

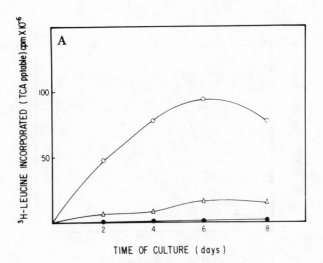

Figure 1A. Differential protein synthesis in B. ternifolia cells
measured by incorporation of ^{3}H-leucine into: (o-o)
membrane fraction; (Δ-Δ) soluble proteins, and (●-●)
secreted proteins.

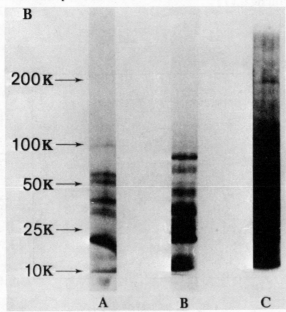

Figure 1B. Slab gel electrophoresis of ^{3}H-leucine-labeled proteins
synthesized by B. ternifolia cells. A) Secreted
proteins, B) Soluble proteins, C) Membrane fraction.

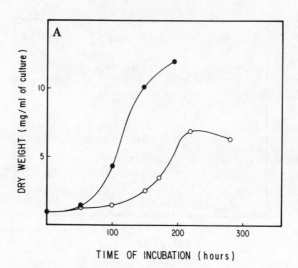

Figure 2A.   Growth curves of B. ternifolia cells cultured in: (●-●)
basal medium and (o-o) 100 mM NaCl containing medium;
growth conditions were as described by Fernandez and
Sanchez (7).

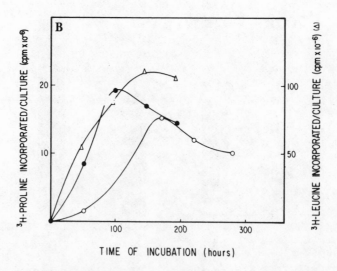

Figure 2B.   Total protein synthesis by B. ternifolia cells cultured
in different conditions; (●-●) Basal medium with
³H-proline ; (o-o) 100 mM NaCl with ³H-proline; (Δ-Δ)
Basal medium with ³H-leucine.

Changes in the intracellular free-proline pools. A 3 fold increase in the levels of free proline was observed in the 100 mM NaCl cultures compared with the basal cells. Although the intracellular pools decreased in both cultures (Table I), the diminution was more pronounced in the basal cultures, whose values at the end of the growth cycle were approximately 5 times lower than those observed in the 100 mM NaCl cultures.

TABLE I: Changes in the Intracellular Free Proline Pools During The Growth Cycle of B. ternifolia Cells and Cultures Grown at 100 mM NaCl[a]

Growth of culture (hours)	Intracellular free-proline[b] (pg/g dry weight)	
	Basal cultures	100 mM NaCl cultures
52	612.90	2125.00
100	469.21	2401.00
148	497.52	-
171	-[c]	983.80
194	225.59	-
219	-	715.36
279	-	1552.00

[a]The values represent the average results obtained from duplicate experiments.

[b]Proline was quantitated by the method of Rojkind and Gonzalez in (8). All determinations were carried out by triplicate.

[c]Proline was only determined in the different cultures to be employed for the labeling experiments.

DISCUSSION

The experiments of incorporation of ^{3}H-leucine in basal cultures showed that protein synthesis reached a maximum at the beginning of the stationary phase and that the membrane fraction had the highest incorporation of the isotope (Fig. 1a). These results can be correlated with an increase in cell division during the preceding growth phase. The incorporations of ^{3}H-proline in basal and 100 mM NaCl cultures when compared, demonstrated a 25% decrease in total protein synthesis in the 100 mM NaCl cells. Although the patterns of labeling were quite similar in both cultures, the decrease in synthesis can be explained by a specific compartmentalization of proline into the vacuoles. Thus, less proline would be available for protein synthesis in the 100 mM NaCl situation. The major changes at 100 mM NaCl occurred in the experiments of differential protein synthesis, where a decrease in the incorporation of ^{3}H-proline into the membrane fraction was observed and this diminution correlated with an increase in the soluble fraction. Although the attempts to evaluate the mRNA activity of the isolated RNAs were unsuccessful, mainly due to starch contamination during the isolation procedure which interfered with the cell-free translation, the variations in total RNA reflect the metabolic changes occurring during the growth phases. In regard to the free-proline pools, although a progressive decline in the values was observed during the growth cycle, the decrease was more pronounced in the basal cultures. The higher levels of intracellular free proline cannot explain the 25% decrease in total protein synthesis in the 100 mM NaCl cultures, because the isotope dilution should be greater if all the proline were available for protein synthesis. In preliminary experiments the conversion of the incorporated ^{3}H-proline into ^{3}H-hydroxyproline was determined. (Results not shown). The Pro/Hyp ratio increased considerably in the 100 mM cultures indicating a decrease in hydroxyproline-containing proteins. This observation could be of relevance, because it has been reported that hydroxyproline-proteins regulate several aspects of morphogenesis (6).

ACKNOWLEDGEMENTS

We wish to thank Dr. E. Sanchez for allowing us to maintain the cultures in our laboratory. The cell line was originally derived at Depto. De Bioquimica, Facultad de Quimica, UNAM.

REFERENCES

1. J. L. Diaz de Leon, H. Soto, M. T. Merchant and L. Diaz
    de Leon, this volume.
2. T. N. Singh, D. Aspinall and L. G. Paleg, Nature New Biology
    236:188-189 (1972).
3. C. R. Stewart and J. A. Lee, Planta 120:279-280 (1974).
4. C. R. Stewart, C. Morris and J. F. Thompson, Plant Physiol.
    41:1585-1590 (1966).
5. C. R. Stewart, S. F. Boggess, D. Aspinall and L. G. Paleg,
    Plant Physiol. 59:930-932 (1977).
6. D. V. Basile, Amer. J. Bot. 66:776-783 (1979).
7. L. S. Fernandez and E. J. Sanchez, Physiol. Plant. (In press).
8. M. Rojkind and E. Gonzalez, Anal. Biochem. 57:1-7 (1974).

BIOCHEMICAL AND ULTRASTRUCTURAL CHANGES INDUCED BY NaCl

IN CELL CULTURES DERIVED FROM Bouvardia ternifolia

J. L. Diaz de Leon, H. Soto, M. T. Merchant,
and L. Diaz de Leon

Departamento de Biologia del Desarrollo,
Instituto de Investigaciones Biologicas,
La Paz, B. C., Mexico

INTRODUCTION

Different experimental systems have been developed to study the various effects of water and salt stress in a variety of plants (1,2). However, the majority of the studies reported in the literature have employed either whole plants or differentiated tissues such as roots, leaves or seeds (3,4). Although in some studies cultures derived from non-halophytic plants have been used, there are only a few reports on the effects of NaCl on plant cell cultures. The possibility to obtain partially resistant or tolerant cultures to NaCl from glycophytes have been explored by other groups (5). The advantages of using more defined conditions in cell cultures to study the effects of NaCl on plant growth, metabolism and morphology were employed to obtain a better knowledge of these effects on a medicinal non-halophytic plant, Bouvardia ternifolia, from which very little is known. The present study describes the biochemical and morphological changes induced by different concentrations of NaCl in cell cultures derived from leaf calluses of B. ternifolia.

RESULTS

Effects of NaCl on growth and cell viability. Basal cultures of B. ternifolia reached the stationary phase about the eighth day of culture (Fig. 1a). In contrast, when NaCl was added a 50% inhibition in the growth of cultures was observed at 50 mM NaCl.

Although cells survived higher salt concentrations growth was poor
and very slow.   Media inoculated with equal number of cells (1 x
10^5 cells) were cultured at different NaCl concentrations for a ten
day period.   Cell viability was determined and compared with basal
cultures.   A 50% decrease in cell viability was observed at 150 mM
NaCl and   it went   down   progressively with   increasing NaCl
concentrations (Fig. 1b).

Changes in free-proline pools and in protein-bound   hydroxyproline. Based on   the results   obtained with cell   suspension cultures, in which a marked decrease in cell growth and cell viability
was observed even at low salt concentrations, it was decided to use
cultures grown in solid media for these experiments.   The determination of dry weight,   free proline and in protein   hydrolysates are
presented   in Table I.   Although cell   growth was   better in solid
media, and considering that the variations in wet   weight accounted
for less than 5% when comparing all the samples taken, the decrease
of about 40% in dry weight more pronounced at 50 mM NaCl is consistent with   the   observed inhibition   in growth   and   cell viability
(Figs. 1a and 1b).   Nevertheless,   the   free-proline pool   showed
almost   a three fold increase at   50 mM NaCl and the   highest value
was observed at 250 mM NaCl, in which free hydroxyproline was also
detected.   When protein-bound   proline and hydroxyproline were
determined in the different samples, it was observed   that protein-
bound proline after an initial increase at 50 mM NaCl did   not show
considerable variations throughout all the NaCl concentrations.   In
contrast, protein-bound hydroxyproline decreased from 15.46 mg/g at
50 mM to less than 1 mg/g at 250 mM NaCl.

Morphological changes induced by NaCl.   Light   microscopy
revealed   an extensive   derangement of intercellular   lamellae with
increasing NaCl concentrations (Fig. 2).   Nuclei became concentric,
round shaped and   poorly stained with surrounding vacuoles.   Lysis
of cells with cytoplasmic contractions was also observed at   200 mM
NaCl.   These observations   were confirmed   by electron microscopy.
In addition, at 50 mM NaCl many mitochondria   could be observed
surrounding the nucleus.   Granules   containing   electron-dense
material appeared at 100 mM NaCl and became prominent at 150 mM
NaCl.

DISCUSSION

Cell   cultures   derived   from   B. ternifolia   showed   a   great
sensitivity to NaCl stress.   This sensitivity was manifested by
decreases in cell growth, cell viability and morphological changes.
The difficulty to culture cells in NaCl was partially overcome by
increasing the initial cell concentration (Results not shown).

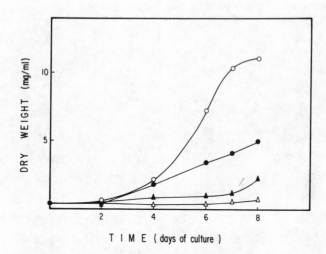

Figure 1A. Growth curves of B. ternifolia cells cultured at
            different concentrations of NaCl. Growth conditions
            were as described by Fernandez and Sanchez (7).

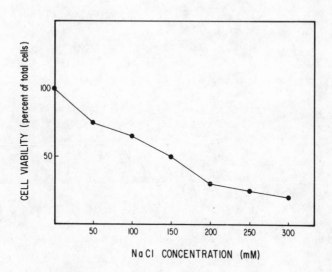

Figure 1B. A plot of cell viability expressed as percent of total
            cells from cultures grown at different concentrations
            of NaCl.

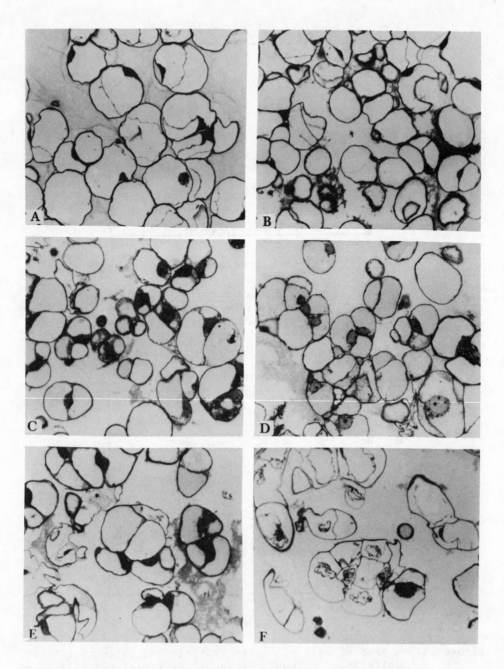

Figure 2.   Light   microscopy   photographs   of   B.   ternifolia   cells
cultured   at: (A) Basal;   (B) 50   mM   NaCl;   (C)   100 mM
NaCl;   (D) 150 mM NaCl;   (E)   200 mM NaCl and (F)   300 mM
NaCl.

TABLE I: Effect of NaCl Concentration on Free Proline Pools and the
          Levels of Hydroxyproline in Total Proteins from B.
          ternifolia Cultures[*]

| NaCl mM | Sample dry weight (mg) | Pool of free proline (μg/g) | Protein hydrolizates[**] | | Pro/Hyp |
			mg Pro/g	mg Hyp/g	
0	68.7	382.5	6.85	6.51	1.06
50	33.6	889.1	15.61	15.46	1.00
100	44.1	704.0	10.39	4.62	2.24
150	56.3	551.5	14.24	2.18	6.53
200	46.1	648.5	12.79	1.06	12.06
250	41.9	1858.1	15.81	0.99	15.96

[*]The values represent the average results from at least three
different experiments.

[**]Proline and hydroxyproline were determined according to the method
described by Rojkind and Gonzalez (8).

Although the growth of "adapted" cells was retarded when compared to
basal cultures, these subcultures showed a slight tendency to regain
their normal morphology. An interesting result was the observed
decrease of protein-bound proline and the decrease in hydroxyproline. The increases in the Pro/Hyp ratio can be interpreted as
consistent with a decrease in hydroxyproline-containing protein (s)
due to salt stress. It has been implied that hydroxyproline-
proteins regulate several aspects of morphogenesis by suppressing
growth and development at critical times (6). The results obtained
in the present study support the use of cell cultures as a better
experimental system to study the effects of salt and water stress in
non-halophytic plants.

ACKNOWLEDGEMENTS

    We wish to thank Dr. E. Sanchez for giving us the inoculum and the Departamento de Bioquimica, Facultad de Quimica, U.N.A.M., for providing us all the tissue culture facilities.

REFERENCES

1. B. Heuer, Z. Plaut and E. Federman, Physiol. Plant. 46:318-323 (1979).
2. M. Tal, I. Rosental, R. Abramovitz and M. Forti, Ann. Bot. 43: 701-708 (1979).
3. N. Bar-Nun and A. Poljakoff-Mayber, Ann. Bot. 44:309-314 (1979).
4. S. Treichel, Ber. Deutsch. Bot. Ges. 92:73-85 (1979).
5. M. W. Nabors, A. Daniels, L. Nadolny and C. Brown, Plant Sci. Lett. 4:155-159 (1975).
6. V. D. Basile, Amer. J. Bot. 66:776-783 (1979).
7. L. S. Fernandez and E. J. Sanchez, Physiol. Plant. (In press).
8. M. Rojkind and E. Gonzalez, Anal. Biochem. 57:1-7 (1974).

SALT TOLERANCE OF DROUGHT-SENSITIVE AND DROUGHT-RESISTANT WHEAT[1]

A. O. Faden

Ministry of Agriculture
and Water
Riyadh, Saudi Arabia

M. B. Kirkham

Evapotranspiration Laboratory
Kansas State University
Manhattan, KS   66506   USA

INTRODUCTION

The objective of this experiment was to determine if a drought resistant wheat cultivar was also salt tolerant.

Materials and Methods

Plants were grown in a growth chamber. Two cultivars of winter wheat (Triticum aestivum L. em Thell.) from the Southern Great Plains in the USA, one drought-resistant ('KanKing') and one drought-sensitive ('Ponca'), were grown along with three cultivars from Saudi Arabia, two spring wheats ('Mexipak' and 'Hinta Medaini') and one durum wheat (T. durum Desf. 'Lokaimy'). Plants were irrigated with three concentrations of salt water. One half of the plants received 80 ml irrigating solution per pot every day. One half of the plants received 240 ml solution per pot every three days. Stomatal resistance was used as an indicator of water stress. Growth and uptake of K and Na were monitored. The specific techniques used have been described in the full length paper[1].

Results

Stomatal resistance. Both with and without salt stress, the drought-resistant cultivar, KanKing tended to have a higher stomatal resistance than did the drought-sensitive cultivar, Ponca (Table 1).

467

TABLE 1:   Stomatal resistance of five cultivars of wheat grown in
           sand or soil with salt-water irrigations daily or every
           three days. Ponca is drought sensitive and KanKing is
           drought resistant. Means in each column followed by the
           same letter are not significantly different at the 0.05
           level, according to Duncan's new multiple-range test.

Cultivar	0 mM NaCl		85 mM NaCl		170 mM NaCl	
	Sand	Soil	Sand	Soil	Sand	Soil

sec/cm
Daily Irrigation

Cultivar	Sand	Soil	Sand	Soil	Sand	Soil
Ponca	12.0b	13.1c	8.5a	8.4a	13.7ab	11.0a
KanKing	13.7b	13.6c	16.7c	9.8a	11.9a	9.4a
Lokaimy	12.7b	6.1b	14.1bc	7.0a	14.6ab	10.9a
Hinta Medaini	13.2b	10.5c	12.8ab	8.3a	17.3b	10.2a
Mexipak	7.9a	3.1a	9.3a	7.8a	16.5ab	12.3a

Irrigation Every 3 Days

Cultivar	Sand	Soil	Sand	Soil	Sand	Soil
Ponca	9.9ab	8.8b	9.6ab	2.9a	13.8a	9.9a
KanKing	11.7b	9.8b	14.3c	8.2c	17.4a	10.9a
Lokaimy	9.6ab	7.7b	8.3a	8.4c	15.1a	9.8a
Hinta Medaini	7.5a	8.8b	12.0bc	8.5c	13.6a	9.0a
Mexipak	8.1ab	4.4a	12.3bc	5.1b	13.6a	11.4a

Growth.   More seeds germinated in sand than in soil (data not
shown).   If grown in sand, the drought-resistant cultivar appeared
also salt tolerant, because dry weight of its shoots and roots
tended to be more than that of other cultivars (Table 2).

Potassium.   For all treatments, concentrations in tops of
plants were within normal concentration ranges (0.5-5% Table 3).
Potassium concentration in grain of plants grown with NaCl was
always more than that in grain of control plants.

Sodium.   With NaCl, KanKing had a higher concentration of Na
in roots grown in sand or soil under both irrigation regimes than
did Ponca, even though the difference was significant at the 5%
level in only one case (0.86 vs. 0.18%, Table 4).

TABLE 2: Dry weight of roots, shoots, and grain of five cultivars of wheat grown in sand or soil with salt-water irrigation daily or every three days. Ponca and KanKing winter wheats did not form grain because they were not vernalized. In each column with five cultivars, measurements of the same plant part followed by the same letter are not significantly different at the 0.05 level, according to Duncan's new multiple-range test.

Cultivar	0 mM NaCl Sand	Soil	85 mM NaCl Sand	Soil	170 mM NaCl Sand	Soil
			g			
		Daily irrigation				
Ponca						
Roots	14.56b	7.45bc	5.42b	1.53ab	3.16ab	0.93ab
Shoots	16.84c	12.37c	13.65c	8.59b	9.08bc	5.22b
KanKing						
Roots	44.55c	8.38c	4.55b	3.14b	3.18ab	0.98ab
Shoots	18.69c	13.87c	14.75c	12.57c	11.63c	4.60b
Lokaimy						
Roots	20.45b	5.42bc	4.21ab	1.95b	4.42b	1.22ab
Shoots	15.16bc	10.53bc	11.08bc	5.66b	7.22b	4.15ab
Grain	1.47a	2.86a	1.42a	1.54a	0.20a	0.27a
Hinta Medaini						
Roots	10.25b	3.87b	2.04a	2.26b	3.31ab	1.75b
Shoots	11.77b	8.64b	8.97b	7.13b	8.09b	5.13b
Grain	7.81b	7.29b	0.95a	3.05b	0.45b	0.75b
Mexipak						
Roots	3.13a	1.37a	2.88b	0.72a	1.86a	0.68a
Shoots	5.79a	4.79a	5.09a	2.65a	5.20a	3.23a
Grain	5.74b	5.44b	2.80b	2.84b	2.89c	1.68c
		Irrigation every 3 days				
Ponca						
Roots	16.51bc	4.23bc	3.62a	1.55a	3.45ab	0.95a
Shoots	12.97b	6.25b	12.60c	7.46bc	8.21bc	4.02ab
KanKing						
Roots	9.27ab	11.84d	6.40ab	3.64bc	5.40bc	2.73b
Shoots	14.40b	10.24c	13.12c	7.82c	9.79c	5.38b
Lokaimy						
Roots	19.85c	6.09cd	11.60b	10.23d	8.07c	1.47ab
Shoots	12.37b	7.42b	8.14b	5.62b	9.07bc	3.14a
Grain	3.89a	2.21a	0.65a	1.58a	0.52b	0.58a
Hinta Medaini						
Roots	6.44a	2.61b	6.37ab	6.09cd	7.73c	2.19b
Shoots	10.86b	7.71bc	9.05b	6.58bc	7.04b	3.17a
Grain	7.50b	1.71a	1.65b	1.69a	0.02a	0.42a
Mexipak						
Roots	4.70a	1.11a	5.29a	1.75ab	2.33a	1.86ab
Shoots	4.67a	3.23a	5.50a	1.57a	3.84a	3.24a
Grain	5.45ab	3.89b	3.04c	1.63a	3.31c	1.91b

TABLE 3: Concentration of potassium in roots, shoots and grain of five cultivars of wheat grown in sand or soil with salt-water irrigations daily or every three days. Ponca and KanKing, winter wheats, did not form grain because they were not vernalized. For letters, see legend of Table 2.

Cultivar	0 mM NaCl Sand	0 mM NaCl Soil	85 mM NaCl Sand	85 mM NaCl Soil	170 mM NaCl Sand	170 mM NaCl Soil
			%			
			Daily irrigation			
Ponca						
Roots	0.32b	0.45a	0.16a	0.33b	0.15a	0.31a
Shoots	2.49a	3.62c	2.28b	2.35c	2.15a	1.89ab
KanKing						
Roots	0.26ab	0.32a	0.24a	0.34bc	0.19a	0.29a
Shoots	2.99ab	2.83bc	2.18b	2.35c	2.16a	1.73a
Lokaimy						
Roots	0.22ab	0.51a	0.16a	0.31a	0.17a	0.27a
Shoots	3.29b	2.76b	2.26b	1.80b	2.25a	1.76ab
Grain	1.58b	1.60b	2.45a	2.69b	NES[†]	2.75a
Hinta Medaini						
Roots	0.25ab	0.36a	0.18a	0.54c	0.13a	0.25a
Shoots	2.81ab	2.91bc	2.09ab	1.28a	2.31a	2.14ab
Grain	0.78a	1.21a	2.29a	2.18ab	2.54a	2.44a
Mexipak						
Roots	0.20a	0.36a	0.23a	0.31a	0.15a	0.26a
Shoots	3.21ab	1.99a	1.52a	2.27bc	1.97a	2.23b
Grain	1.64b	1.26ab	2.48a	1.83a	2.31a	2.22a
			Irrigation every 3 days			
Ponca						
Roots	0.40b	0.39a	0.22b	0.40a	0.15a	0.25a
Shoots	3.29b	3.74c	2.12a	1.80a	2.32a	1.87a
KanKing						
Roots	0.39b	0.55a	0.23b	0.40a	0.19a	0.32a
Shoots	2.26a	2.81b	2.12a	1.86a	2.06a	1.83a
Lokaimy						
Roots	0.47b	0.49a	0.19b	0.49a	0.19a	0.30a
Shoots	2.46a	2.37b	2.97b	1.66a	1.87a	1.74a
Grain	1.26a	1.39a	2.34a	2.06b	1.93a	2.23a
Hinta Medaini						
Roots	0.30ab	0.40a	0.21b	0.35a	0.31b	0.28a
Shoots	2.72ab	2.39b	2.27a	1.90a	2.37a	1.64a
Grain	1.22a	1.33a	2.17a	1.95ab	NES	2.28a
Mexipak						
Roots	0.23a	0.56a	0.12a	0.32a	0.13a	0.33a
Shoots	3.13a	0.76a	2.17a	1.71a	2.08a	1.93a
Grain	1.66b	1.18a	2.09a	1.57a	2.01a	1.79a

[†] Not enough sample for analysis.

TABLE 4: Concentration of sodium in roots, shoots, and grain of five cultivars of wheat grown in sand or soil with salt-water irrigations daily or every three days. Ponca and KanKing, winter wheats, did not form grain because they were not vernalized. For letters, see legend of Table 2.

Cultivar	0 mM NaCl		85 mM NaCl		170 mM NaCl	
	Sand	Soil	Sand	Soil	Sand	Soil
			%			
Daily irrigation						
Ponca						
Roots	0.17b	0.08ab	0.48ab	0.50ab	0.43a	0.36a
Shoots	0.39c	0.36b	11.51bc	11.32b	9.28a	9.20a
KanKing						
Roots	0.04a	0.04a	1.07b	1.02b	0.52a	0.40a
Shoots	0.13b	0.54b	14.46c	7.61ab	8.57a	6.23a
Lokaimy						
Roots	0.04a	0.10b	0.55ab	0.33a	0.53a	0.46a
Shoots	0.31b	0.65b	12.02bc	9.58ab	6.79a	6.12a
Grain	0.11b	0.05a	6.87a	7.53a	NES[†]	8.48b
Hinta Medaini						
Roots	0.08ab	0.12b	0.57ab	0.31a	0.39a	0.42a
Shoots	0.16a	0.17a	5.89ab	4.36a	12.26a	10.25a
Grain	0.03a	0.10b	5.82a	5.96a	7.71b	3.16a
Mexipak						
Roots	0.04a	0.10b	0.39a	0.59ab	0.64a	0.45a
Shoots	0.41b	0.35ab	4.63a	6.48ab	10.31a	10.76a
Grain	0.09b	0.08ab	5.83a	4.57a	3.44a	4.57ab
Irrigation every 3 days						
Ponca						
Roots	0.10a	0.09ab	0.77b	0.49bc	0.18a	0.41a
Shoots	0.35ab	1.09c	4.20a	4.45a	6.32b	3.75a
KanKing						
Roots	0.10a	0.12ab	0.85b	0.92c	0.86cd	0.47a
Shoots	0.34ab	0.44b	6.77ab	7.24a	1.79a	4.83ab
Lokaimy						
Roots	0.08a	0.10ab	0.69b	0.26ab	1.00d	0.40a
Shoots	0.24a	0.47b	7.30ab	6.06a	1.24a	9.07b
Grain	0.26b	0.10b	5.00a	4.01a	7.87b	7.82a
Hinta Medaini						
Roots	0.08a	0.17b	0.52b	0.23ab	0.28ab	1.47b
Shoots	0.56bc	0.53bc	12.16b	6.97a	7.88b	8.13ab
Grain	0.17b	0.03a	5.71a	4.00a	NES	6.43a
Mexipak						
Roots	0.05a	0.07a	0.10a	0.17a	0.46bc	0.37a
Shoots	0.84c	0.15a	5.96ab	5.01a	6.16b	4.22ab
Grain	0.06a	0.06ab	4.51a	3.27a	3.52a	7.00a

[†] Not enough sample for analysis.

Discussion

Under saline conditions in a sandy medium, the drought-resistant winter wheat cultivar, KanKing, tended to grow better than did the drought-sensitive winter wheat cultivar, Ponca. The better growth of KanKing compared to Ponca appeared to be due, in part, to its higher stomatal resistance, its larger amount of roots, and its capacity to bind Na in the roots and limit transfer to the shoots. KanKing often produced more dry matter with daily irrigations than with irrigations every three days. Daily irrigations apparently were more effective in leaching salt out of the root zone than the less frequent ones.

In summary, the results suggested that, for maximum growth of wheat under saline conditions, the following steps should be taken:

1. Select a drought-resistant cultivar.
2. Plant seeds in sand.
3. Irrigate frequently.

Acknowledgements

We thank Mr. Ziad Adham, Director General of Research and Development, Plant Protection Division, Ministry of Agriculture and Water, Riyadh, Saudi Arabia, for seeds from Saudi Arabia, and Dr. E. L. Smith, Department of Agronomy, Oklahoma State University, for seeds from Oklahoma.

Footnote

Contribution No. 81-152-A, Dept. of Agronomy, Evapotranspiration Laboratory, Kansas Agric. Expt. Sta., Manhattan, KS 66506. Partial support for work was provided by National Science Foundation Problem Focussed Research Applications, Grant No. NSF ENV77-04092. This is an abbreviated form of the paper. The full-length paper has been deposited in the National Technical Information Service, U. S. Dept. of Commerce, Springfield, Virginia 22161 USA.

HALOPHYTES: NEW SOURCES OF NUTRITION

R. S. Felger[1] and J. C. Mota-Urbina[2]

[1]Arizona-Sonora Desert Museum and
    Environmental Research Laboratory
University of Arizona
Tucson, Arizona   USA

[2]Vaso de Ex-Lago Texcoco
Centro de Investigaciones Forestales Region Central
   (CIFREC)
Instituto Nacional de Investigaciones Forestales
(INIF - SARH), Mexico, D. F.

Our basic premise is the development of agricultural crops and practices to fit the environment rather than modifying the environment to fit the crops. This can largely be accomplished by choosing species already adapted to the diverse environments of the world. Recent investigations in the Sonoran Desert indicate there is indeed a great diversity of species available for development as new crops (1).

It is not known how many species on earth are palatable by man, how many might be adapted to modern agriculture, or how many might be halophytes. However, we can make some crude predictions and indicate areas for investigation. Using the Sonoran Desert as a model, we restrict discussion mostly to food plants with emphasis on species generally not covered by other investigators.

The Sonoran Desert, covering 310,000 km^2 in northwestern Mexico and southwestern United States, has a flora of about 2,500 species of seed plants (2). More than 20% of these have been used by man for food, and about 1.6% of the total flora served as major food resources (3). As a means of comparison, on the order of 131 species (about 5% of the total flora) are halophytes, and at least 38 have been used by man for food (Table 1). Seven species in this list served indigenous peoples as major food resources [Distichlis

TABLE 1:   Halophytes from the Sonoran Desert

MONOCOTS

CYMODOCEACEAE
Halodule wrightii
CYPERACEAE
Eleocharis rostellata
Scirpus americanus
†S. paludosus
GRAMINEAE
*Arundo donax
Cenchrus palmeri
*Cynodon dactylon
†Distichlis palmeri
D. spicata
†Echinochloa crusgalli
Jouvea pilosa
Monanthochloe littoralis
†Phragmites australis
Spartina foliosa
†Sporobolus airoides
S. pyramidatus
†S. virginicus
JUNCACEAE
Juncus acutus
J. cooperi
J. mexicanus
PHOENICACEAE
†Brahea armata
†Sabal uresana
RUPPIACEAE
Ruppia maritima
SCHEUCHZERIACEAE
Triglochin concinna
TYPHACEAE
†Typha latifolia
ZOSTERIACEAE
?Phyllospadix torreyi
†Zostera marina

DICOTS

AIZOACEAE
†*Mesembryanthemum spp.
†Sesuvium verrucosum
†Trianthema portulacastrum
BATACEAE
†Batis maritima
BORAGINACEAE
Heliotropium curassavicum
H. procumbens
CAPARACEAE
Cleomella obtusifolia
†Wislizenia refracta
CARYOPHYLLACEAE
Spergularia macrotheca
S. marina
CELASTRACEAE
†Maytenus phyllanthoides

CHENOPODIACEAE
†Allenrolfea occidentalis
Aphanisma blitoides
Atriplex barclayana
†A. californica
†A. canescens
†A. elegans
A. hymenelytra
A. julacea
†A. lentiformis
?A. linearis
A. magdalenae
A. pacifica
A. parishii
†A. polycarpa
*A. rosea
†*A. semibaccata
†A. serenana
†A. wrightii
Chenopodium album
C. arizonicum
†C. californicum
C. flavellifolium
C. lanceolatum
†C. leptophyllum
†*C. murale
†Monolepis nuttalliana
Nitrophila occidentalis
†Salicornia europea
†S. pacifica
†S. subterminalis
†Suaeda californica
S. fruticosa
S. taxifolia
S. torreyana
COMBRETACEAE
Conocarpus erecta
COMPOSITAE
Alvordia fruticosa
Aster intricatus
Baccharis sarothroides
Cirsium mohavense
Coreocarpus johnstonii
C. parthenioides
C. sonoranus
Encelia lacinata
E. ventorum
Helianthus niveus
Hemizonia pungens
Iva ascerosa
I. hayesiana
Jaumea carnosa
Pluchea purpurascens
†P. sericea
Porophyllum crassifolium
P. maritimum
P. tridentatum

CONVOLVULACEAE
?Cressa truxillensis
†Ipomoea pes-carpae
CUCURBITACEAE
Vaseyanthus brandegei
V. insularis
FRANKENIACEAE
Frankenia grandifolia
F. palmeri
EUPHORBIACEAE
Croton californicus
Ditaxis brandegeei
Euphorbia incerta
E. leucophylla
E. platysperma
LEGUMINOSAE
Astragalus hornii
†Prosopis articulata
†P. torreyana
†P. velutina
MALVACEAE
Gossypium sp.
Sida hederacea
S. lepidota
NYCTAGINACEAE
Abronia maritima
OLACACEAE
Schoepfia californica
S. shreveana
PHYTOLACCACEAE
Phaulothamnus spinescens
Stegnosperma halimifolium
PLANTAGINACEAE
?Plantago heterophylla
PLUMBAGINACEAE
Limonium californicum
PORTULACACEAE
?Calandrina ambigua
?C. maritima
RESEDACEAE
†Oligomeris linifolia
RHIZOPHORACEAE
†Rhizophora mangle
SOLANACEAE
†Lycium brevipes
L. californicum
SAURACEAE
Anemopsis californica
TAMARICACEAE
*Tamarix spp.
*T. aphylla
UMBELLIFERAE
?Cymopterus bulbosus
VERBENACEAE
†Avicennia germinans
ZYGOPHYLLACEAE
Viscainoa geniculata

---

† Species documented to have been utilized by humans for food.
* Non-native, introduced or adventive naturalized species.
? Species probably edible by man.

palmeri, Brahea armata, Zostera marina, Prosopis (3 species) and Lycium brevipes] and seem to have agronomic potential (1). Species listed in Table 1 are those observed by us or reported in the literature to tolerate or grow in saline or alkaline environments. The list includes a range of halophytes, from species restricted to pure sea water, or even hypersaline desert conditions, to those which only range into brackish water or saline/alkaline conditions. Further investigations may show that some are not halophytes, as well as add others to the list.

Several discoveries reported at this workshop will probably greatly influence development of halophyte crops: 1) increasing salinity decreases productivity; 2) salts in vegetative parts can be reduced by soaking them in fresh water, and 3) at least some halophytes do not accumulate or concentrate salts in their seeds (4).   The decrease in productivity with increasing salinity could be a major factor limiting saline or sea water agriculture. We suggest optimizing productivity by choosing from the diversity of halophytic species and culivating them on less than maximum salinity levels, rather than attempting to develop high-yielding salt-tolerant crops from traditional non-halophyte crops. This problem becomes even more acute when non-desert crops are attempted in the desert.

New food crops will generally find best acceptance when developed for animal feed or fodder prior to, or simultaneous with development as new foods for people.   In many cases the vegetative parts may be valuable as fodder and not suitable for human consumption, while the seeds may be desirable food for humans. Furthermore, multiple products will probably become increasingly important for most agronomic crops.

The chenopods, composites, and grasses are the largest families of halophytes in the Sonoran Desert. Certain of these chenopods are well known, valuable forage species, and it is within this family that significant new halophytic crops can be expected to be developed. Most of the composites in Table 1 are only moderately halophytic and do not seem to hold much agronomic potential.   Legumes are important in the desert ecosystem and some show major agronomic potential.   However, only four in the Sonoran Desert are even marginally halophytic. The mesquites (Prosopis ssp) often grow at the edges of mangroves and on upper beaches.

The spurges (Euphorbiaceae) are prominent in tropical and semi-arid regions, but poorly represented in saline environments. Euphorbia incerta grows on upper beaches and is often tidally inundated.   E. platysperma, apparently closely related, extends into upper beaches (5).   These plants may be of value because of the high concentrations of hydrocarbons characteristic of the

genus. Lycium brevipes, in the Solanaceae, produces prodigious
quantities of small tomato-like edible fruit (6).   It often grows
in alkaline or saline coastal soils.

Among the grasses Distichlis and related saltgrasses show
considerable agronomic potential.   D. spicata is being cultivated
experimentally at Lago Texcoco in the Valley of Mexico under highly
saline conditions as a forage crop for cattle and sheep (7).
Weight gains for the animals are significant, although the forage
is most palatable during months of rain or dew.

Salt excretions on saltgrasses and various other halophytes
are hygroscopic and can be an important fresh water source for the
plants.   Another species, D. palmeri, with grain about the size of
wheat (6.5 mm long), served Cocopa Indians at the Colorado Delta as
a major food resource.   They may have been harvesting an improved,
selected form which now may be extinct (1).

D. palmeri is being grown experimentally under hypersaline
conditions (8).   With halophytes such as these, which certainly
lend themselves to domestication, we urge the implementation of
genetic improvement including hybridization and selection.   For
example, a closely related saltgrass, Jouvea pilosa, a beach
species from the Gulf of California, has a grain 20 mm long and
might be hybridized with Distichlis.

The dream of farming the sea might someday be realized with
seagrasses.   Four seagrasses occur in coastal waters along the
Sonoran Desert: Halodule, Phyllospadix (surfgrass), Ruppia (ditch-
grass), and Zostera (eelgrass).   There is considerable inter-and
intra-specific variation among seagrasses, indicating the possibility for genetic improvement.

Eelgrass seed served the Seri Indians of Sonora as a major
food resource (9).   Using flour prepared by Seri women, the first
loaf of bread from the sea was prepared for the Environmental
Research Laboratory by Hazel Fontana and Mahina Drees.   The
vegetative parts are also edible, and the discovery of salt removal
by washing opens a new spectrum of agronomic potential. Vegetative
productivity of seagrasses can be very high (300-600 mg dry
wt/m^2/yr), approaching that of sugar cane (10).   It is noteworthy
that this occurs in full strength seawater.

Although commercial production might be realized in managed,
semi-natural marine fields, we envision seagrass cultivation on
rafts in protected coastal waters.   Most of the seagrasses grow in
sandy-muddy substrates, which was presumed to be required for
successful growth.   However, Tom Backman at the University of
Washington discovered that eelgrass can be grown with its roots

suspended in water without soil.   Conceivably, the seagrass rafts
might have roosts for marine birds in order to provide fertilizer.
Commercially significant fish might be attracted to the nutrient-
rich water beneath the rafts, offering a means to finance the long
term research and development for seagrass farming.   The rafts
might ride a meter or so below the surface, be secured with
multiple anchor clumps, and be comprised of modular trays. Neutral
buoyancy can be provided by constructing the rafts of PVC pipe,
perhaps filled with styrofoam.   To avoid destruction from storms,
the rafts could be temporarily lowered to the seafloor (Wayne
Collins, pers. comm.)

        There is obviously a great diversity of halophytes available
for agronomic development, even in deserts and in the sea.

References

1. R. S. Felger, in: "New Agricultural Crops," (G. A. Ritchie,
        ed.) pp. 5-20, Westview Press, Boulder, CO (1979).
2. I. L. Wiggins, in: "Vegetation and Flora of the Sonoran
        Desert," (F. Shreve and I. L. Wiggins, eds.) 2 vols.,
        Stanford Univesity Press, Stanford, CA (1964).
3. R. S. Felger and G. P. Nabhan, in: "Social and Technological
        Management in Dry Lands," (N. Gonzales, ed.) pp. 129-149,
        Westview Press, Boulder, CO (1978).
4. E. Glenn, M. Fontes, S. Katzen and B. Colvin, in: "Proceedings
        2nd International Workshop on Biosaline Research,"
        (A. San Pietro, ed.) Plenum Press, (1981).
5. R. S. Felger, Desert Plants, 2(2):87-114 (1980).
6. R. S. Felger and M. B. Moser, Ecology of Food and Nutrition
        5:13-27 (1976).
7. J. C. Mota-Urbina, Ciencia Forestal 4(22):21-44 (1979).
8. E. Glenn, M. Fontes and N. Yensen, this volume.
9. R. S. Felger and M. B. Moser, Science 181:355-356 (1973).
10. C. P. McRoy and C. McMillan, in: "Seagrass Ecosystems:
        A Scientific Perspective," (C. P. McRoy and C. Helfferich,
        eds.) pp. 53 - 81, Marcel Dekker, N. Y. and Basil (1977).

MANGROVE SYSTEMS OF THE BAY OF LA PAZ AS EXPLOITABLE RESOURCES

J. P. Gallo, A. Maeda and O. Maravilla

Centro de Investigaciones Biologicas
de Baja California
La Paz, Baja California Sur, Mexico

Mangroves, one of the most productive ecosystems of the world, have a very important ecological influence on subtropical coastal fisheries as shown by Odum (1), Heald (2) and Lugo (3). However, these ecosystems continue being altered and destroyed, in order to satisfy the needs of the fast developing population.

Due to the climatic conditions that prevail in the Bay of La Paz, the Californian mangroves receive only occasional fresh water. Thus, these mangroves present a very different condition in hydrology, geomorphology and function to those found in other regions of Mexico (4). The present interdisciplinary study attempts to give an integrated view of the mangrove ecosystem in Baja California.

The study zones are three coastal lagoons ("Esteros"), which are found in the Bay of La Paz.

The coastal lagoon of Puerto Balandra is found at 110° 18' 45" L.W. and 24° 18' 30" L.N., with an area of 10,600 m² (Map 1).

The tidal channel of Zacatecas, is found at 110° 25' 30" L.W. and 24° 10' 10" L.N., with an area of 5,000 m² (Map 2).

The coastal lagoon of Enfermeria, at 110° 18' L.W. and 24° 14' L. N., with an area of 2,500 m² (Maps 3 and 4).

The present study includes plankton, benthos, necton and Geology (sediments). The methods and collecting procedures are described by Espinosa, et al. (5). Each area is characterized with respect to the relation between species, their abundance and their relation with environmental factors such as temperature, salinity, dissolved oxygen, pH and sediment.

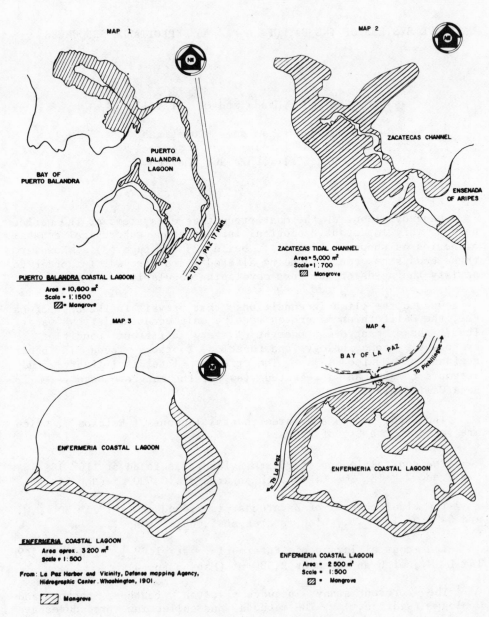

Fig. 1.   Studied Zones

In Balandra lagoon a high species diversity occurs on all biotic levels, probably due to the oceanic influence, the magnitude of the area and the heterogeneity of substrate (6). These factors permit the implantation of many organisms of commercial interest (Table 1).

Zacatecas channel presents a somewhat similar species abundance and diversity to Balandra's but with a minor production. This is most likely due to the shape of the tidal channel and smaller area (thin but long) with very fast tidal currents. Such conditions do not permit the development of planktonic and nectonic organisms typical of the coastal lagoons. The zone is directly influenced by the water masses of the Ensenada de Aripes and similar phytoplankton (7), fishes and fauna are observed in both (Table 1).

TABLE 1: Commercially important species present in the studied areas.

Species	Balandra	Zacatecas	Enfermeria
Eucinostomus gracillis	+	+	+
Eucinostomus argenteus	+	+	+
Lutjanus aratus	+	+	
Lutjanus argentiventris	+	+	+
Lutjanus novemfasciatus	+		+
Anchoa ischana	+		
Mugil curema	+	+	+
Paralabrax maculatofasciatus	+		
Opisthonema libertate	+		
Holopagrus guntheri	+		
Gerres aureolus	+		+
Alubula vulpes	+		
Bairdiella icistia	+		
Diapterus peruvianus		+	+
Eugerres axillaris		+	+
Haemulon scuderii		+	
Harengula thrissina			+
Penaeus californiensis	+		+
Callinectes bellicosus	+	+	+
Megapitaria squalida	+		
Ostrea palmula	+	+	
Andara sp	+	+	

Enfermeria lagoon is the most disturbed system. It presents the highest relative abundance, but the lowest species diversity. Referring to phytoplankton, Enfermeria presents ticoplanktonic species typical of eutrophic zones (8). The same occurs with the benthic fauna (5). No nectonic predators of fishes have been found, thus allowing their development in juvenile stages. The species of commercial importance found in the zone are shown in Table 1.

With respect to abiotic factors, they do not have a determinant influence over the planktonic, benthonic and nectonic species in this coastal lagoon.

Concerning the substrate in Enfermeria, this is homogeneous constituted principally by gross lime, which confirms the eutrophic conditions of this coastal lagoon (6). The orientation and aperture of the mouth of the lagoon has been altered with the construction of the La Paz - Pichilingue Highway. As a result the communication of the lagoon with the sea has been restricted, being constantly closed by the accumulation of sand (9). Occasionally, an effective communication with the sea can be attained at high tides.

In spite of the low water flow, these disturbances are not sufficient to kill the lagoon, rather they act as catalyzers of eutrophication and destroy the productivity of the system (Maps 3 and 4).

The fact that the Puerto Balandra coastal lagoon system has not been substantially disturbed permits the establishment of aquaculture of some of the same species found in the Bay, such as Mullet (Mugil curema and M. cephalus), clam (Anadara sp.), mangrove oyster (Ostrea palmula), chocolato clam (Megapitaria squalida), and brown shrimp (Penaeus californiensis).

In Zacatecas lagoon, some commercial species are present but it is not feasible to carry out any type of aquaculture practice on a commercial level due to the small area and physiography.

With respect to the Enfermeria coastal lagoon, the reduction and misorientation of the mouth has resulted in negative changes in the flux of water. Therefore, a great homogeneous organic sedimentation has occurred within the area promoting the selectivity of species adapted to eutrophic conditions, thus yielding few species but a greater number of individuals per species. Nevertheless, this coastal lagoon is yet amenable to aquaculture on a smaller scale, since some of the species found are commercially important and occur in high abundance at certain stages of their biological cycle.

REFERENCES

1. E. P. Odum, (ed.), Ecologia Interamericana (1972).
2. E. Heald, "The Production of Organic Detrius in a South
   Florida Estuary," Univ. Miami, Sea Grant Tech. Bull. 6:110
   (1971).
3. A. I. Lugo and S. C. Snedaker, Ann. Rev. Ecol. & Syst. 5:36-64,
   (1974).
4. J. P. Gallo, "Classificacion de las Lagunas Costeras del
   Litoral Sureste de la Peninsula de Baja California,"
   (in press). Centro de Investigaciones Biologicas, La Paz,
   B.C.S., Mexico.
5. M. Espinosa, et al., Informe General de Labores, Centro
   de Investigationes Biologicas de Baja California, A.C.,
   La Paz, B.C.S. (1979).
6. E. Diaz, et al., Informe General de Labores, Centro de
   Investigaciones Biologicas de Baja California, A.C.,
   La Paz, B.C.S. (1979).
7. H. Nienhuis, Some Aspects of the Phytoplankton Ecology in the
   Ensenada of La Paz, B.C.S., Mexico, CICMAR-IPN (1979).
8. J. J. Bustillos, et al., in: CIBCASIO, Vol. 5, La Paz, B.C.S.
   (1979).
9. R. Mendoza, et al., "Determinacion del flujo a traves de 24
   horas, con analisis de Parametro Fisicoquimicos y Nutrientes
   en el Estero de Enfermeria, Bajia de La Paz, B.C.S., Univ.
   Autonoma de Baja California Sur, (1978).

NUTRITIONAL VALUE OF HALOPHYTES GROWN ON HYPERSALINE SEAWATER

E. P. Glenn, M. R. Fontes, S. Katzen and L. B. Colvin

Environmental Research Laboratory
Tucson   International Airport
Tucson, Arizona   85706   USA

Irrigation of terrestrial crops with undiluted seawater has been advocated as a way to increase agricultural productivity along coastal deserts (1).   At present, no conventional grain or forage crop can survive seawater irrigation in a desert environment, but wild, salt-tolerant plants (halophytes) can be grown in relatively high yield in the desert, using even hypersaline seawater for irrrigation (Glenn et al, this volume).   The usefulness of these plants for human and animal nutrition remains to be determined.

A. barclayana, A. lentiformis and S. europaea were grown in open-field plots irrigated twice daily with nutrient-enriched, hypersaline (40,000 ppm) seawater at Puerto Penasco, Sonora, Mexico (Glenn et al, this volume).   Fruits of D. palmeri and C. truxillensis were collected from wild plants growing in a negative estuary (no fresh water inflow) near El Golfo, Sonora, Mexico.   Harvested material was sun-dried and stored at room temperature.   Seeds were separated from fruits by hand and with the aid of a commercial seed cleaner.

Plant material was dried at 75 C and ground in a Wiley Mill (20-mesh screen) prior to analyses.   Chemical analyses were performed in house and by a commercial laboratory (EFCO, Inc.) Tucson, Arizona), using standard analytical procedures (2).

Leaching experiments to remove salts were conducted with 5g (dry wt) samples suspended in 50 ml dH_2O), and gently mixed on a rotary shaker (50 rpm) at room temperature.   Plant material was recovered by squeezing the suspension through 3-layers of wet cheesecloth.

Animal feeding trials were conducted at a commercial feedlot (Mesquit del Oro, Hermosillo, Sonora, Mexico), and at the Animal Sciences Department, University of Arizona. The chemical composition of Atriplex and Salicornia spp. varied according to the developmental stage of the plants at the time of harvest. Immature A. lentiformis plants (aerial parts only) were 34% lower in ash, 43% lower in fiber, and 20% higher in protein than mature plants. Results were similar for A. barclayana except that ash contents were equal in immature and mature tissues. Ash was not determined for immature S. europaea plants; however, protein was 62% higher in immature than in mature plants. Protein contents of immature A. lentiformis, A. barclayana and S. europaea plants were 18.7%, 12.4%, and 14.4%, respectively.

Ash and cation contents of the above species were lowered significantly by soaking ground material in fresh water (Table 1). This treatment removed primarily monovalent cations (Na and K). Time-course experiments with A. lentiformis showed that salt (i.e., ash) was removed more readily than protein, resulting in enhanced levels of protein at optimal soaking times. However, soaking also resulted in losses of organic matter. In experiments with all three test species, protein losses amounted to 15-56% of the total, even though percentage protein in final products was approximately equal to the starting materials.

A relatively high percentage of the essential amino acids were present in both A. lentiformis and A. barclayana proteins, although supplementation would be necessary to provide adequate levels of the sulfur amino acids for chicks and rabbits.

Un-desalted A. barclayana ground leaves and stems, incorporated at up to 15% of the diet, supported the growth of mice and chickens (Table 2), and the somewhat lower weight gains at the higher levels could be attributed to the lower energy value of the halophyte material due to its high salt content. On the other hand, A. lentiformis tissue was inhibitory to the growth of both mice and chickens and, in addition, caused 94% mortality of chicks at the 15% level in the diet (Table 3).

The toxicity of A. lentiformis to chicks was substantially alleviated by prior soaking of the plant material for 24 hr in a saturated CaOH solution. In this experiment, chicks fed untreated A. lentiformis at 10% of the diet for four weeks, had 23% mortality and a weight gain of only 219 g/bird, compared to 0% mortality and a weight gain of 669 g/bird for controls. However, when switched to CaOH-treated A. lentiformis for weeks 4 to 6, the former group suffered no further mortality, and had a weight gain of 326 g/bird compared to 409 g/bird for controls during weeks 4 to 6. The percent weight gain during weeks 4 to 6 was actually higher for

TABLE 1. Ash, P and cation content of soaked and unsoaked halophytes. Dry material was ground in a Wiley Mill (10-mesh screen) and soaked for 16 hrs in 10 volumes of dH_2O.

| | A. lentiformis | | A. barclayana | | S. europaea | |
	Unsoaked	Soaked	Unsoaked	Soaked	Unsoaked	soaked
Ash (%)	20.2	6.7	33.5	15.4	42.4	6.5
Na (ppm)	105,000	11,000	111,000	50,000	130,000	35,000
K (ppm)	80,000	3,000	75,000	3,000	10,500	500
Ca (ppm)	4,150	5,500	10,950	10,700	4,950	6,100
Mg (ppm)	4,000	3,000	9,000	13,000	7,000	2,500
Cu (ppm)	5	5	10	5	10	7.5
Zn (ppm)	30	25	27	40	33	40
Fe (ppm)	60	35	105	110	85	65
P (%)	0.27	0.06	0.20	0.06	0.18	0.06

TABLE 2. Performance of mice and chicks fed A. lentiformis and A. barclayana. 18-20 animals/treatment were fed basal diets plus halophytes for 4 weeks. Diets balanced for protein, Ca, & P.

Diet	Chicks			Mice	
	Wt. Gain (g)	F.C.R.	Mortality (%)	Wt. Gain (g)	F. C. R.
Basal	574.5	1.723	0	17.4	5.07
5% A. barclayana	607.9	1.766	0	17.0	5.21
10% A. barclayana	555.5	1.796	5.6	17.2	4.90
15% A. barclayana	467.5	1.928	5.6	15.8	4.83
5% A. lentiformis	332.6	2.104	0	15.0	5.05
10% A. lentiformis	98.1	3.005	72.2	12.2	5.24
15% A. lentiformis	55.7	4.785	94.4	9,8	6,30

TABLE 3. Nutritional Content of seeds of Distichlis palmeri and Cressa truxillensis

	Distichlis palmeri	Cressa truxillensis
%Cell Wall	26.7	53.3
% Cell Soluble	73.4	46.6
% Ash	1.3	1.6
% Protein	9.0	13.0

birds fed CaOH-treated A. lentiformis than for control brids, because of the smaller starting weights of the former birds at week 4.

We hypothesized that treatment with CaOH may render oxalate insoluble in the animals digestive system, thereby reducing toxicity. However, A. barclayana, which was not toxic to mice or to chicks, proved to have a higher oxalate content than A. lentiformis (5.90% vs 3.57%). Both Atriplex spp. had higher oxalate levels than S. europaea (1.53%).

D. palmeri and C. truxillensis seeds had low ash contents (Table 3). The fiber content of C. truxillensis seeds was higher than that of D. palmeri, reflecting the hard seed coat of the former plant, but protein and cell soluble constituents were sufficiently high in both plants to indicate potential value as grain crops.

The present results support Goodin's (3) conclusion that immature Atriplex plants can provide high-quality forage, offering levels of protein and soluble carbohydrates equal to alfalfa and other high-protein conventional forage crops. Immature S. europaea also appear to be high in nutritional value (i.e., 14% protein).

The major effect of seawater irrigation is to increase the ash content of the harvested product, and this will limit the amount that can be incorporated into a livestock diet. On the other hand, desalted material could be incorporated at much higher levels.

Utilization of seawater-grown halophytes for forage and fodder would appear to have promise if economic procedures for removing excess salts and toxins are developed. Halophyte grains do not accumulate salts and may be used directly for animal or human food.

We would like to thank Drs. C. Weber, B. Reid, S. Swingle and W. Brown, University of Arizona, for supplying data on mice and chick feeding trials and protein quality indices.

References

1. H. Boyko (ed). "Salinity and Aridity: A new approach to old problems." W. Junk, The Hague (1966).
2. Associations of Analytical Chemists, " Official Methods of Analysis," 12th edition, Washington, D. C. (1975).
3. J. R. Goodin, Atriplex as a forage crop for arid lands, in: "New Agricultural Crops." (G. A. Richie, ed.), AAAS Selected Symposium 38, pp. 133-148 (1979).

PRODUCTIVITY OF HALOPHYTES IRRIGATED WITH HYPERSALINE SEAWATER IN THE SONORAN DESERT

E. P.Glenn, M. R. Fontes and N. P. Yensen

Environmental Research Laboratory
Tucson, Arizona 85706 USA

The use of highly saline irrigation water for salt-tolerant terrestrial crops has been advocated as part of a biological solution to agricultural salinity problems (1-3).   If undiluted seawater could be used, potentially 32,000 km of desert coastline are available for agriculture, and millions of hectares of inland desert overlie saline aquifers (1). Most plants are harmed by salt concentrations greater than one-tenth the salt content of seawater (35 ppt) (4), but suitable crops may be developed by improving salt tolerant strains of conventional crops (2) or domesticating existing salt tolerant wild plants (halophytes; 3). The problem of salt build-up in the soil may be solved by irrigating in excess of plant requirements on highly permeable soils to leach salts below the root zone (4).

An important question is the extent to which terrestrial plants are productive on highly saline water (5). Known mechanisms of salt tolerance require expenditures of metabolic energy and hence even the most salt tolerant halophytes suffer growth reductions with increased salinity of irrigation water (5).   Halophytes grown on seawater may produce only one-third as much dry matter as fresh water controls (6, 7) in greenhouse experiments. Estimates of halophyte productivity under natural conditions range from low values 10-400 gDW $m^{-2}yr^{-1}$) to very high values for salt marsh species receiving both tidal innundation and fresh water influx (4000 gDW $m^{-2}yr^{-1}$; 5).   In controlled field experiments in Israel (8), Atriplex spp. drip irrigated with 42 ppt seawater yielded 470 gDW $m^{-2}yr^{-1}$, while in Delaware (9) Spartina alterniflora grown on two-thirds to full strength (32 ppt) seawater, yielded 400 gDW $m^{-2}yr^{-1}$.

We tested the yield potential of halophytes under agricultural conditions in an extreme desert environment at Puerto Penasco, Sonora, Mexico, at the northern extremity of the Gulf of California. Irrigation water was hypersaline (40 ppt) seawater discharged from a shrimp aquaculture facility, which added 30-70 ppm of organic matter and 0.01-0.03 ppm of dissolved ammonia to the water. Precipitation at the study site is 9 cm/yr with 250 cm of potential evaporation; the mean temperature is 20.2 C with highs of 43 C; humidity is normally low, even though dew may form on cold nights. Field trials were conducted on beach sand and sandy desert soil, with saturated infiltration rates of 2.5-3.0 and 1.3-2.0 cm/hr, respectively. Leveled plots (12.3 x 18.5 m) were flooded at intervals to a depth of 6-10 cm. Addition of inorganic fertilizers to some of the plots did not affect plant yields.

Except for <u>Salicornia europaea</u>, which was direct-seeded into field plots, seedlings were established in the greenhouse on fresh water and conditioned to seawater irrigation prior to transplanting to the field. The survival rate was over 80% for approximately 30,000 seedlings transplanted. Fresh water usage was held to 500 ml per plant by recycling and planting in compact silviculture trays.

Nine candidate species, chosen from an initial screening of 32 species in 19 genera (6), were evaluated for total above ground yield (Table 1). The most productive species, irrigated at 12-hour intervals, yielded from 895-1365 gDW $m^{-2}yr^{-1}$. These values compare favorably with total yields from such conventional fresh water crops as alfalfa (760 gDW $m^{-2}yr^{-1}$ U. S. average 1978; 10), corn (790 gDW $m^{-2}yr^{-1}$ average in areas of high yield; 11) and oats, (926 gDW $m^{-2}yr^{-1}$ average in areas of high yield; 11), but are considerably less than yields from sugarcane (3430 gDW $m^{-2}yr^{-1}$ Hawaii average; 11).

We previously found that 4 of the test species, <u>A. barclayana</u>, <u>A. glauca</u>, <u>A. rependa</u> and <u>Batis maritima</u> (6) yielded only one-third as much dry matter on seawater as on fresh water in the greenhouse. While field productivities of <u>S. europaea</u> and <u>B. maritima</u> on fresh water are unknown, <u>Atriplex</u> spp. yield up to 1,000 gDW $m^{-2}yr^{-1}$ on fresh water (12). Therefore, the respectable yields obtained on seawater in the present experiments may be due to high intrinsic productivities of some of the test species.

As the time between irrigations increased, salt accumulation increased in the soil surface and the moisture content of the soil decreased (Table 2). This resulted in high soil-water salinities at the surface of plots watered every 72 hours compared to plots watered every 12 or 24 hours. Plant yields were reduced on the less frequently flooded plots (Table 1).

TABLE 1: Above/Ground Yield (gDW m^{-2}) of Halophytes on 0.61 m Centers Irrigated Every 12, 24 and 72 hours with 40 ppt Seawater.

Water Frequency (hrs.):	Above Ground[1] Yield (gDW m^{-2})		
	12	24	72
Salicornia europaea	1356	nd	nd
Batis maritima	1137	nd	nd
Atriplex linearis	927	1134	245
Atriplex barclayana	901	733	431
Atriplex lentiformis	895	620	230
Atriplex glauca	413	348	130
Atriplex canescans	178	165	111
Atriplex polycarpa	143	61	31
Atriplex rependa	22	15	9

[1]Atriplex spp. were transplanted from greenhouse to eighteen desert plots (12.3 x 18.5 m) on 23 April 1979. Alternate plants were harvested at t= 217 days from transplanting. Batis maritima was planted in beach plots on 31 July 1979 and harvested at t= 355 days. Salicornia europaea was sown on 16 February 1979 and harvested at 5 = 265 days.

[2]0.31m centers.

TABLE 2: Moisture and Salt Content (mg/g soil) in Desert Plots Watered Every 12, 24 and 72 Hours, with 40 ppt Seawater. Measurements were made on soil samples from the surface and 15 cm depth, immediately before a scheduled irrigation, using techniques in Richards (4) (1:5 soil extract for measurement of salinity by electrical conductivity). Each entry is the average of four determinations.

Watering Frequency (hrs.)	Moisture (mg/g)		Salt (mg/g)	
	Surface	15cm	Surface	15cm
12	73.6	68.2	4.0	2.6
24	77.6	72.0	5.3	2.9
72	18.4	54.9	10.3	2.3

In the present experiments, frequent irrigation increased productivity, apparently by removing salts from the soil surface and by providing nutrients in the irrigation water. O'Leary has calculated that the metabolic cost of salt tolerance in a higher plant irrigated with seawater is approximately equal to the amount of energy partitioned into fruit production in conventional grain crops (5); for this reason, he is pessimistic that they can never be adapted to high yields on highly saline water. On the other hand, the present results show that the highest yielding halophyte species can produce acceptable yields of forage material on seawater, despite the inhibitory effect of salt on growth.

References

1. H. Boyko, "Salinity and Aridity: New Approaches to Old Problems," Junk, the Hague (1966).
2. E. Epstein and J. Norlyn, Science 197:249 (1977).
3. G. Sommers, in: "The Biosaline Concept: An Approach to the Utilization of Unexploited Resources, (A. Hollaender, ed.) pp. 101-115, Plenum Press, N. Y. (1979).
4. L. Richards, "Diagnosis and Improvement of Saline and Alkali Soils" USDA Handbook 60 (1954).
5. J. W. O'Leary, "Arid Land Plant Resources," (J. R. Goodin and D. K. Worthington, eds.) pp. 574-581, Texas Tech. University Lubbock, TX (1979).
6. E. P. Glenn, N. P. Yensen and M. R. Fontes, "Symposium on the Gulf of California" (In Press).
7. N. J. Chatterton and C. M. McKell, Agronomy Jour 61:448 (1969).
8. D. Pasternak, J. Ben Dov and M. Forti, Annual Research Rpt., R & D Authority, Ben Gurion University of the Negev, Rpt. No. BGUN-RDA 222-79 (1979).
9. G. F. Somers, M. Fontes and D. M. Grant, "Arid Land Plant Resources," (J. R. Goodin and D. K. Northington, eds.) pp. 402-417, Texas Tech. Univ. Lubbock, TX (1979).
10. Agricultural Statistics, USDA (1977).
11. E. P. Odum, "Fundamentals of Ecology," p. 73, W. B. Saunders Co., (1959).
12. J. Goodin, "New Agricultural Crops," (G. Ritchie, ed.) pp. 133-148, AAAS Selected Symposium 38 (1979).
13. C. H. Hubbell, Feedstuffs 52(3):42 (1980).

NUCLEAR MAGNETIC RESONANCE STUDIES OF ION MOBILITY AND BINDING

IN A DEAD SEA BACTERIUM Ba$_1$ AND MODEL SYSTEMS

M. Goldberg[1], H. Gilboa[1], R. Gelber[1] and M. Risk[2]

[1] Physical Chemistry
Technion
Haifa, Israel

[2] Division of Biochemistry
University of Texas Medical Branch
Galveston, Texas 77550 USA

SUMMARY

Pulsed ion-specific N.M.R. analysis was performed on pelleted Ba$_1$ bacteria grown at 0.5 M or 3.0 M salt containing 50% lithium, and on various ion exchange resins as model systems. In low salt medium, all the lithium was observable by N.M.R., but in high salt grown bacteria, only about 2/3 of the cation was discernable by N.M.R. These observations suggest restricted mobility of ions in high salt bacteria, indicating ion binding. Relaxation time studies (T_1 and T_2) showed 10 to 15-fold reduction for lithium in Ba$_1$ grown or washed in high salt, in comparison with E. coli washed in this medium. These data show unusual binding in Ba$_1$ in comparison with other microorganisms, somewhat comparable with that observed with ion exchange resins.

INTRODUCTION

Nuclear magnetic resonance probe studies can be used to investigate the physical state of atoms of interest. Since the microenvironment alters easily determined parameters of these nuclei, this technique is an appropriate one for ascertaining binding. This question is especially pertinent with regard to intracellular water and its associated cations. This topic has been associated with much controversy recently. The current study was undertaken to examine mobility of intracellular sodium and lithium in a bacterium capable of growing in a high salt medium and over a wide range of osmotic values. Ion exchange resins were used as a physical model for the presumptive binding observed in this bacterium.

MATERIALS AND METHODS

Ba$_1$ was grown in one liter suspension cultures at 36°C in two liter Erlenmeyer flasks at 125 strokes/min in Difco nutrient broth (8 g/liter) 0.05M KCl, 0.1 M MgCl$_2$, NaCl to supplement to 0.05 low or 3.0 M (high) salt. In some cases, 50% of NaCl was replaced by LiCl. Plateau phase cells were harvested by pelleting at 10 x 10^3 g, followed by two washes in isotonic medium without nutrient broth, transferred to glass N.M.R. tubes, and spun at 14 x 10^3 g for 30 min to remove excess medium, all at 4°C.

E. coli W3110 was grown in Eijkman medium, suspended in appropriate wash solution, and prepared as for Ba$_1$. A-15 and Dowex 50-8x ion exchange resin were washed in the same solutions and pelleted similarly. N.M.R. measurements of total ion intensity, T$_1$, and T$_2$ were performed by a modified Bruckner pulsed N.M.R. spectrometer at 60 M Hz with tunable radio frequency generator and on-line Nicollet data processing. Total ion intensity in prepared sample was compared with ashed samples resuspended in distilled water to equal volume of the original sample.

RESULTS

Pelleted Ba$_1$ bacteria grown at 3.0 M and 0.5 M total salt (1.4 M and 0.175 M LiCl, respectively) were examined by N.M.R. as fresh and ashed preparations. Ion intensity was calculated from aqueous LiCl standard solutions. Observed concentrations reflect degree of intracellular accumulation, binding, bulk exclusion of solution, and immobilized ions. Data are expressed for fresh sample ion concentration as percent of ashed sample concentration (equated to 100%) (Table I). In low salt Ba$_1$, all the lithium is detectable in fresh sample, whereas in high salt bacteria, only about 2/3 was discernable. For the two ion exchange resins tested, about 1/6 and 1/3 of the total ions were detectable in fresh samples.

A more sensitive measurement of ion state is relaxation time. In this technique, nuclei are stimulated by imposed radio frequency irradiation, and the time required to return to ground state is determined for the longitudinal (T$_1$) and transverse (T$_2$) components. Escherichia coli cells washed in Ba$_1$ medium showed relaxation values for lithium similar to those observed in free solution. Ba$_1$ grown in 50% LiCl replacing NaCl gave values similar to those grown in lithium free medium and washed in lithium wash solutions (i.e., treated as E. coli). Ba$_1$ in 0.5 M salt showed slightly reduced T$_1$ and about 2x reduction in T$_2$. When grown in 3.0 M, T$_1$ was about 10% that of E. coli, and T$_2$ about 15% observed for this ion in E. coli (Table 2). Conversely, sodium values examined in presence or absence of lithium did not differ significantly

TABLE I

Observable N.M.R. Li$^+$ Signal in Fresh Bacteria and Model Systems as Fraction of Total Lithium Concentration

Sample	% Detectable
A-15 Resin	14.5
Dowex 50-8x	34.2
Ba₁ 0.5M Salt (0.175M Li$^+$)	100
Ba₁ 3.0M Salt (1.4M Li$^+$)	63.6

TABLE II

Relaxation Times of Excited Li$^+$ Nuclei in Bacterial Pellets

Sample	T_1 msec	T_2 msec	Ba₁ Time / E. coli Time T_1	T_2
E. coli 0.5M Salt	1620	86	100%	100%
E. coli 3.0M Salt	1740	100	100%	100%
Ba₁ 0.5M Salt	1380	45	85%	52%
Ba₁ 3.0M Salt	160	5.9	9.9%	6.9%

All values determined at 17.45 MHz

from free solution in either Ba_1 or E. coli for both salt concentrations.

DISCUSSION

Ba_1 is an interesting euryhaline mesohalophilic bacterium with growth optimum from 0.5 M to 2.0 M salt, and capable of growth over a 20x salt range (0.2 M to 4.0 M). There are many similarities with the newly described genus Halomonas (1) and with Ps. hadestorga (2). The advantage of using this organism for these experiments measuring physical correlates of intracellular ion mobility is the intense signal in this system. This is due to the ability of this organism to grow in high total salt and also to withstand unusually high lithium concentrations.

It has been shown previously that for ^{23}Na N.M.R. studies in many types of bacteria, the observed signal was appreciably different from aqueous solution (3). This is apparently due to ion anisotropy at cell-medium interface. These effects are especially pronounced in Gram negative species; they seem to be due in large part to the lipopolysacharide component isolated from the membrane-wall complex, since this extract alone shows sodium line width increase virtually identical with intact bacterial suspensions.

It has been suggested that in Ba_1 three types of sodium binding sites exist, reflecting exchange between free and bound sodium (4). Equations relating the properties of the extracellular ion with free and bound introcellular ions determined strong binding in a significant segment of the ions in bulk phase pelleted bacteria (5).

In sarcoplasmic reticulum preparations, it has been suggested that such ion mobility reduction (reflected in changes in relaxation rate) may be due to binding at specific (hydrophobic?) sites on membrane molecules (6). Relaxation time of water in biological systems is much less than in pure water. This is due to binding to macromolecules; isolated proteins and protein-rich components have a strong effect, but lipid-rich components show greater effect (7). Since Ba_1 has unusual surface properties and lipid composition, these components are being investigated as the sites responsible for the anomolous ion properties observed by N.M.R. signals.

REFERENCES

1. R. H. Vreeland and E. L. Martin, Can. J. Microbiol. 6:746-52 (1980).
2. R. S. Breed, ed., "Bergey's Manual of Determinational Bacteriology," 7th ed., p. 120, Williams and Wilkins publisher (1957).

3. R. C. Lyon, N. S. Magnuson and J. A. Magnuson, in: "Extreme
    Environments", M. R. Heinrich, ed., pp 305-320,
    Academic Press (1976).
4. M. Goldberg and H. Gilboa, in: "Nuclear Magnetic Resonance
    Spectroscopy in Molecular Biology," B. Pullman, ed., pp.
    481-491, Reidel publisher (1978).
5. M. Goldberg and H. Gilboa, Biochim. Biophys. Acta.
    538:268-283 (1978).
6. E. M. Stephens and C. M. Grisham, Biochem. 18:4876-85 (1979).
7. R. M. De Vre, Prog. Biophys. Molec. Biol. 35:103-34 (1979).

ARSENIC DETOXICATION IN <u>MACROCYSTIS</u> <u>PYRIFERA</u>

J. M. Herrera-Lasso and A. A. Benson

Scripps Institution of Oceanography
La Jolla, CA 92093 USA

SUMMARY

Arsenic accumulates in the giant kelp <u>Macrocystis</u> <u>pyrifera</u> up to 60 ppm. In surface waters where phosphate concentrations are low, arsenate enters the cell via a transport system unable to discriminate between phosphate and arsenate. In vitro experiments revealed migration of trimethylarsoniumlactate from the kelp to the surrounding seawater. This relatively nontoxic organoarsenic compound is a degradation product of an arsenolipid synthesized by the cells through a series of reactions which involve arsenate reduction and successive methylation. The excretion of trimethyl-arsoniumlactate constitutes a detoxication mechanism developed by many algae to avoid accumulation of toxic forms of arsenic.

INTRODUCTION

Arsenic occurs in seawater in concentrations of 2 ppb (1). The study of its distribution and speciation has attracted the attention of biologists and chemists in connection with the toxicity of arsenic compounds and the ability of certain organisms to accumulate it. Although it is accumulated in almost all marine organisms to some extent (2, 3), lower members of the food chain often contain the highest concentrations; biomagnification through the food chain is not the general case (4). In surface waters where phosphate concentrations are low, arsenate enters the cell via a transport system unable to discriminate between phosphate and arsenate. In green and red algae the range of arsenic concentra-tions is about 1-10 ppm, whereas in brown algae higher concentrations have been detected (5). The highest concentrations have been found in seaweeds where values range from 10 to 70 ppm and values of up to 139 ppm have been recorded (Table 1) (6). Further concentrations of arsenic by direct grazers (sea urchin), does not occur (4). One of the reasons for this might be the ability of algae to metabolize the arsenate to form organo-arsenic compounds which are excreted by the alga and the animal.

Methanearsonic acid and dimethylarsinic acid (cacodylate), which have been detected in seawater, are known to be excreted by some algae under continuous culture (8). In this paper we report even greater the excretion of a novel organo-arsenic compound, trimethylarsoniumlactate, by the giant kelp <u>Macrocystis</u> <u>pyrifera</u>. This compound is a component of a membrane-associated arseno-phospholipid in many algae (9, 10).

For in vitro experiments, fresh <u>Macrocystis</u> was collected from kelp beds near La Jolla. Lamina pieces were illuminated in 250 ml flasks containing filter-sterilized seawater with 250 mg/l strepto-mycin, 250 mg/l penicillin G, and 20 Ci of ^{74}As carrier-free arsenic acid at 16° under constant illumination. Samples of the medium were removed over a 48-hour period and analyzed (Fig. 1). For in situ labelling of the kelp, a double plastic bag system was used, adding the radioisotope to the inner bag containing 3 l of seawater and the last 30 cm of a stripe which had several young laminae and using the second bag for protection. After three hours the seawater from the first bag was frozen for further analysis and the still radioactive laminae were rinsed thoroughly and put into a new plastic bag containing fresh seawater for 20 hours.

After incubation, the seawater was evaporated to dryness in a Rotavapor and salts were extracted with ethanol. Radioactive components were separated and identified by two-dimensional paper chomatography (11) and electrophoresis.

Results and Discussion

Two dimensional chromatograms revealed the presence of three arsenic compounds, A, B and C in the seawater after 4 days incubation in the laboratory. These three compounds correspond to arsenate, cacodylic acid and trimethylarsoniumlactic acid (TMAL). The chromatographic R_f values and elctrophoretic R_m values for these three compounds extracted from the seawater, correspond to those obtained with standards of arsenate, cacodylic acid and TMAL under the same conditions (Table 2).

Figure 1 shows the results of a time series experiment, which indicates that the excretion of cacodylic acid and TMAL begins 6

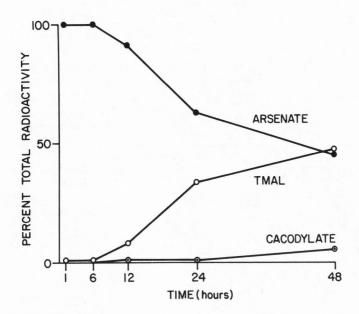

Fig. 1. Excretion of ^{74}As-organoarsenic compounds and consumption of ^{74}-arsenate by excised Macrocystis laminae in seawater containing streptomycin and penicillin G.

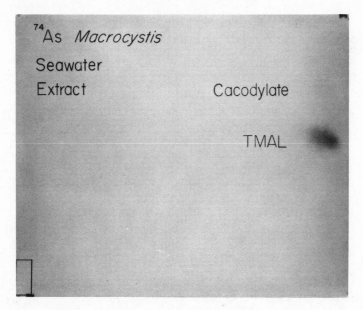

Fig. 2. Radiochromatogram of ^{74}As-labeled compounds excreted to the seawater during 20 hours by labeled Macrocystis laminae in situ.

TABLE 1:   Concentration of Arsenic in Seaweeds.

Seaweed	ppm	Reference	
Laminaria hyperborea	139	Lunde	(6)
Laminaria digitata	109	Lunde	(3)
Sargassum vulgare	68	Dandhukia	(5)
Sargassum cinctum	60	Dandhukia	(5)
Macrocystis pyrifera	60	Wilson	(7)

TABLE 2:   Chromatographic R_f and Electrophoretic R_m values for compounds A, B, C and Standards.

	R_f		R_m*	
	phenol/water	But/prop	pH 3.0	pH 6.0
A	0.06	0.25	-56	-64
B	0.79	0.64	-2	-15
C	0.80	0.50	+22	0
Arsenate	0.10	0.18	-49	-62
Cacodylate	0.77	0.69	-.2	-16
TMAL	0.84	0.54	+23	0

+ Cation; - Anion

* R_m compared to indigotetrasulfonate x 100

hours after the addition of the radiolabelled arsenic acid, and also shows that TMAL accounts for most of the arsenic excreted to the seawater medium.

In situ experiments confirmed the presence of both cacodylic acid and TMAL in the Seawater after three hours exposure to the radioarsenate. Figure 2 shows that after rinsing the laminae exposed to the radioisotope during three hours most of the arsenic released by the cells is in the form of TMAL.

The biosynthesis of TMAL and its excretion to the surrounding seawater represents a detoxication mechanism for Macrocystis pyrifera.

Acknowledgements

Collaboration of Dr. C. F. Phleger with the experiments made in situ is greatfully acknowledged. This work was supported by a grant from the National Science Foundation OCE 78 25707 and by the Consejo Nacional de Ciencia y Technologia of Mexico.

References

1. H. Onishi, "Handbook of Geochemistry," (K. H. Wedepohl, ed.) Vol. II-2, Chapter 33 (1969).
2. A. P. Vinogradov, "The Elementary Chemical Composition of Marine Organisms," Memoir-Sears Foundation for Marine Research, No. II, pp. 110-152, New Haven, CN (1953).
3. G. Lunde, Env. Health Perspectives 19:47-52 (1977).
4. W. R. Penrose, H. B. S. Conacher, R. Black, J. C. Meranger, W. Miles, H. M. Cunningham and W. R. Squires, Ibid. 19: 53-59 (1977).
5. M. M. Dandhukia, Phykos 8:108-111 (1969).
6. G. Lunde, Jour. Sci. Food Agri. 21:416 (1970).
7. S. H. Wildon and M. Fieldes, N. Z. J. Sci. Tech. B 23:47-48 (1971).
8. M. O. Andreae, Deep-Sea Research 25:391-402 (1978).
9. R. V. Cooney, R. O. Mumma and A. A. Benson, Proc. Natl. Acad. Sci. 75(9):4262-4264 (1978).
10. R. V. Cooney, J. M. Herrera-Lasso and A. A. Benson, IInd Intl. Workshop on Biosaline Research, Memoirs, Plenum Press (1981).
11. A. A. Benson, J. A. Bassham, M. Calvin, T. C. Goodale, V. A. Haas and W. Stepkh, Am. Chem. Soc. 72:1710-1718 (1950).

SALT-AFFECTED RANGELANDS: POTENTIAL FOR PRODUCTIVITY AND MANAGEMENT

David B. Kelley
Dept. of Land, Air and Water Resources
University of California
Davis, CA   95616   USA

Rangelands are productive, smoothly functioning ecosystems which with proper management can remain productive without the enormous economic inputs that might be required to convert them to irrigated agricultural systems. The wise manipulation of the World's range resources, particularly in terms of the intensity of their use by man and his livestock, could lead to an increase in the productivity of these lands. This is unfortunately not their history. I wish to focus the attention of this workshop on rangelands of the United States and Mexico in particular. The term "rangelands," as I will use it here, refers to the extensive and usually arid or semi-arid shrublands and grasslands which are grazed by and provide habitat for domestic livestock (cattle, sheep, goats) and wildlife.

Rangelands can be quite diverse in character. Their topography may be steep and rough or flat and smooth; their vegetational cover may be sparse or rich; the density of their native fauna may be high or low. Saline rangelands may be of geologically-recent development and may exist for relatively short time spans, though the complexity of their vegetational systems may be of a high order.

Saline rangelands are predominantly shrublands whose productivity is derived chiefly from halophytic and xerophytic shrubs. The salinity of their soils and waters is a function primarily of two factors: aridity and geology. Their high evapotranspiration rates allow the concentration of unleached salts in their soils and high concentrations of salt in groundwater supplies. The most prominent saline rangelands occur in interior drainage basins (eg. parts of the Great Basin Desert such as those near Great Salt Lake, and the southern portion of California's San Joaquin Valley). They may be associated with outcrops of salt-laden strata of marine origin such as the Mancos shale deposits of the upper Colorado River watershed and may, in certain cases, be subject to windborne salt deposition from maritime sources.

In the western United States, where rangelands dominate the landscape, there are over 300 million hectares of rangelands, forest rangelands, and grasslands (1). Over 80% of the total land area of the 11 westernmost states is rangeland (2). Rangelands of Mexico cover over 100 million hectares and support millions of cattle, sheep, goats, horses, mules, and non-domestic grazing animals (3). Worldwide, nearly one-half of the total land area, representing the largest portion of the arid and semi-arid zones, is classified as rangeland and provides most of the meat and other animal products consumed by people who inhabit the arid zones (4). Over 30% of these arid lands are salt-affected (5, 6) and of comparatively low biomass productivity.

Rangelands are most important as watersheds, especially in the arid western U.S. and northern Mexico. The management of rangelands must have water production and quality as foremost criteria. The uses to which rangelands are put (grazing, recreation, hunting, mining, etc.) must be compatible with water management and are necessarily limited by considerations thereof. The contributions of saline rangelands to water quality problems can be significant when those lands are mismanaged and the consequences of that mismanagement are ignored. Many of the uses of rangelands are compatible with water production, but the potential for water quality degradation (by salts, eroded soil, agricultural residues, and mining wastes) must be recognized. These problems are compounded as population centers develop along the water systems of rangelands.

Topography, water availability, and proximity to agricultural, mining, and recreational resources are factors which have led to the location in saline rangelands of some of the largest cities of the western U.S. and Mexico. Several of these cities - Salt Lake City, Boise, Bakersfield, Phoenix, El Paso, Ciudad Juarez, Mexicali - developed where before only salt-desert shrubs grew. The impact of these population centers on surrounding saline rangelands and their water systems has been severe, as has the development of irrigation agriculture. These may be the highest and best uses of saline rangeland resources, for there is no question that the irrigated lands are highly productive and that the cities provide highly desirable living conditions. But when agricultural and domestic water supplies become limiting, and when residents of these arid lands demand that more land be devoted to

housing, transportation corridors, and other intensive (and entrap-
ping) land uses, conflicts arise. Extension of the illogic of
wholesale development to lands better suited to water production,
grazing, recreation, and wildness leads to these conflicts and they
must be addressed.

Values can be assigned to the uses of these rangelands -
gallons of water produced by this watershed, beef cattle exports
from that valley, tourist dollars generated by this canyon, hunting
license fees generated by that mountain range - but doing so
skirts the issues, in my opinion, of the ultimate fragility of
these rangeland ecosystems and the management options available to
us.

Saline rangelands and their flora are unique resources.
Certain native shrubs and grasses have very high productivity,
palatability, and nutritiuonal characteristics even under very
stressful conditions (cf. 1, 2, 7, 8).   The web of nutrient and
water cycling in these lands is complex and need not be compromised.
by man's manipulation of the resources available therein.

Management alternatives in these delicately balanced ecosystems are keynoted by the importance of determining, and adhering
to, the carrying capacity of the land.   Techniques and technology
for the appropriate management of water resources are available.
Grazing management, combining objectives suitable for the production of meat and the maintenance of vegetation cover and wildlife
habitat, is a well-developed science and, by now, part of the
heritage of the West.   Principles of multi-uses/multi-benefits are
part of the body of law governing all public and most private
rangeland resources, at least in the United States. The methods of
proper management is of wisdom and time, not necessarily of
economics.

I do not propose here that salt-affected rangelands be left
untouched or undeveloped. A poet of the American West, Gary
Snyder, suggested to me, while speaking of irrigation agriculture
in arid lands, that "...an importation of water is an importation
of culture..." I would extend this thought to rangelands of
western North America and say that our development of them is an
importation of culture and a revision of their biological and
sociological order.   We must be wise when conducting these revi-
sions.

Wallace Stegner wrote that "..the West's ultimate unity..[is]
aridity..." (9).   Our management of rangeland resources of the
West would do well to key on aridity and on the nature of arid
lands: fragility, balance, and a certain peculiar order of life
and landscape.

Finally, there are those who look upon our range resources as goods available for the taking, and have proceeded to take without thought of the massive changes that taking promotes.   For them, I would echo Daniel Axelrod, who said, while speaking of arid lands in general and the Sonoran Desert in particular, "Stay the hell out" (D. I. Axelrod, pers. comm.)

References

1. A. P. Plummer, in: "Wildland Shrubs, Their Biology and Utilization," USDA Forest Service, Gen. Tech. Rpt. INT-1, pp. 121-137 (1972).
2. C. M. McKell and J. R. Goodin, in: "Arid Shrublands, Proc. of the Third Workshop of the US/Australia Rangelands Panel, Tucson, AZ, pp. 12-18 (1973).
3. M. H. Gonzales, in: "Wildland Shrubs, Their Biology and Utilization," USDA Forest Service, Gen. Tech. Rpt. INT-1, pp. 429-434 (1973).
4. L. A. Sharp and K. D. Sanders, "Rangeland Resources of Idaho. A Basis for Development and Improvement. Univ. Idaho Forest, Wildlife and Range Expt. Sta. Contribution No. 141, Moscow (1978).
5. D. B. Kelley, J. D. Norlyn and E. Epstein, in: "Arid Land Plant Resources," ICASALS, Texas Tech. University,Lubbock, TX. 326-334 (1979).
6. E. Epstein, J. D. Norlyn, D. W. Rush, R. W. Kingsbury, D. B. Kelley, G. A. Cunningham and A. F. Wrona, Science 210: 399-404 (1980).
7. D. B. Kelley, J. R. Goodin and D. R. Miller, in: "Contributions to the Ecology of Halophytes," (D. N. Sen and K. S. Rajpurohit, eds.) W. Junk, The Hague (1981).
8. R. S. Felger, in: "New Agricultural Crops," (G. A. Ritchie, ed.) pp. 5-20, Amer. Assoc. Advancement Sci., Washington, DC (1979).
9. W. Stegner, "The Sound of Mountain Water," Ballantine Books, NY (1972).

SALT TOLERANCE OF COTTON GENOTYPES IN RELATION TO K/Na-SELECTIVITY

André Läuchli[1]

Wilfried Stelter[2]

Dept. of Land, Air
  and Water Resources
University of California
Davis, CA 95616 USA

Botanitches Institut
Tierärztliche Hochschule
Hannover, D-3000
Hannover, 71, West Germany

Most halophytes accumulate ions, particularly in the leaves (1, 2). Many salt sensitive species, however, exclude Na^+ and/or Cl^- from the shoot (3). The regulation of K/Na-selectivity in higher plants under salinity is complex, and there is no uniform response to salinity stress. Many salt tolerant plants show a decrease in K^+ when exposed to salinity (4) but, as reviewed by Epstein (2), selective uptake of K+ by halophytes is maintained under saline conditions. Potassium is an osmotically active solute and important for generation of turgor in growing plant cells (5, 6).

Cotton is known to be fairly salt tolerant. It translocates K^+ and Na^+ to the shoot (7) and therefore does not appear to be a salt excluder. However, the significance of K/Na-selectivity in salt tolerance of cotton has not been studied.

Results

1. An experiment with Gossypium hirsutum grown in the presence of 1 mM K^+ and 10 mM Na^+ (Table 1) showed high K/Na ratios in the plant, particularly in the growing regions (tips of roots and shoot). Thus, cotton exhibits high K/Na-selectivity at low salinity.

2. In a long-term experiment, the effect was tested of a much higher salinity level (144 mM NaCl, 6 mM KCl) on growth of 4 cotton genotypes (G. arboreum cv. Nanking 2558, G. barbadenese PS-4, G hirsutum cv. Deltapine 16, G. hirsutum cv. Tamcot SP 37). Growth

511

TABLE 1:   Ratios of K/Na in cotton (G. hirsutum)

Solution culture: nutrient solution, 1 mM KNO_3 + 10 mM NaCl
                    32 days of growth.

Apical 15 mm	Root Other Parts	Hypo-cotyl	Coty-ledons	Inter-node 1	Leaf 1 Petiole	Leaf 1 Blade	Shoot Tip
8.1	2.3	0.75	0.70	1.8	5.3	2.6	10

TABLE 2:   Na/K-ratios in two cotton genotypes grown under salt
           stress.

G. hirsutum cv. Tamcot SP 37:   salt tolerant.

G. hirsutum cv. Deltapine 16:   more salt sensitive.

Solution culture: nutrient solution, 6 mM KCl + 144 mM NaCl,
                    40 days of growth.

		Tamcot SP 37	Deltapine 16
Solution		24	24
Apical root		0.71	0.62
Hypocotyl		1.5	1.2
	1	1.9	1.3
	2	1.8	1.0
Internodes	3	1.5	0.80
	4	2.1	0.44
	5	2.0	0.37
	1 + 2	1.8	2.7
Leaves	3 + 4	4.2	1.0
	5	4.0	0.76

of all 4 genotypes was reduced, but growth reduction was least in Tamcot SP 37. Table 2 shows Na/K ratios in Tamcot SP 37 and the more salt sensitive Deltapine 16; Tamcot had higher ratios in most parts of the shoot, particularly in the upper leaves. These high Na/K ratios in Tamcot were due to accumulation of Na^+, and Cl^- was also accumulated in the shoot of this cultivar. In addition, Tamcot had a relatively uniform K^+ content of about 100 μmoles/g fresh weight throughout the plant, but K^+ levels were much more variable in Deltapine.

3. The significance of K/Na-selectivity in Tamcot was further studied at low and high equimolar K^+ and Na^+ concentrations (Table 3). The highest K/Na ratios (indicating K/Na selectivity) were found in the apical root and the first leaf. Discrimination between K^+ and Na^+ was less pronounced at the higher salt level, at which Na^+ accumulated in hypocotyl.

TABLE 3: Ratios of K/Na in cotton (G. hirsutum cv. Tamcot SP 37)

Solution culture: nutrient solution, equimolar K^+ + Na^+ concentrations, 14 days of growth					
K^+, Na^+ in solution, mM	root apical	proximal	hypocotyl	cotyledons	leaf 1
5	19	7.6	5.6	5.8	37
50	5.3	3.8	1.2	2.6	7.9

4. In excised low-salt roots of Tamcot, Na^+--in contrast to K^+--was not transported into the xylem vessels over a period of 24 hours, indicating Na^+ exclusion from the shoot in low-salt cotton roots.

Conclusions and Outlook

Salt tolerance in cotton appears related to accumulation of Na^+ and Cl^- in the shoot and to K/Na-selectivity which is set up in the root. In low-salt roots, K/Na-selectivity is also documented by a lack of Na^+ transport to the xylem. Presently, we are studying 7 cultivars of G. hirsutum to test whether salt tolerance of cotton is correlated with accumulation of Na^+ and Cl^- in the expanded regions of the shoot and with maintenance of adequate K^+ levels in the growing tissues.

References

1. T. J. Flowers, F. Troke and A. R. Yeo, Ann. Rev. Plant Physiol. 28:89-121 (1977).
2. E. Epstein, in: "Genetic Engineering of Osmoregulation: Impact on Plant Productivity for Food, Chemicals and Energy", (D. W. Rains, R. C. Valentine and A. Hollaender, eds.) pp. 7-21 (1980)
3. A. Läuchli, Ber. Deutsch, Bot. Ges. 92:87-94 (1979).
4. H. Greenway and R. Munns, Ann. Rev. Plant Physiol., 31:149-190 (1980).
5. A. Läuchli and R. Pfluger in: "Potassium Research - Review and Trends", pp. 111-163, International Potash Institute, Bern, Switzerland, (1979).
6. R. G. Wyn Jones, C. J. Brady and J. Speirs, in: "Recent Advances in the Biochemistry of Cereals", (D. C. Laidman and R. G. Wyn Jones, eds.), pp. 63-103, Academic Press (1979).
7. D. W. Rains, Experientia, 25:215 (1969).

CRUSTACEAN DIVERSITY RELATED TO THE SUBSTRATE IN TWO COASTAL LAGOONS IN BAJA CALIFORNIA SUR, MEXICO

J. Llinas, E. Diaz, E. Amador, M. Espinoza

Centro de Investigaciones Biologicas de
Baja California, A. C.
La Paz, Baja California Sur, MEXICO

Benthic communities, especially those in shallow water, are seasonally affected by different factors such as substrate, organic matter, climatic changes and physicochemical parameters (1, 2). A direct relation exists between the diversity of macrobenthic species and the type of substrate (3, 4). Although these studies are of great importance, very little research has been done in Mexico concerning interactions between benthic organisms and the environment. The crustaceans of coastal lagoons may be of commercial value and exploitation may need protection against human impact. The present study was initiated in 1979 to establish the relationship between the type of sediment and benthic crustacean diversity in two coastal lagoons.

The study areas were Enfermeria and Balandra, located in the Bay of La Paz. Enfermeria is a small shallow water lagoon, about 2500 m^2. It is bordered by Rhisophora mangle, Avicennia germinans, and Laguncularia racemosa and communicates with the Bay through a narrow mouth. Balandra is about 10,575 m^2 and communicates with the sea via an open mouth approximately 180 m wide. In addition to the mangrove species mentioned above, Conocarpus erectus is found in some places.

Sediments in Enfermeria correspond to thick limes; in contrast, Balandra varies from thick limes to medium sands. Pelecypods and gasteropod shells are abundant in the sandy sediments.

515

Table 1. Crustacean Diversity (Genera) in <u>Enfermeria</u> Summer 1979.

STATION	JUNE		JULY		AUGUST	
	ORGANISM	No.	ORGANISM	No.	ORGANISM	No
2	+ Callinectes sp.	2	+ Callinectes sp	1	+ Callinectes sp.	1
4	+ +		+ +		+ +	
5	Callinectes sp.	1	+ +		Eurytium sp. Callinectes sp.	1 1
8	+ +		Penopeus sp.	1	+ +	
9	+ +		+ +		+ +	
V	Balanus sp.	1	Eurytium sp Penopeus sp	1 1	Euritium sp	1
VII	+ +		Penopeus sp	1	+ +	

+ Observed in the body of water, but not appearing in the sample.
+ + No organisms found.

Table 2. Crustacean Diversity (Genera) in Balandra Summer 1979

STATION	JUNE		JULY		AUGUST	
	ORGANISM	No.	ORGANISM	No.	ORANISM	No.
I	* *		* *		* *	
2	Callinectes sp. Palaemon sp	1 1	Eurytium sp. Penaeus sp.	3 3	Callinectes Penaeus sp Palaemon sp	1 1 1
5	Eurytium sp. Penaeus sp.	1 1	* *		* *	
7	Callinectes sp. Eurytium sp. Alpheus sp. Squilla sp. Herbstia sp.	2 9 2 3	Callinectes sp. Eurytium sp Alpheus sp. Penaeus sp	1 13 1 2	Eurytium sp. Herbstia sp. Alpheus sp.	8 2 1
8	Eurytium sp. Squilla sp. Alpheus sp. Callinectes sp.	2 1 1 1	Eurytium sp. Squilla sp. Alpheus sp. Penaeus sp.	4 1 1 2	Eurytium sp Penaeus sp.	2 1
9	* *		Eurytium sp.	3	Squilla sp.	1
II	Balanus sp Eurytium sp.	10 2	Balanus sp. Eurytium sp.	3 3	Balanus sp Eurytium sp.	7 3
IV	Balanus sp. Eurytium sp.	8 1	Balanus sp.	23	Balanus sp Eurytium sp.	12 2
V	Balanus sp.	3	* *		Balanus sp.	2
VI	* *		* *		* *	

* NO ORGANISMS FOUND.

The study was done during the summer months of 1979. Seven sampling stations were established in Enfermeria (Map 1) and ten for Balandra (Map 2). Monthly samples were taken in June, July and August. Benthos samples were collected with a suction apparatus for soft bottoms and with a shovel for hard substrates. In both cases the sample covered an area of 0.2 m^2. All samples were sieved through 1 mm, 5 mm and 10 mm meshes and the organisms collected were preserved in 10% formalin. Physicochemical parameters were measured in situ with a Monitor Martek Mark V and organic matter content was determined by the Walkly and Black method (5). The shannon and Weaver index was used to determine diversity values.

Physiochemical parameters in Balandra were more or less constant during the study; however, dissolved oxygen changed during the last month (Figs. 1 and 2). The slight variation in pH (7.9 - 8.8) was probably due to changes in dissolved oxygen or in organic matter.

Organic matter varied from 1.2% to 21% in June, from 3% to 37.5% in July and 0.79% to 20% in August. The highest values are for the sampling stations located between the mangrove roots of Rhizophora mangle.

Enfermeria also showed a constant pattern in the physico-chemical parameters. As in Balandra, pH probably reflected dissolved oxygen and organic matter content and ranged from 8.1 to 8.8. Organic matter was again highest between the roots and where there is a predominance of fine grain sediments; ranges were 1.94% to 8.82% in June, 1.95% to 9.70% in July and 4.33 to 9.64% in August.

Abundance was low in both areas; however, Balandra showed a higher diversity (Tables 1 and 2). The highest diversity was found at stations 7 and 8; diversity indexes of 0.48 and 0.60, respectively.

Physicochemical parameters seemed to be constant in both areas, although for Balandra there were variations in August, probably due to the summer-autumn transition. Ranges in temperature, salinity and dissolved oxygen may be a result of evaporation and water exchange by tidal currents. The constancy of these factors suggests that they are not determinant for species divesity.

Organic matter limits the diversity and abundance of organisms (2), possibly in combination with the grain size of the sediment. The high values detected between roots is explained by the constant supply of mangrove leaves. Further, the diversity of benthic

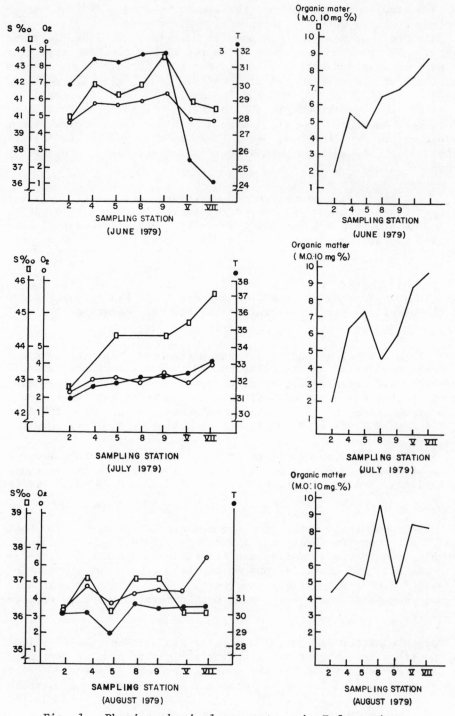

Fig. 1.   Physico-chemical parameters in <u>Enfermeria</u>

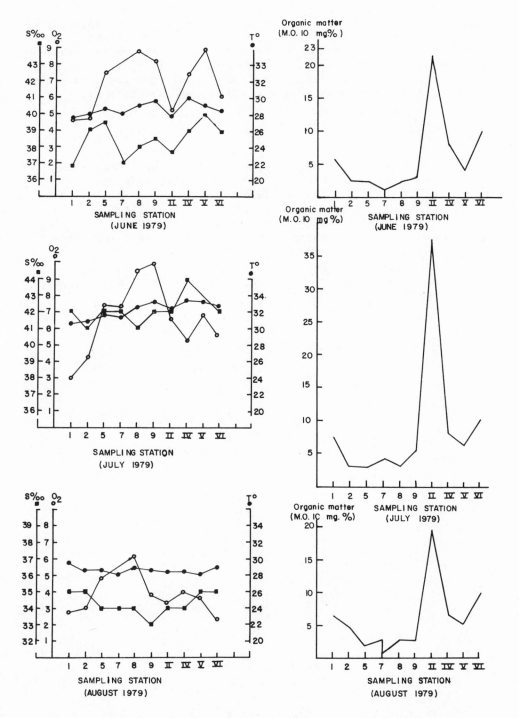

Fig. 2. Physico-chemical parameters in Balandra

organisms is directly related to the size of the interstitial space. In fine sediments, the interstitial space is small and the diversity is low. In medium and thick sands, and where corals and mollusc shells are present in the substrate, there is a consider-able increase in diversity. Increased oxygenation supports a wider variety of better morphologic and physiologic adaptations.

References

1. Espinoza, et al., "Informe General de Labores", Centro de
       Investigaciones Biologicas de Baja California, A. C.,
       La Paz, B.C.S., Mexico (1979).
2. R. G. Bader, Jour, Mar, Res. 13(1):32-47 (1954).
3. J. S. Gray, "Animal sediment relationships", Univ. Leeds,
       Wellcome Marine Laboratory, Robin Hood's Bay, Yorkshire,
       England (1974).
4. B. G. Thom, Jour. Ecol. 55:307-343 (1967).
5. H. A. Holme and A. D. McIntyre, I.B.P. Handbook No. 16, pp.
       30-52 (1971).

CALLUS INDUCTION AND CELL SUSPENSION CULTURES OF Ipomea pes-caprae

(L.) FOR SALT TOLERANCE STUDIES

G. F. Lopez, A. E. Castro and F. Cordoba Alva

Centro de Investigaciones Biologicas de Baja California
A. C. La Paz, Baja California Sur
Mexico

It is advisable that a vigorous research program on the mechanisms of salt resistance in halophytes should be part of any effort devoted to the establishment of crop culture under saline conditions (1). Recently, a survey of plants with halophytic characteristics in the neighboring areas of La Paz, B.C.S., Mexico, was made with the aim of studying such properties at a cellular level. I. pes-caprae has been selected for two main reasons: Firstly, it was found that this creeper plant is capable of growing in fresh water. This fact suggests that I. pes-caprae is a facultative salt tolerant plant and not an obligate halophyte (2). However, little is known about its salt tolerance capacity. Secondly, it was also found that I. pes-caprae possesses a particularly favorable response to plant tissue culture techniques.

The callus induction from leaves and stem segments was obtained using 20 ml of a basal media of MS (3) supplemented with 2, 4-Dichlorophenoxyacetic acid (2,4-D) and 6-Benzylaminopurine (6-BAP), both from Sigma Company, St. Louis, Missouri, USA, in 200 ml flasks (10x5 cm). After a one month period under continuous light, the callis were transferred and incubated in similar media increasing monthly the salt concentration: 0.1, 0.4, 0.5, and 0.08% NaCl, at 24°C.

The cell suspension culture grown in dark and/or under continuous light, were obtained by disintegration of 5 g of callus cells in 50 ml of basal media supplemented with 2,4-D and 6-BAP in 250 ml Erlenmeyer flasks. The cell suspension was maintained under constant shaking in a gyratory shaker (Eberbach Mod. 6140) at 24°C, for about 2 weeks. Five ml aliquots of suspended cells were

then transferred to 250 ml Erlenmeyer flasks containing 50 ml of
basal media at different NaCl concentrations, and incubated for a
month as indicated above. Cel growth was determined by dry weight
estimation.

Different hormone combinations (Table 1) were prepared to
study their effect in callus induction in leaves and stem segments.
The experiments showed that for callus production, stems responded
much better to the hormone treatment than leaves. Therefore, only
stems were used in the subsequent studies. In Table 1, it is also
shown that the optimal hormone concentration for callus induction
in stems, incubated under continuous light at $24^{\circ}C$, is 10^{-8} M of
6-BAP and 10^{-5} M of 2,4D. Hence, all cultures were carried out
using these hormone concentrations.

A significant light effect in the capacity of I. pes-caprae
callus cells or cells in suspension for growing in saline media has
been observed (Fig. 1). Cells cultured in dark showed always a
higher resistence to salinity than those incubated under continuous
light. Nevertheless, the cells seem to be unable to withstand
drastic changes in salt concentration, indicating that adaptation
to high salinity is only possible through stepwise increments of
salt in the media. These observations are in agreement with the
results of previous works on cultures of isolated tissues of glycophytes and halophytes (4). Thus, it appears that the salt
tolerance of callus tissues is indeed unrelated to the natural salt
tolerance of the plant.

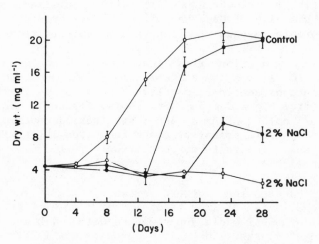

Figure 1. Light effect on the growth of I. pes-caprae cell
cultures in 2% NaCl. (o—o), Light; (•—•) Dark.

TABLE 1:   Hormonal effect on callus induction in $\underline{I}$. $\underline{pes\ caprae}$ stem segments.

Benzylaminopurine (M)	2,4-Dichlorophenoxyacetic acid (M)						
	0	10^{-3}	10^{-4}	10^{-5}	10^{-6}	10^{-7}	10^{-8}
0	−	−	−	−	−	−	−
1×10^{-3}	−	−	−	−	−	±	−
1×10^{-4}	−	−	−	±	−	−	−
1×10^{-5}	−	−	±	−	−	+	−
1×10^{-6}	−	−	+	+	+	±	−
1×10^{-7}	−	−	+	++	+	+	−
1×10^{-8}	−	−	+	+++	+	+	+

It remains to be answered whether the salt tolerance in callus is maintained throughout its different development stages up to the one considered as an adult plant. Work in this direction is in progress at our laboratory.

Acknowledgement

The authors express their appreciation to Biol. Jose Leon de la Luz, for his assistance in the identification of the creeper plant $\underline{I}$. $\underline{pes-caprae}$ (L).

References

1.   E. Epstein, "Crop Production in arid and semiarid regions using saline water," University of California, Davis (1978).
2.   Y. Waisel, Marine Halophytes, in: "Biology of Halophytes," pp. 261-291, Academic Press, Inc., New York and London (1972).
3.   T. Murashige and F. Skoog, Physiol. Plant 15:473-497 (1962).
4.   B. Golleck, (ed.) Salt Tolerance in Isolated Tissues and Cells, in: "Structure and Function of Plant Cells in Saline Habitats, pp. 19-35, John Wiley and Sons, Inc., N.Y. (1973).

BARLEY PRODUCTION: IRRIGATION WITH SEAWATER ON COASTAL SOIL

J.D. Norlyn and E. Epstein

Department of Land, Air and Water Resources
University of California
Davis, California 95616

The world's coastal deserts constitute a substantial natural resource that is presently underexploited (1). In view of the increasing problem of providing food for the world's population these areas ought to be utilized for some form of agricultural production if it is at all possible. Seawater, however, with the heavy load of salt it carries, has been considered the very antithesis of a resource for agriculture. If seawater, or some dilution of it, could be utilized on a portion of the vast expanses of coastal deserts for some form of agricultural production, a contribution toward alleviating the world's food dilemma could be made.

The first step in a project to utilize these resources would be to find a plant that would be tolerant of extremely high salinity and also provide some useful product. There are two basic types of plants that could be studied: the halophytes, salt tolerant wild plants, and the more salt sensitive crop plants. We elected to begin our studies with crop plants. Somers (2) has reviewed the work that he and others have done with halophytes; so this approach is also being studied.

Search for Maximum Salt Tolerance

The variation among crop plants in their ability to tolerate soil salinity has been summarized by Ayers and Westcott (3). Among the most tolerant of the crop plants is barley. The U.S. Department of Agriculture maintains a large and diverse collection of barley, some 22,000 different genotypes, that might include some very salt tolerant types. Until recently, however, no large-scale

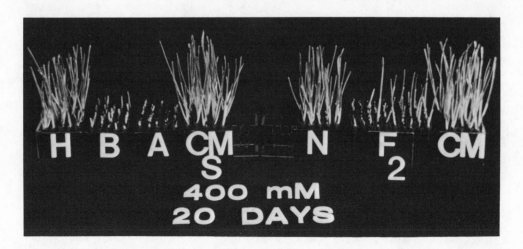

Figure 1. Germination paper and plastic rack seed support system
            used to treat barley seed with various levels of salt
            solutions in containers in the laboratory. Entry "H" is
            'U. C. Signal,' "B" is 'Briggs,' "A" is 'Arivat,' "CMS"
            is a selection from 'California Mariout,'   "N" is
            'Numar,' "F_2" is the F_2 generation from a cross between
            Arivat and the selection from California Mariout, and
            "CM" is 'California Mariout.'   The treatment is 400 mM
            NaCl seed imbibition for a period of 20 days.

testing for this trait in barley had been conducted. For these two
reasons, and the additional one that barley is a very well understood experimental organism (4), this crop was our first choice to
initiate studies of salt tolerance. If we could find high
tolerance to salinity in any of the barley strains, or could
develop it through a breeding program, it could be utilized to
improve existing cultivars by crop breeders.

     In our preliminary studies, we compared the ability of various
cultivars to germinate and establish seedlings under increasing
levels of salinity.   It became apparent that there was a rather
wide variation in the performance among the cultivars.   The
cultivar Arivat, "A", in Figure 1, is clearly unable to establish
seedlings, whereas the selection from California Mariout, "CMS",
appears rather successful.   The entry labeled "F_2", is the seed of
the second generation from a cross between Arivat and California
Mariout selection.   It appears that this F_2 population is segregating for the ability to germinate and establish seedlings when
stressed with this level of salinity, suggesting that this ability
is under some form of genetic control.

In subsequent studies it was observed that the ability to germinate during high salinity stress did not necessarily indicate that a particular cultivar or strain of barley was also tolerant of high salinity throughout the vegetative and reproductive stages of the life cycle. This change in performance during the life cycle has been observed in intermediate wheatgrass by Hunt (5), and in other crops by other workers.

A cross was made between Arivat and California Mariout. The F_1 seeds were planted and grown out producing the F_2 generation of seed. One hundred of these F_2 seeds were planted and grown out to maturity in the greenhouse to provide individual seed lots representing each individual F_2 seed.

The resulting seed lots were compared for yield when they were subjected to salt stress in the nutrient solution from the time the plants had reached the fourth leaf stage until maturity. This test was conducted again in the next generation to provide yield perfor-mance under stress during the vegetative and reproduction stages for both the F_2 and F_4 generations. The results of the comparison indicated that the ability to yield, under the salt stress imposed in these tests, is under genetic control as the coefficient of correlation (r) was 0.586, between the two generations, and was significant at the 5% level.

In another comparison of the F_4 progeny of this cross, the seeds of each lot were compared for the ability to germinate and establish seedlings under a salt stress of 350 mM NaCl. The results showed a considerable variation for this ability, ranging from types like Arivat that were unable to germinate, to those similar to California Mariout which gave almost 100% germination. When these results were compared with the yields as tested in the previous experiment, in which stress was not imposed until the vegetative stage, and was then continued through the life cycle of the plants, the correlation coefficient (r) was 0.12 and not signi-ficant. This finding supports the previous observations that the ability to germinate and establish a seedling under salt stress is not necessarily related to the ability of the plant to cope with salt stress during the remainder of the life cycle.

Field Trials at Bodega Marine Laboratory

Various selections that have been isolated from different screening programs that we have conducted were field tested on dune sand near the Bodega Marine Laboratory of the University of California, about 80 km north of San Francisco. Early results were reported by Epstein and Norlyn (6) and Epstein et al. (7).

The soil on which the field trials were planted had an analysis of 90% sand, 4% silt and 6% clay, and a cation exchange capacity of 10.8 me/100g. In addition this sand has a deep and open profile, allowing a very rapid internal drainage, well suited to this type of experimentation.

The first two years of these field tests were a period of drought in California, so that there was minimum interference from rain. In the next two seasons, the normal rainfall pattern returned and interfered with our experiments. During the 1979-80 season we erected a 24 by 30 meter plastic canopy to prevent rain from contaminating our experiments. The ends of the canopy were not covered and the sides were only partially covered to provide adequate air circulation.

We provided freshwater for the first irrigation to get the plants established, and then stepped up the proportion of seawater in the irrigation applied to selected plots as rapidly as possible until the desired treatments were being imposed: control (fresh water), 1/3, 2/3, and undiluted seawater.

The grain yields for the 1979-80 crop year are 3,102 kg/ha for fresh water, 2,390 kg/ha for 1/3 seawater, 1,436 kg/ha for 2/3 seawater, and 458 kg/ha for full seawater irrigation. These values are the average of all of the entries in each treatment.

Where we irrigated with undiluted seawater for nearly the entire season, the average yield of 458 kg/ha is clearly not too impressive. When, however, the two main resources (seawater and sand) are considered, this yield does gain some measure of significance. The grain yields at the lower concentrations of seawater are near the levels that are normally observed for the average yield of barley in the world which was estimated at 1710 kg/ha for 1975, despite the fact that these were still very highly saline irrigations (2/3 seawater corresponds to 23,000 ppm salt, and 1/3 to over 11,000).

Conclusions

The evidence presented suggests that irrigation of barley with up to at least 2/3 seawater is feasible and may result in yields the economic significance of which will depend on local conditions. Selection and breeding are likely to improve the performance of barley under these highly saline regimes. The work should be extended to coastal areas more likely to be put to this use than is the coast of Northern California, in terms of climate, extent, and economic and social conditions.

Acknowledgements

This work was supported by the Office of Sea Grant, U.S. Department of Commerce. We are indebted to International Plant Research Institute, Inc., for making available the plastic canopy.

References

1. P. Meigs, "Geography of Coastal Deserts," UNESCO, Paris (1966).
2. G. F. Somers, in:"The Biosaline Concept: An Approach to the Utilization of Under-exploited Resources," A. Hollaender, ed., Plenum, New York (1979).
3. R. S. Ayers and D. W. Westcot, "Water Quality for Agriculture," Food and Agriculture Organization of the United Nations, Rome (1976).
4. D. E. Briggs, "Barley," John Wiley and Sons, New York (1978).
5. O. J. Hunt, Crop Sci. 5:407-409 (1965).
6. E. Epstein and J. D. Norlyn, Science 197:249-251 (1977).
7. E. Epstein, J. D. Norlyn, D. W. Rush, R. W. Kingsbury, D. B. Kelley, G. A. Cunningham and A. F. Wrona, Science 210:399-404 (1980).

INDUCTION OF RESISTANCE TO SALINITY IN THE FRESHWATER ALGA

CHLAMYDOMONAS REINHARDTII

G. T. Reynoso, B. A. de Gamboa and S. R. Mendoza

Centro de Investigaciones Biologicas
  de Baja California
Baja California Sur, Mexico

The unicellular green flagellate Chalmydomonas reinhardtii (Chlorophyta) has been widely used for investigations both at the cellular and the molecular level (1). Although C. reinhardtii is a freshwater organism, it has a close phylogenetic relationship with some halotolerant organisms like Carteria and Dunaliella (2). The mechanism of osmoregulation in such microorganisms is yet unclear (3). Some organisms show variations in the biosynthesis of compounds considered to be osmoregulatory, in response to hyper-osmotic stress (4). Thus, attempting to shed some light on this problem, we report in the present paper the effect of exogenous proline and taurine during NaCl hyperosmotic stress in Chalmy-domonas reinhardii.

The clonal cell line (11/32 C wild Type Mt) used was isolated from a strain obtained from the Culture of Algae and Protozoa Center of Cambridge, England. Asparagine, glutamic acid, gluta-mine, histidine, hydroxyproline, methionine, phenylalanine, proline and taurine (L-aminoacids) were purchased from Sigma Co., St. Louis, MO, USA.

The cells were grown in liquid cultures in 250 ml Erlenmeyer flasks, containing 100 ml of basic culture media MIA (5) at different salt concentrations and 25 C under constant illumination (10,000 Lux). Cell growth was measured by turbidimetry in a Linson 3 spectrophotometer using a 480 nm filter.

The salt concentration effect on cell growth is shown in Figure 1. Table 1 shows the effect of different aminoacids in the

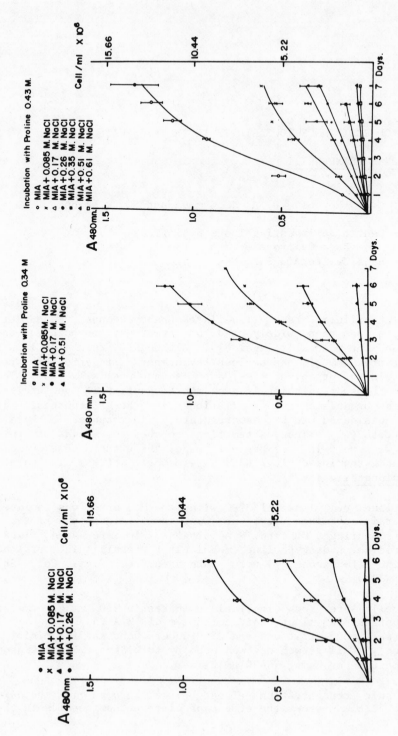

Fig. 1. Effect of NaCl concentrations on the growth of *C. reinhardtii*. Mean values of about 9 determinations are indicated together with standard deviation.

Fig. 2. Effect of salt concentration on *Chlamydomonas* cells preincubated in proline. Mean values of more than 6 determinations are indicated together with standard deviation.

TABLE 1:   Effect of different Amino Acids on <u>Chlamydomonas</u>
<u>reinhardtii</u> algae.*

	% of Growth
Control (no salt)	100
0.05 M NaCl	49
Pro plus 0.05 M NaCl	111
Tau "	99
Glu "	76
Gln "	72
Asp "	54
Hyp "	53
Met "	50
His "	49
Phe "	47

*Amino Acid concentration 10^{-2} M.

growth of <u>Chlamydomonas</u> under salinity stress (0.085 M NaCl). Of all amino acids tested, proline and taurine shows the most significant effect on the cell growth at 0.085 M NaCl. Glutamine and glutamic acid also increase cell growth under similar conditions, but less efficiently. Others show no effect at all. The function of these aminoacids in osmoregulatory processes is a matter of controversy (6, 7). However, our results suggest that the presence of exogenous taurine and proline might confer NaCl resistance. The effect is less pronounced with taurine than with proline.

The effect of proline and taurine is more evident when the algae are cultured for several days in an isosmotic medium before being transfered to an amino acid-free media with different NaCl concentrations. If proline concentration in the culture is 0.34 M, the algae are able to grow at 0.51 M NaCl. Moreover, if the concentration of this amino acid is raised up to 0.43M, the algae can still grow at 0.61 M NaCl, which is the marine salt concentration (Fig. 2).

It is a remarkable fact that <u>Chlamydomonas</u> incubated in a medium with proline resists high <u>salinites</u>. This suggests an uptake of the amino acid into the cell and therefore mechanisms of proline action can be suggested:

a)   As an osmoregulator. This is probably not the case since
     cell division will cause dilution of the proline pool.

b)   As an activator of some complex mechanism which assures
     cell viability. This idea is supported by the maintenance
     of halotolerance during succeeding generations.

The induction of salinity resistance in an obligate freshwater
alga provides a new approach to the study the phenomenon of halotolerance.

ACKNOWLEDGEMENTS

The assistance of Dr. Felix Cordoba Alva and Dr. Jose Luis
Ochoa in the preparation of this manuscript is greatly appreciated.

REFERENCES

1.   G. A. Hudock and H. Rosen, in: "The Genetics of Algae,"
        Vol. XII 29-48,
        Blackwell Scientific Publications (1976).
2.   A. D. Brown and J. L. Borowitzka, in: "Biochemistry and
        Physiology of Protozoa," Vol. I,
        139-190, Academic Press, Inc. (1979).
3.   J. A. Hellebust, Can. J. Bot., 54:1735-1741 (1976).
4.   B. Gollek, (ed.) "New Trends in the Study of Salt
        Tolerance," Halsted Press, NY, Toronto, Canada (1973).
5.   R. Sager, J. Gen. Physiol. 37:729 (1954).
6.   R. Gilles and A. Pequeux, Biochem. Physiol. Vol. 57:183-185
        (1977).
7.   B. Schobert, Physiol. Pflanzen 175:91-103 (1980).

SELECTION FOR YIELD IN CEREALS FOR SALT-AFFECTED CROPLANDS

R.A. Richards, C.W. Dennett, C.W. Schaller,
C.O. Qualset and E. Epstein

Departments of Agronomy and Range Science and
Land, Air, and Water Resources
University of California, Davis, CA 95616

The development of crops with genetic resistance to salt in the soil has sometimes been proposed as a partial solution to the increasing problems of salt buildup in agricultural lands. Despite the fact that this approach may partially offset the detrimental effects of saline soils on crop yields, data on genetic relationships of crop growth and yield in saline soils are meager, so that specific breeding objectives for crops to be grown on these soils are not yet clear.

As a first step in a breeding program to improve the yield of wheat and barley in salt-affected soils, we have set out to obtain some of this basic information. The results presented here are part of a larger study to characterize the performance of many wheat and barley genotypes when grown on a highly variable, but typical salt-affected field in the San Joaquin Valley of California.

The experimental area was part of an irrigated wheat field. The entire area was managed by the grower using standard cropping practices. At crop maturity a sampling procedure was adopted whereby, for every genotype, total aboveground biomass, total grain yield, and soil salinity were measured from areas with low salt concentrations to areas with high salt concentrations in the soil.

Results

For all genotypes there was a linear relationship between salt concentration in the soil, and biomass and grain yield. Comparisons among genotypes over all salt concentrations indicated three major findings. These are listed below, and are further

TABLE 1. Grain yield, biomass and harvest index (HI) of some wheat
and barley cultivars at an ECe of 5 and 15 mmhos/cm.

	ECe (mmhos/cm)					
	5			15		
Cultivar	Yield	Biomass	HI	Yield	Biomass	HI
	g/m^2			g/m^2		
Wheat						
Baart	278	1216	0.23	104	458	0.23
Yecora Rojo	395	946	0.42	244	583	0.42
Barley						
Arivat	529	1317	0.40	227	575	0.39
WPB-20-16	382	985	0.39	226	637	0.35

illustrated in Table 1 using examples of two cultivars of both
wheat and barley.

1. Significant differences between genotypes over all
    salinity levels.

2. The presence of substantial genotype x soil salinity
    interactions in both wheat and barley. These inter-
    actions were such that genotypes yielding most biomass
    on the most saline soils often yielded comparatively
    less on soils with lower salt concentrations and vice
    versa.

3. Intrinsic differences in harvest index (the ratio of
    grain to total above ground biomass) among the wheat
    genotypes altered their relative performance in terms
    of grain yield compared with biomass. Thus wheat geno-
    types with the highest grain yield on the more saline
    soils were not necessarily those that produced the most
    biomass. Wheats containing the Norin 10 dwarfing genes
    are known to have a higher harvest index than taller
    wheats without these genes and it was the former that
    yielded most grain in these saline soils. Variation
    in harvest index among the barley genotypes tested was
    not great and did not influence yield variation among
    genotypes in the salt-affected soils.

Conclusions

Breeding wheat and barley for salt-affected areas must not only take the above findings into consideration but also must consider the nature of salt-affected fields. Fields are typically highly variable, having areas where salt concentrations are so low that crop yield is not affected and areas where not even halophytes will grow. The predominant area, however, has salt concentrations in the soil with an ECe below 10 mmhos/cm. Areas with concentration above this are fewer and contribute comparatively little to the total grain yield of a salt-affected region.

Considering the nature of salt-affected soils, and the presence of substantial genotype x salinity level interactions, we suggest that breeding emphasis should be placed on the production of cultivars with good yield potential on soils with salt concentrations below an ECe of 10 where growth and yield of presently available varieties are significantly, but not disastrously reduced.

OSMOREGULATORY MECHANISMS OF THE MESOHALOPHILIC

DEAD SEA BACTERIUM BA$_1$

M. Risk[1], R. Gelber[2]. S. Knock[3], and R.C. Wood[4]

[1]Biochemistry, Univ. of Texas Medical Branch, Galveston,
TX 77550, USA; [2]Physical Chemistry, Technion
Haifa, Israel; [3]M.B.I., U.T.M.B., Galveston, TX;
[4] Microbiology, U.T.M.B., Galveston, TX

SUMMARY

Ba$_1$ is a motile, negative, obligately aerobic bacterium
from the Dead Sea salt flats. Under standard growth conditions,
Ba$_1$ grows in the range of 0.2 to 4.0 M total salts. When grown
over a wide range of NaCl concentration, betaine was synthesized as
a linear function of osmolarity of the growth medium, presumably as
an internal osmotic regulator. However, betaine also changes five-
fold within a growth curve at a single molarity. Cyclopropane
fatty acids constitute a large proportion of the total cellular
fatty acids, up to 55% during plateau phase growth. These unusual
lipids may aid in forming a barrier to salt entry.

INTRODUCTION

Ba$_1$ is a motile bacilliform nonspore-forming gram negative
bacterium isolated in the mid 1960's from a salt drying pond at the
southern end of the Dead Sea (Raefeli-Eshkol, 1968). Although this
organism can grow in a minimal medium with glucose as the sole
carbon source, the majority of work on it has been in nutrient
broth medium, in which Ba$_1$ grows well in suspension in the range
0.2 M to 4.0 M total salts. Since this is an unusually wide range
of osmotic tolerance, this study was undertaken to ascertain the
mechanisms permitting such growth.

Since there are precedents for accumulation of low molecular
weight hydrophilic metabolites within cells in proportion to external osmotic pressure, several such components were examined. In

addition, lipid composition was examined as a possible adaptive mechanism permitting high salt tolerance.

MATERIALS AND METHODS

Ba$_1$ was grown in Difco nutrient broth (8 g/l) in 0.05 M KCl, 0.1 M MgCl$_2$, and NaCl added to give the desired molarity. Suspension cultures were grown in a water bath with a shaker or New Brunswick Shaking incubator at 37°C, 150 strokes/min.

Growth was measured by absorbance at 600 nm. Harvested cells were pelleted at 9000 g for 10 minutes at 25°C, washed twice in isotonic medium without nurient broth, suspended in distilled water, and an aliquant removed for protein measurement (Sedmak and Grossberg, 1977). Cells were ruptured at ice temperature in water/acetone/chloroform 1:1:0.5 with a Polytron homogenizer for 10 seconds, and centrifuged at 2000 g, 5 minutes. Suitable volumes of the upper (aqueous) phases were used for betaine determination (see below). The remainder was dried under nitrogen gas and saponified; free fatty acids were methylated, extracted, and analyzed by gas liquid chromatography as described previously (Jungkind and Wood, 1974), using a Varian model 1200 gas chromatograph with 5% diethylene glycol succinate coating Chromosorb W; the column was run at 190°C. Retention times were compared with those of known standards. Identity of the peaks was verified by catalytic hydrogenation (removing unsaturated lipid peaks) and bromination (which cleaves cyclopropane derivatives at the ring substituent site, resulting in the disappearance of these peaks) (Brian and Gardner, 1968).

Betaine was separated from nonacid congeners by a two step precipitation method and quantified by a spectrophotometric assay of the iodine complex, modified from previous techniques (Story and Wyn-Jones, 1977; Barak and Tuma, 1979). Concentrations were determined from a standard curve and expressed per mg of protein. Verification of the precipitated metabolite as betaine was accomplished by comparison with authentic iodide salt of betaine, using a Finnigan 4000 mass spectrometer. The sample was introduced on a solid probe and volatilized by gradual heating. In the electron impact mode, a 250°C ion source at 70 ev was used to analyze peaks from 20 to 400 m/z. Isobutane chemical ionization was carried out at 200°C.

RESULTS

Ba$_1$ was grown in suspension, in media from 0.2 to 3.0 M salt to approximately the same cell density. Betaine synthesized by these cells expressed per mg protein was plotted as a function of growth medium salt concentration. A linear function of betaine vs. molarity was obtained (Fig. 1).

Alternatively, cells were grown in 3.0 M total salts, and harvested periodically. Endogenous betaine per mg protein was determined throughout the growth curve from early log to plateau phase (Fig. 2); a sharp peak was observed in mid log phase, followed by a decline. The betaine declining segment of the curve can be fitted by a power function ($r^2=0.989$).

The residue of the homogenized cells after betaine determination was analyzed for total fatty acids; six GLC peaks were observed consisting of C_{16}, C_{17}, C_{18}, and C_{19} components; two unidentified peaks comprised 5 to 10% of the total. The two cyclopropane fatty acid (C_{17} and C_{19}) peak areas were totaled and expressed as a percentage of total fatty acids; this value was plotted vs. time of harvest. This component increased monotonically from about 20 to 55% of total fatty acids; a power function described this relationship with good fit ($r^2=0.938$) (Fig. 2).

Betaine identification was performed by mass spectrometry. Sample applied to the solid probe was heated from ambient temperature to 360°C; ion intensities as a function of temperature were compared for standard and cell isolated material. Both showed similar pyrolysis peaks. The molecular ion was not observed, but characteristic marker ions of quaternary amines were found for both substances (Table I).

DISCUSSION

Various microorganisms living in high salt concentrations have been shown to osmoregulate by synthesizing aqueous soluble intracellular metabolites. This includes glycerol in dunaliella (Brown, 1978), and amino acids in bacteria (Stanley and Brown, 1974). Betaine accumulates in vascular halophytes (Story and Wyn-Jones, 1977) but apparently has not been observed previously in procaryotes in a similar habitat. We have now identified this compound as a major metabolite in Ba$_1$, a euryhaline bacterium isolated from a saturated salt lake, the Dead Sea (Neev and Emery, 1967). Thus an additional strategy permitting life in extreme environments has been discovered. It appears that betaine is synthesized as a direct response to external salt levels (Fig. 1). Since betaine concentration varies within the growth curve for cells grown at 3.0 M, it will be necessary to quantitate betaine synthesis over a time course for each molarity to validate this observation. This method may be utilized by other mesohalophilic bacteria from the same source, especially Pseudomonos halestorga which shows many similarities with Ba$_1$ (Vulcani, 1940).

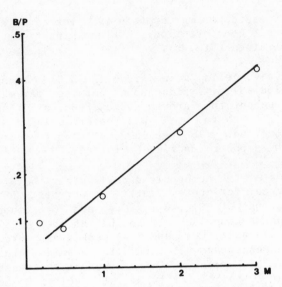

Figure 1. Ba₁ cells were grown at the designated molarities as
described. Cells of 0.6 to 1.0 absorbance units were analyzed for
betaine iodide at 365 nm; betaine concentration per mg protein was
plotted vs. the molarity of the growth medium (r=0.967).

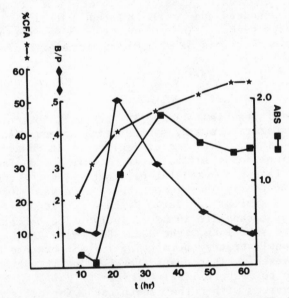

Figure 2. Ba₁ cells were grown at 3.0 M as described. Cells were
analyzed for growth (Abs. 600 nm) at indicated times (squares) and
protein determined on washed pellets. The same samples were
analyzed for betaine (diamonds) and fatty acids. The cyclopropane
fatty acids were expressed as percent of total fatty acids
(stars).

TABLE I

Positive Ions Observed by Electron Impact of Betaine Iodide Standard and Presumptive Betaine Iodide from Ba$_1$

m/z	Identity
44	$\overset{\displaystyle H}{\underset{+}{\overset{\displaystyle \mid}{CH_3-N=CH_2}}}$
58	$\underset{+}{CH_2N\ (CH_3)_2}$
72	$\underset{+}{(CH_3)_2N(CH_3)_2}$
127	I^+
142	$CH_3\ I^+$
149	Pthalate contaminate

The N-demethylated ion at 58 m/z is the characteristic marker of quaternary amines in E.I.M.S. (Karlanden et al., 1973). In chemical ionization (isobutane reagent gas), both standard and cell extract exhibit the protonated N-demethylated betaine molecular ion (m/z=104). This confirms the identity of the Ba$_1$ quaternary amine as betaine.

Cyclopropane fatty acids have been shown in high levels in Ba$_1$ previously (Peleg and Tietz, 1973); however, only a single determination on the growth curve was reported. At this stage (presumably a plateau phase), they constituted about 60% of the total. The C$_{17}$ cyclopropane fatty acid was found in low levels, whereas in our study it constitutes a much higher proportion (data not shown).

It is well known that cyclopropane fatty-acid synthesis increases during the later growth stages of cultured bacteria (O'Leary, 1967). Since this is the case with Ba$_1$ as well, the relative decrease of betaine concentration from mid log phase, concomitant with the rise of cyclopropane fatty acids suggests an additional mechanism for coping with osmotic stress. These unusual lipid components are characteristic of bacteria living in acid medium (Levin, 1977) and extreme temperature (Chan et al., 1971; Marr and Ingraham, 1962).

ACKNOWLEDGEMENTS

We would like to express our appreciation to Prof. Y. Avidor, Technion University, for Ba_1, to Prof. B. Volcani, Scripps Institute of Oceanography, for Ps. halestorga, and Dr. E. W. Czerwinski for support.

REFERENCES

1. D. Raefali-Eslkol, Biochem. J. 109:670-85 (1968).
2. J. J. Sedmak and S. E. Grossberg, Anal. Biochem. 79:544-52 (1977).
3. D. L. Jungkind and R. C. Wood, Biochim. Biophys. Acta. 357:286-97 (1974).
4. B. L. Brian and E. W. Gardner, Appl. Microbiol. 16:549-52 (1968).
5. R. Story and R. G. Wyn Jones, Phytochem. 16:447-53 (1977).
6. A. J. Barak and D. J. Tuma, Lipids 14:860-4 (1979).
7. A. D. Brown, in: "Energetics and Structure of Halophillic Microorganisms," S. R. Caplan and M. Ginzberg, eds., pp. 625-40, Elsevier/North Holland (1978).
8. S. O. Stanley and C. N. Brown, in: "Effect of the Ocean Environment on Microbial Activities," R. R. Colwell and R. Y. Morita, eds., pp. 92-103, University Park Press (1974).
9. D. Neev and K. O. Emery, "The Dead Sea," Geological Survey No. 41, Minister of Development, Jerusalem, Israel (1967).
10. B. Volcani, "Studies on the Microflora of the Dead Sea," Ph.D. dissertation, Hebrew University (1940).
11. E. Peleg and A. Teitz, Biochim. Biophys. Acta. 306:368-79 (1973).
12. W. M. O'Leary, in:"The Chemistry of Metabolism of Microbial Lipids, p. 15, World Pub. Co. (1967).
13. R. A. Levin, J. Bacteriol. 108:992-5 (1971).
14. M. Chan, R. H. Hines, and J. M. Akagi, J. Bacteriol. 106:876-1 (1971).
15. A. G. Mar and J. L. Ingraham, J. Bacteriol. 84:1260-7 (1962).
16. S. G. Karlanden, K. A. Karlsson, and I. Paschea, Biochim. Biophys. Acta. 326:174-83 (1973).

PROBLEMS OF SPECIES INTRODUCTION WITH SEAWEEDS

O.W. Terry

Marine Sciences Research Center
State University of New York
Stony Brook, New York, USA

The biosaline concept emphasizes the development of saline plant agriculture through a variety of approaches, presumably including the use of introduced species. Introduction of such "exotic species" of either plants or animals has, however, been a frequent target of concerned ecologists --and with some reason. The record of many past introductions is such as to justify their concern. All of us know examples, ranging from the rabbits of Australia to the water hyacinth of Florida, where introductions have become serious and expensive pests. Caution is certainly indicated.

Some introductions have, on the other hand, been apparently quite successful. Pacific salmon and striped bass were moved across the United States, in opposite directions, with good results. The Pacific oyster may have brought some of its pest species to American and European shores but, on balance, seems to be considered an asset there. The problem clearly is to know which species to introduce. Unfortunately there is no sure way of making such decisions, no matter how complete amd methodical the prior investigation. The introduced species will often react in unexpected ways with its new environment, sometimes occupying a different niche from the one previously filled. Careful study reduces the risk of damage from unwanted proliferation but cannot eliminate it. This is a significant risk but one which can and should, under appropriate controls, be accepted. The real problem is that of choosing an appropriate control mechanism.

Previous considerations of this problem have usually proposed that it be dealt with by some form of a committee of experts. The committee's membership would be either professional ecologists or officials of the appropriate regulatory agency, in the United States probably a state environmental agency or department. Species introductions would be approved or rejected, after due study and consideration, by this committee and would be allowed only upon approval.

This seemingly rational approach contains, from the point of view of the prospective mariculturist and also that of the general public, a significant flaw. Such a committee's decisions will almost inevitably lean in the direction of excessive caution and especially so in the case of seaweed introductions. Seaweeds, even more than marine animal introductions, are almost certain to escape from an open water culture system and therefore present, from the viewpoint of the authorizing group, a special and very likely an unacceptable risk. My home state of New York has a ready example of the dangers involved. We have there a non-native green seaweed, Codium fragile, which was apparently introduced by accident under wartime conditions and has now spread widely and become a nuisance on beaches and a pest in shellfish beds. Though actual losses have probably been exaggerated, the result of this and other similar examples is that seaweed introductions are under tight control by a state agency.

Although no one has, to my knowledge, even applied as yet for authorization to culture an exotic seaweed in New York waters the possibility is not as remote as it might seem. For several years, since the energy problem became acute, there has been a substantial research project directed toward the large-scale culture of giant kelp, Macrocystis spp, as a source of biomass energy. The project has been conducted off the California coast where giant kelp grows naturally. During the past year a companion project was initiated at Stony Brook to investigate the potential for seaweed biomass production on the East Coast. Our goal was to develop a suitable culture system, comparable in productivity and efficiency to the giant kelp system, using native East Coast species. It will be several years, perhaps more, before the results of this research are known but it is fairly apparent already that this goal will not be easily achieved. The Macrocystis plant has several important characteristics, including size and surface-floating growth habit, which appear to give it significant advantages over any East Coast species we might choose to work with. Assuming that an introduction of giant kelp to the East Coast would succeed biologically, such an introduction might well prove a much easier alternative to the current project plan. Let me emphasize that no such introduction has been seriously proposed at this time or is likely to be proposed in the near future. Let us suppose, however, that it is proposed, perhaps at some future time when alternative energy sources are desperately needed.

Whatever group has the authority to approve introductions --I assume that some possibility of approval will exist --would, properly, consider the risks involved in a <u>Macrocystis</u> introduction. They would almost certainly conclude that the risks are substantial. Giant kelp would inevitably escape, if not as spores then as drift plants, from whatever culture structure is used. Many square miles of habitat similar to that occupied by <u>Macrocystis</u> in California waters exists on the East Coast. If it became established there on a large scale the impact would not be minor; drift weed would accumulate on beaches, fishing and boating activities would be interfered with, and powerful political interest groups would be offended. With this prospect in view, and without even considering possible harmful impact on local marine ecosystems, it would be a bold committee indeed that would take the responsibility for approving introduction. Yet it is quite possible that the public interest, and actual preference if consulted, would support introduction even under these circumstances. The social and economic values of a successful culture industry might far overbalance any nuisance problems associated with the culture, or in any case the <u>risk</u> of such problems. No committee of scientists or agency officials can be expected to weigh potential benefits as heavily as they do risks in this kind of situation.

The regulation of seaweed introductions is only one instance of a more general environmental problem, that of insuring a properly balanced regulatory mechanism. Mariculture is particularly vulnerable to over-regulation simply because it involves unfamiliar concepts. The risk/benefit analysis, never easy under any circumstances, is almost inevitably biased toward risks, especially if the risks are to large and well-established interest groups such as those that now utilize the marine environment. It has become commonplace to note that mariculture's problems are not so much biological as social and legal. This being so it becomes necessary for mariculture as an industry, and a research enterprise, to make sure that its efforts are directed to where the problems are.

Again as with environmental problems generally, the basic solution appears to lie in the direction of greater public participation in the decision-making process. Unless this involvement is somehow facilitated the demands of (perfectly legitimate) established interests will almost inevitably receive undue weight when balanced against the potential values of a new enterprise, especially one as novel as mariculture. It is essential that a broader perspective be brought into the decision-making process, essentially a planning perspective.

The need for better-planned use of the limited space and resources of coastal ocean is not a particularly novel discovery,

and efforts to meet the need have made progress, though at a sometimes agonizingly slow pace. Coastal Zone Management (CZM) in the
United States is an attempt to institute, against some very well-
established interests, a more rational and broadly based planning
and management structure. Essentially it is an appeal to the
political process to correct perceived abuses such as pollution and
mismanagement of resources under customary procedures. Each state
is encouraged to develop its own unique approach to CZM, based on
its own existing institutions and regulations. Usually the result
is a shifting of responsibility away from municipal government
levels to regional and statewide groups, which are likely to have
the advantages of access to planning expertise and lessened vulner-
ability to special interests. How much actual public participation
is achieved in the process depends on many factors and in large
degree remains to be seen.

A related but different recent initiative toward more
effective planning through increased public participation was
introduced on Long Island in response to Section 208 of the Federal
Water Pollution control Act Amendments of 1972 (PL 92-500). Long
Island's groundwater is its only source of fresh water supply, one
which is ultimately limited by the rate of precipitation but also
by increasingly severe pollution problems. The objective of
Section 208 planning was to develop comprehensive area-wide plans
for optimum use and protection of this limited water resource from
further pollution. A novel feature of the Long Island 208 project
was the extensive involvement of a large and active volunteer
"Citizen Advisory Committee" in the entire process, from the
establishment of goals and objectives to the presentation of the
completed plant to the general public. Section 208 has only
peripheral involvement with the marine environment (though a very
important effect through control of pollution) but may be a
suggestive model. It demonstrates one route to greater public
participation in planning.

Summing up, the exotic species problem promises to become an
increasingly important issue as marine agronomy, and all of mariculture, expand toward industry status. Control over seaweed introductions to minimize ecological risk is necessary and inevitable
but it is crucial that this control be exercised in a reasoned and
appropriate manner. The control methods most often suggested in
the past are unlikely to provide the kind of balanced decisions
that are needed. It therefore is important for marine agronomists
and mariculturists to be aware of the exotic species issue, as a
major facet of the whole regulation problem in their industry, and
to exert their influence to bring about more rational ways to deal
with it.

PHOTOSYNTHESIS IN SALT STRESSED WOODY PERENNIALS

R. R. Walker

CSIRO, Div. of
Horticultural Research
Merbein, Victoria, 3505
Australia

W. J. S. Downton

CSIRO, Div. of
Horticultural Research
Glen Osmond, Adelaide, SA, 5001
Australia

In this paper we summarize our recent work on the photosynthetic responses to NaCl stress of three horticulturally important crop plants - grapevine, citrus and guava. Special attention is given to the physiological and biochemical changes occurring in grapevine and citrus leaves during stress and following cessation of salt-treatment. The changes are also related to leaf water status and chloride concentrations.

Comparative effects of NaCl on photosynthesis

Photosynthesis was measured by infrared gas analysis in mature leaves of citrus (Citrus medica cv. Etrog. Citron), guava (Psidium guajava L. cvs 11-56 and GA-37) and grapevines (Vitis vinifera L. cv. Sultana syn. Thompson Seedless) at various times after the commencement of NaCl treatments. Treatments ranged from 0-50, 0-75 and 0-90mM for citrus, guava and grapevine, respectively. In all cases, photosynthesis was progressively reduced by salinity. The initial decline was largely a result of increased stomatal resistance to CO_2 transfer (Table 1). With increasing duration of exposure to salinity increased internal resistances to CO_2 fixation became evident (Table 2). A stimulation of photorespiration relative to net CO_2 fixation was apparent by comparing rates of photosynthesis in 20% and 2% O_2. This effect occurred initially in the absence of a marked increase in internal resistance (Table 1), but with further duration of salt treatment was associated with both higher stomatal and internal resistances (Table 2). These data indicate alterations in the relative activities of carboxylation and oxygenation in salt-stressed plants. This was checked by

549

TABLE 1. Leaf chloride concentrations, photosynthesis, stomatal and internal resistances and the enhancement of photosynthesis between 20% and 2% O_2 in control and salt-stressed leaves of guava, grapevine and citrus. Measurements were made 12, 3 and 8 weeks respectively after the beginning of salt-treatments. Values are means $\pm$s.e. of 8 replicates (guava), 3 replicates (grapevine, citrus control) or 2 replicates (citrus, 50mM NaCl).

Plant	Treatment	Leaf Chloride (mM)[1]	Net Photo-synthesis (ngCO$_2$cm^{-2}s^{-1})	Stomatal Resistance (s.cm^{-1})	Internal Resistance (s.cm^{-1})	% Enhancement of photosynthesis between 20% and 2% O_2
Guava	Control	77+5	57.1+2.6	0.61+0.06	3.38+0.27	48.1+5.9
	75mM NaCl	152+9	36.8+2.2	3.01+0.34	4.11+0.75	106.3+15.5
Grapevine	Control	19+1	45.5+2.5	1.13+0.13	4.81+0.45	55.6+5.3
	90mM NaCl	99+9	26.1+1.2	7.88+3.16	4.73+0.93	98.0+12.7
Citrus	Control	49+4	37.0+3.1	1.96+0.06	5.10+0.69	63.8+4.8
	50mM NaCl	339+30	22.8+2.4	5.34+1.70	6.50+0.52	98.5+19.3

[1]Tissue water basis

TABLE 2. Leaf chloride concentrations, photosynthesis, stomatal and internal resistances and the enhancement of photosynthesis between 20% and 2% O_2 in leaves of control and salt-treated grapevine and citrus at the time of stress-relief (5 and 10 weeks respectively after the beginning of salt-treatment) and either 40 or 14 days after stress-relief for grapevines and citrus, respectively. Values are means +s.e. of 3 replicates (grape-vines, citrus control) or 2 replicates (citrus salt-treated). Salt-treatments were 90mM NaCl and 50mM NaCl for grapevine and citrus, respectively.

Plant	Time	Treat-ment	Leaf chloride (mM)[1]	Net Photo-synthesis (ngCO$_2$cm^{-2}s^{-1})	Stomatal Resistance (s.cm^{-1})	Internal Resistance (s.cm^{-1})	% enhancement of photosynthesis between 20% and 2% O$_2$
Grape-vine	Stress-relief	Control	5±0.2	45.0±1.4	1.2±0.1	6.2±1.3	40.6±3.1
		Salt-treated	206±2	20.4±3.0	5.0±0.6	11.2±1.3	72.9±9.2
Grape-vine	40 days after Stress-relief	Control	11±0.4	38.4±0.5	1.6±0.3	6.6±0.2	48.4±12.3
		Salt-treated	118±13	41.0±5.4	1.7±0.5	6.2±0.04	47.3±2.2
Citrus	Stress-relief	Control	46±12	35.4±2.2	2.3±0.3	5.9±0.6	55.1±4.0
		Salt-treated	335±34	19.1±0.2	4.8±1.1	8.5±0.3	100.3±4.2
Citrus	14 days after Stress-relief	Control	48±6	39.2±2.9	1.3±0.4	5.4±0.4	68.2±5.3
		Salt-Treated	324±2	33.7±2.0	0.8±0.2	6.8±5.3	65.3±6.4

tracing the flow of carbon through the photosynthetic pathway with isotopically labelled CO_2 (2). Increased leaf chloride concentrations were associated with a decrease in the labelling of sugar monophosphates and 3-phosphoglyceric acid and an increase in labelled ribulose-1, 5-bisphosphate (RuBP), glycolate, glycine and serine which again indicated an increase in photorespiratory activity in salt affected vine leaves.

Each of these effects occurred in the absence of visible leaf damage. The thickness of expanded grapevine leaves also did not appear to be altered by salinity stress (5,cf. ref. 3).

Capacity for photosynthetic recovery following stress-relief

Cessation of salt treatment reversed the photosynthetic decline and eventually there was a complete recovery in photosynthetic rates in leaves of both grapevines and citrus (Table 2). Stomatal and internal resistances to CO_2 fixation and the relative stimulation of photosynthesis by low O_2 returned to near control values. In grapevine leaves, these changes were independent of changes in levels of Fraction I protein and specific activity of RuBP carboxylase (5).

The physiological recovery shown by grapevine leaves was associated with a marked decrease in leaf chloride (Table 2). The recovery did not appear to be turgor dependent. In contrast, mean turgor potential in salt-stressed citrus leaves increased from 0.16 ± 0.04 MPa to 0.40 ± 0.08 MPa during the 14 day period following stress-relief, compared to turgor potential in control leaves which remained between 0.43 and 0.47 MPa over this time interval.

Continued salt treatment in both vines and citrus eventually led to leaf death. Chloride was the principal ion accumulated in grapevines (5) and caused symptoms of "leaf burn" when concentrations reached approx. 300mM. In Etrog citron there was a high incidence of leaf abscission when chloride concentrations exceeded approx. 400mM. Sodium concentrations did not exceed 40mM in any of the citrus leaves measured.

Discussion

Salinity stress caused a reduction in photosynthesis and a stimulation of photorespiration in leaves of grapevine, citrus and guava. On cessation of salt treatment, these effects were reversed. The increased oxygen enhancement of photosynthesis, initially apparent in the absence of a change in internal resistance, may relate to the partial closure of stomata which would have lowered intercellular concentrations of CO_2 and led to the stimulation of oxygenase activity. Alternatively, differential effects of accumulated ions on carboxylase and oxygenase activity in situ are also possible, e.g. a salinity-induced stimulation of oxygenation relative to carboxylation could raise intercellular CO_2 levels and result in increased stomatal resistance to CO_2 diffusion (6).

Grapevine and citrus leaves differed during their photosynthetic recovery in that a reduction in foliar chloride levels rather than a marked change in turgor potential occurred in grapevine leaves, while the converse was evident for citrus. The reason for this is unknown, but may be associated with differences in the location of accumulated chloride. For example, high ion concentrations in the cell walls of citrus leaves may have caused a reduction in turgor. A stress-relief induced redistribution of accumulated solute, possibly into vacuoles, would have allowed the restoration of turgor. The absence of a marked increase in turgor in grapevine leaves following stress-relief might indicate that most of the accumulated solute was located within cellular compartments.

The accumulation of salt in cell walls or other cell compartments could alter the cellular CO_2/O_2 levels somewhat through differential effects on the solubilities of CO_2 and O_2. More importantly, however, such an accumulation, by shifting the first apparent dissociation constant for carbonic acid downwards, e.g. from approx. 6.4 in pure water to approx. 6.0 in seawater, could significantly lower the proportion of CO_2 in equilibrium with HCO_3^- at physiological pH (1). Consequently photorespiration could be stimulated relative to photosynthesis in the absence of any stomatal changes. This may explain the stimulation of photorespiration which was associated with increased internal resistances of CO_2 fixation under conditions of salinity stress (Table 2; ref 2, 5).

Leaves of grapevine (cv. Sultana) and citrus (cv. Etrog citron) grown under glasshouse conditions tolerated at least 200mM (2.0% dry wt.) and 350mM (2.6% dry wt.) chloride, respectively, without sustaining permanent damage to the photosynthetic apparatus. Other grapevine and citrus cultivars may be more or less tolerant. It should be borne in mind, however, that tolerance to chloride will likely be reduced under field conditions where water may be limiting and evaporative demand higher.

References

1. J. A. Berry and W. J. S. Downton, in: "Photosynthesis: CO_2 Assimilation and Plant Productivity" (Govindjee, ed.) Academic Press, New York, (1981).
2. W. J. S. Downton, Aust. J. Plant Physiol. 4:1-10 (1977).

3. D. J. Longstreth and P. S. Nobel, Plant Physiol. 63:700-703
    (1979).
4. R. R. Walker, P. E. Kriedemann and D. H. Maggs, Aust. J. Agric.
    Res. 30:477-488 (1979).
5. R. R. Walker, E. Torokfalvy, N. S. Scott and P. E. Kriedemann,
    Aust. J. Plant Physiol, (In Press).
6. S. C. Wong, I. R. Cowan and G. D. Farquhar, Nature 282:424-425
    (1979).

MECHANISM OF SALT TOLERANCE IN SALICORNIA PACIFICA VAR UTAHENSIS

D. J. Weber

Department of Botany
 and Range Science
Brigham Young University
Provo, UT 84602 USA

The halophytes in the genus Salicornia (Chenopodiaceae) are among the most salt-resistant higher plants (1). The mechanism by which Salicornia species survive high saline environments is not clearly known. This paper will propose a general mechanism of salt tolerance in Salicornia pacifica var. utahensis.

Environmental Factors

Salicornia pacifica var utahensis (Tidestrom) Munz and soil from three sites in Goshen, Utah, were analyzed for soil moisture, pH and ion content over a growing season. While the moisture in the top two inches of soil varied during the season, the subsurface soil moisture (6 to 10 inch zone) remained relatively constant (25 to 35%) (2). The pH of subsurface soil changes little (pH 7.5 to 8.0) during the growing season. The osmotic potential of the surface soil (upper 2 inches) as determined by freezing point depression reached as high as 135 atm in September. Sodium and Cl^- were the major ions in the soil. The Na^+, K^+, Ca^{++} and Cl^- were in much lower concentrations in the subsurface soil (2).

Anatomical Structure of Salicornia

DeFraine (3) concluded that the succulent internode of Salicornia was of foliar origin and formed by the fusion of two small opposite leaves. However Fahn and Arzee (4) concluded that the cortex was not foliar in origin and that the fleshy tissue

555

external to the central vascular cylinder was true cortex tissue.
Each internode appears to function as an independent unit connected
to the central vascular system. The outer cells (palisade) contain
chloroplasts which photosynthesize. The inner region (cortex) lacks
chloroplasts.   Isolated tracheoids (5) in the palisade region have
thickening in the walls and plasma membranes (6).   They appear to
function as a water reserve in the succulent tissue.

Osmotic Potentials in Salicornia

Freezing point depression measurements indicated that osmotic
potential values increased as measurements were made from the base
of the stem shoot (75 to 90 atms range) to the top (110 to 170 atms
range) (2).   The osmotic measurements across a stem indicated that
two regions (an inner and outer region) exist.   A logical division
would be the palisade and cortex regions.   The three major ions,
Na^+, K^+, and Cl^-, in S. pacifica tissue were analyzed over the
growing season.   The chloride ion concentration remained constant
throughout most of the growing season.   The sodium ion concentration gradually increased to 16.1% by the end of the growing season,
whereas the potassium ion decreased slightly through the growing
season. The evidence indicates that Salicornia is capable of maintaining high concentrations of salts in its tissue.

Salt Sensitivity of Salicornia Enzymes

Weber et al. (7) found that RuBPCase from S. pacifica var.
utahensis and S. rubra was almost as salt sensitive as the RuBPCase
from tomato. Greenway and Osmond (8) compared malic dehydrogenase,
glucose-6-phosphate dehydrogenase and isocitrate dehydrogenase from
halophytes to the same enzymes from glycophytes and found that the
enzymes from both were very similar in salt sensitivity.   These
results suggest that some type of compartmentalization must occur
if photosynthesis is to occur normally in Salicornia without interference from high salt concentrations.

Distribution of Ions in Salicornia

Weber, et al (9) using wavelength dispersive x-ray microanalysis to determine the distribution of ions in a cross section of S.
pacifica var utahensis found the Na^+, K^+, and Cl^- concentrations
were very low in the palisade region but high in the cortex area.
This suggests a type of compartmentalization in the nonphotosyn-
thetic cortex cells. When an older shoot was scanned, the ion concentration was still high in the cortex region (9).

Energy dispersive x-ray microanalysis was used to determine concentration of ions in <u>Salicornia</u> tissues. Fresh, freeze-dried and frozen shoots were analyzed in an AMR 1000/1000A scanning electron microscope equipped with an EDAX x-ray spectrometer. The shoots were frozen in place in the field with liquid nitrogen and then transported to the laboratory in liquid nitrogen. The ion concentrations were low for palisade and high for cortex regions which was similar to the wave length dispersive analysis.

Localization of Ions in Individual Cells

One of the major problems in determining the ion concentration in specific locations in a cell is the possible movement during sample preparation for the electron microscope. In order to reduce ion movement, Hess et al (6) precipitated the chloride ions with silver to form silver chloride. Silver chloride precipitations were present in the palisade and cortex cells. In the palisade cells, the silver chloride precipitate was common in the vacuole, present in the cytoplasm but low in organelles such as chloroplasts (6). These results suggest that organelle membranes and perhaps an active transport system keep the high salt concentrations out of the chloroplasts where it could interfere with photosynthesis.

ATPase in Salicornia

Weber et al. (10) detected ATPase in sections of <u>S. pacifica</u> var. <u>utahensis</u> using lead to precipitate the phosphate released from ATP due to the action of ATPase. The ATPase was found along plasma membranes in the palisade and cortex cells of shoots (10). Energy dispersive x-ray microanalysis was used to verify that the precipitate was lead phosphate (10).

Kim and Weber (11) isolated ATPase from S. <u>pacifica</u> var. <u>utahensis</u> by sucrose density gradient centrifugation. The isolated ATPases were salt tolerant and functioned in 3M NaCl. Thus, the ATPase could function in the presence of high salt concentrations in the cortex cells. L'Roy and Hendrix (12) evaluated the cell membrane potential of roots of <u>Salicornia bigelovii</u>. They concluded that chloride was accumulated at the expense of metabolic energy, but that accumulation of Na^+ and K^+ into the roots did not need metabolic energy to explain the movement into the roots. They did not analyze the membranes of the palisade or cortex cells in stems (12).

General Theory for the Mechanisms of Salt Tolerance in Salicornia Pacifica var. Utahensis

The general concept of salt tolerance in S. pacifica var. utahensis is based on the assumption that salt tolerance (of S. pacifica var. utahensis) reduced competition from other plants which cannot survive in higher salt concentrations. Also the increased salt concentration in the shoot permits greater osmotic potentials to be developed in the shoot which would help water movement from the soil into the plant. As the young shoots develop, the ATPase maintains an ion pumping action to keep ion concentration in the palisade cells low. In contrast the salts accumulate in the cortex cells. In addition, ATPases in the chloroplast membranes maintain a low salt concentration in the chloroplasts which permits photosynthesis to function normally. As a shoot grows older, the amount of salt in the cortex cells becomes very high and moves into the palisade cells. Eventually the salt concentration becomes too high in the palisade cells and the cells are killed. The tissue shrivels, but the vascular system continues to function for internodes above the shriveled internode.

In summary, the mechanism of salt tolerance is a combination of compartmentalization, active ion transport, and selective death of independent internodes.

References

1. Y. Waisel, "Biology of Halophytes", Academic Press, New York (1972).
2. D. J. Hansen and D. J. Weber, Great Basin Nat. 35:86-96 (1975).
3. DeFraine, E., Linn. J. Bot. 41:317-348 (1912).
4. A. Fahn and T. Arzee, Amer. J. Bot. 46:330-338 (1959).
5. N. Lersten and C. G. Bender, Proc. Iowa Acad. Sci. 82:158-162 (1976).
6. W. M. Hess, D. J. Hansen and D. J. Weber, Can. J. Bot. 53:1176-1887 (1975).
7. D. J. Weber, W. R. Anderson, S. Hess, D. J. Hansen and M. Gunasekaran, Plant and Cell Physiol. 18:693-699 (1979).
8. H. Greenway and C. B. Osmond, Plant Physiol. 18:693-699 (1972).
9. D. J. Weber, H. P. Rasmussen and W. M. Hess, Can. J. Bot. 55:1516-1523 (1977).
10. D. J. Weber, W. M. Hess and C. Kim, New Phytol. 84:285-291 (1980).
11. C. Kim and D. J. Weber, Plant and Cell Physiol. 2:263-272 (1980).
12. A. L'Roy and D. L. Hendrix, Plant Physiol. 65:544-549 (1980).

SCREENING FOR SALT TOLERANCE IN PLANTS: AN ECOLOGICAL APPROACH

Anne F. Wrona and Emanuel Epstein

Department of Land, Air and Water Resources
University of California
Davis, CA 95616 USA

INTRODUCTION

Although plant responses to climatic variables such as light, temperature, humidity and the length of the growing season have been extensively investigated, comparatively little ecophysiological work has been devoted to mineral nutrition and salt relations as factors explaining plant distribution (1, 2). Studies of salt effects on plants are, however, becoming increasingly important, particularly in light of the encroachment of salt on much of irrigated agriculture (3). Crop yields have decreased in many areas as a result of continued irrigation in arid and semi-arid lands and concomitant salt accumulation. One way of fighting this trend may be through breeding and selection of salt-tolerant crops. The goal of improved crop yields on salt-affected soils requires complementary research into the mechanisms imparting salt tolerance to plants.

A wild halophyte, Distichlis spicata (Gramineae, tribe Festuceae) was chosen for this study because of its wide distribution. It can be found growing on desert salt flats, submerged in seawater in coastal salt marshes, and in more hospitable environments such as fields and pastures where it is occasionally grazed when better forage is unavailable. Such a diversity of habitats made it seem likely that salinity ecotypes, or populations differentiated in their responses to salt, may exist. The purpose of this project was to see if salt-tolerant and salt-sensitive ecotypes of Distichlis spicata could be identified, and thereby make possible future physiological research into the mechanisms of salt tolerance.

Materials and Methods

Soils and _Distichlis_ plants were sampled from six sites in Northern California that ranged from inland (experiencing hot, dry summers and cold, wet winters) to coastal where year-around temperatures are more moderate, but plants are periodically inundated with seawater. Electrical conductivity, pH, K^+, Na^+, Cl^-, Ca^{2+} and Mg^{2+} were measured in extracts of soil saturation pastes of site samples. Levels of K^+, Na^+, Cl^-, Ca^{2+} and Mg^{2+} were also measured in plant samples from the sites. On the basis of these measurements we selected four sites as potentially yielding ecotypes differing in salt tolerance. Two of these sites, Tomales (38° 6' N, 122° 50' W) and Bodega (38° 20' N, 123° 4' W) are along the coast roughly 80 km north of San Francisco, and two, Putah (38° 31' N, 122° 4' W) and Dixon (38° 17' N, 121° 48' W) are inland and removed from the influence of seawater. The work reported here deals with the coastal Bodega and the inland Putah populations.

In order to screen salinity ecotypes, we brought _Distichlis_ plants from these varied sites to the more uniform environment of the greenhouse in Davis. There they were vegetatively propagated and planted into plastic grids over 100 liter tanks (Fig. 1) of half-concentration modified Hoagland solution (1). Roots were slipped through a square in the grid and the plants were held in place with a wrapping of Dacron batting around their bases. Black polyethylene kept the solutions dark. Using Rila mix, a synthetic sea salt mix (Rila Products, P. O. Box 114, Teaneck, N. J. 07666), we salinized the solutions in the tanks in increments of 10% of seawater salinity per week. Final treatments were 0.0, 0.8, and 2.0 times the salinity of seawater. Plants were grown six weeks at the final concentrations before they were harvested, dried, and analyzed. Atomic absorption spectophotometry was used for determinations of cations and titrations on a Buchler-Cotlove chloridometer for assaying Cl^-.

Results and Discussion

Plants from the two sites responded differently throughout the experiment. At a salinity level equaling twice seawater salinity, the differences were quite apparent visually (Fig. 2). The coastal Bodega plants (left) appeared reasonably healthy at 2.0 seawater while the plants from the inland Putah site (right) showed extensive salt injury.

Analyses of the ion contents in the roots and shoots of the plants also indicated the existence of different ecotypes. When K^+ contents of the tissues grown at 0.0, 0.8, and 2.0 times seawater are compared, an interesting difference in response of the two

Fig. 1. System used for culturing <u>Distichlis</u> plants in seawater and nutrient solution.

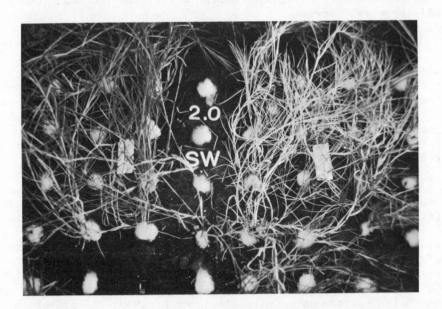

Fig. 2. Marked difference in appearance of two ecotypes grown at 2.0 seawater for six weeks. Coastal Bodega plants (left) appear reasonably healthy whereas the inland Putah plants (right) show significant salt damage.

ecotypes becomes apparent. At 2.0 seawater the K^+ level dropped in the roots of both ecotypes to roughly 0.6% of tissue dry weight, and the shoots of the coastal plants had more than double the K^+ concentration of the inland ones.

When Na^+ content is expressed in a similar fashion as a function of salinity, the Na^+ level increased over the levels found at 0.0 seawater, particularly so in the roots of the Putah ecotype which accumulated 2.4% Na^+ on a dry weight basis when exposed to twice seawater salinity for six weeks.

When plants were grown in seawater, Cl^- also increased in the roots and shoots of both ecotypes to levels well above those found in the controls. Again, the greatest increase occurred in the roots of the Putah ecotype grown at 2.0 seawater.

The content of Ca^{2+} was much lower in the roots and shoots of both ecotypes when they were grown in seawater than when no seawater was in the culture medium. The greatest drop from the control levels occurred in the roots of the Putah ecotype.

Levels of Mg^{2+} increased over control values in the roots and shoots of the Bodega ecotype and in the roots of the Putah ecotype when Distichlis was grown in seawater. The shoots of the Putah ecotype, however, maintained approximately the same Mg^{2+} content at all three salt levels.

Increases in tissue Na^+, Cl^- and Mg^{2+} reflect the high levels of these ions found in seawater. The observed decrease in Ca^{2+} in all tissues and the accompanying increase in Mg^{2+} noted in all tissues except for the shoots of the Putah ecotype seemingly correspond to the low Ca^{2+}/Mg^{2+} ratio (1:5.6) of seawater. Although the Ca^{2+} content of seawater is high at roughly 10 mM (1), favorable agricultural soils have much more Ca^{2+} present than Mg^{2+}. The greater abundance of Mg^{2+} in the surrounding seawater might result in a substitution of Mg^{2+} for Ca^{2+} in the plants.

The Putah ecotype would appear to have the least control over its internal ion content when exposed to increasing salt levels (Fig. 3). Sodium and chloride -- ions toxic to many plants in high concentrations -- were at considerably higher levels in the roots of Putah plants than in those of Bodega plants. Furthermore, the K^+ levels in the shoots of the Putah plants were possibly lower than necessary for healthy metabolism in the presence of elevated Na^+ and Cl^- levels. The coastal plants, adapted to high Na^+ levels in their environment, maintained high shoot K^+ levels. This differential response may indicate the existence of a tolerance mechanism in the coastal ecotype.

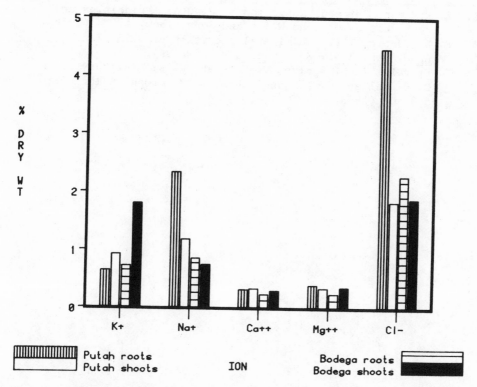

Fig. 3. Comparison of ion contents of inland Putah and Coastal
Bodega ecotypes after six weeks of growth at twice
seawater salinity.

Now that distinct salinity ecotypes have been shown to exist
in _Distichlis_ _spicata_, comparative physiological studies may elucidate some of the mechanisms of salt tolerance in this species.

Work supported by the National Science Foundation and the
Institute of Ecology, University of California, Davis, CA 95616 USA
and done in partial fulfullment of the Ph. D. degree by the senior
author.

References

1. E. Epstein, "Mineral Nutrition of Plants: Principles and
      Perspecitves." John Wiley and Sons, Inc., N. Y. (1972).
2. F. S. Chapin, III, The mineral nutrition of wild plants,
      Annu. Rev. Ecol. Systematics (In Press).
3. E. Epstein, J. D. Norlyn, D. W. Rush, R. W. Kingsbury,
      D. B. Kelley, G. A. Cunningham, A. F. Wrona, Science 210:
      399-404 (1980).

LIST OF PARTICIPANTS

Aaronson, S., Biology Dept., Queen's College, City University of
New York, 65-30 Kissena Blvd., Flushing, NY 11367, USA

Ahmad, R., Plant Physiology/Botany, University of Karachi,
Karachi-32, PAKISTAN

Alcaraz Melendez, L., Depto. de Biologia Terrestre, Centro de
Investigaciones Biologicas, de Baja California, Guillermo
Prieto 1042, La Paz, B.C.S., MEXICO

Aller, J. C., National Science Foundation, 1800 'G' St. N.W.,
Washington, D. C. 20550, USA

Ascencio, F., Depto. de Biologia Terrestre, Centro de
Investigaciones Biologicas, de Baja California, Guillermo
Prieto 1042, La Paz, B.C.S., MEXICO

Ben-Amotz, A., Israel Oceanographic & Limnological Research,
Tel-Shikmona, P. O. Box 8030, Haifa, ISRAEL

Benson, A. A., Scripps Institute of Oceanography, University of
California, La Jolla, CA 92093, USA

Brand, T., Depto. de Biologia Marina. Centro de Investigaciones
Biologicas, de Baja California, 5 de May No. 9, La Paz,
B.C.S., MEXICO

Brosseau, G., Office of Problem Analysis, National Science
Foundation, 1800 'G' Street, N.W., Washington, D. C.
20550, USA

Brunet Martins, C., Centro de Ciencias Agrarias, C. P. 35-A,
Fortaleza, CEARA, BRAZIL, S. A.

Carlson, W. T., Agricultural & Applied Biological Science,
Smithsonian Science Information Exchange,
1730 M Street, N. W., Washington D.C. 20036, USA

Castellanos Cervantez, T., Depto. de Biologia Celular, Centro de
    Investigaciones Biologicas, de Baja California, Guillermo
    Prieto 1042, La Paz, B.C.S., MEXICO

Castro Acevedo, R.E., Depto. de Biologia Celular, Centro de
    Investigaciones Biologicas, de Baja California, Guillermo
    Prieto 1042, La Paz, B.C.S., MEXICO

Cendejas, Consejo Nacional de Ciencia, & Tecnologia, Insurjentes
    Sur 1677, Mexico D. F.   MEXICO

Chapman, V. J., Botany Dept., Aukland University, Aukland C. I.,
    NEW ZEALAND

Cooney, R. V., Scripps Institute of Oceanography, University of
    California, La Jolla, CA 92093, USA

Cordoba Alva, F., Centro de Investigaciones Biologicas, de Baja
    California, Jalisco Y Madero, APDO Postal No. 128, La Paz,
    B.C.S., MEXICO

Cota de La Pena, A., Centro de Investigaciones Biologicas, de Baja
    California, Ocampo 312, La Paz, B.C.S. MEXICO

Cunningham, G. A., Dept. of Land, Air & Water Resources,
    University of California, Davis, CA 95616, USA

Cunningham, G. L., Biology Dept., New Mexico State University,
    Las Cruces, New Mexico, USA 88001

Dantas, P., Conshelo Nacional de Desenvolvimiento, Cientifico e
    Tecnologico, Eficio CNPQ., AV W3 Norte, Brasilia, D. F.,
    BRASIL, S.A.

De Alba Perez, C., Universidad Nacional Autonoma, de Baja
    California Sur, La Paz, B.C.S., MEXICO

De Cunha, J.B., Conshelo Nacional de Desenvolvimiento, Cientifico e
    Tecnologico, Edificio CNPQ., AV W3 Norte, Q. 507/B, Brasilia,
    D.F., BRASIL S.A.

Dennett, C., Dept. of Agronomy, University of California, Davis, CA
    95616, USA

Devey, G., Middle East Section, National Science Foundation, 1800
    'G' St., Washington D.C. 20550, USA

Diaz de Leon, L., Instituto de Investigaciones, Biomedicas UNAM,
    Mexico 20, D. F. MEXICO

Diaz de Leon, J. L., Depto. de Biologia Celular, Centro de
    Investigaciones Biologicas, de Baja California, Guillermo
    Prieto 1042, La Paz, B.C.S., MEXICO

Diaz Rivera, E., Depto. de Biologia Marina, Centro de
    Investigaciones Biologicas, de Baja California, 5 de Mayo No.
    9, La Paz, B.C.S., MEXICO

Dimayuga, R. E., Depto. de Biotecnologia, Centro de Investigaciones
    Biologicas, de Baja California, Ocampo 312, La Paz, B.C.S.,
    MEXICO

Duarte, M. R., Conshelo Nacional de Desenvolvimiento, Cientifico e
    Tecnolgico, Edificio CNPQ, AV W3 Norte Q 507/B, Brasilia, D.
    F., BRASIL, S.A.

Dubinsky, Z., Dept. of Life Sciences, Bar-Ilan Unversity,
    Ramat-Gan, ISRAEL

Epstein, E., Dept. of Land, Air & Water Resources, University of
    California, Davis, CA 95616, USA

Espinosa Garduno, M., Centro de Investigaciones Biologicas, de Baja
    California, 5 de Mayo No. 9, APDO Postal No. 128, La Paz,
    B.C.S., MEXICO

Faden, A. O., 14-3 N. University Place, Oklahoma State University,
    Stillwater, OK 74074, USA

Felger, R., Arizona-Sonora Desert Museum, Rte. 9, Box 900, Tucson,
    AZ 85704, USA

Fenical, W., Scripps Institute of Oceanography, University of
    California, San Diego, CA 92093, USA

Fontes, M., Environmental Research Lab., University of Arizona,
    Tucson International Airport, Tucson, AZ 85706, USA

Fuentes Solis, H. R., Universidad Autonoma, de Baja California, La
    Paz, B.C.S., MEXICO

Gale, J., Dept. of Botany, Institute of Life Sciences, Hebrew
    University, Jerusalem, ISRAEL

Gallagher, J. L., Marine Studies, Robinson Hall, University of
    Delaware, Newark, DE 19711, USA

Gallo, J. P., Depto. de Biologia Marina, Centro de Investigaciones
    Biologicas, de Baja California, 5 de Mayo No. 9, La Paz,
    B.C.S., MEXICO

Gamboa de Buen, A., Depto. de Biologia Celular, Centro de
    Investigaciones Biologicas, de Baja California, Guillermo
    Prieto 1042, La Paz, B.C.S., MEXICO

Gillespie, D. M., P. O. Box 13687, University of Georgia, Savannah,
    GA 31406, USA

Glenn, E. P., Environmental Research Lab., University of Arizona,
    Tucson International Airport, Tucson, AZ 85706, USA

Gonzales de Alba, M. A., Depto. de Biologia Marina, Centro de
    Investigaciones Biologicas, de Baja California, 5 de Mayo No.
    9, La Paz, B.C.S., MEXICO

Guerrero Godinez, R., Depto. de Biotecnologia, Centro de
    Investigaciones Biologicas, de Baja California, APDO Postal
    No. 128, La Paz, B. C. S., MEXICO

Gutierrez Fuentes, J. A., Depto. de Biologia Terrestre, Centro de
    Investigaciones Biologicas, de Baja California, Guillermo
    Prieto 1042, La Paz, B.C.S., MEXICO

Herrera-Lasso, J., Scripps Institute of Oceanography, University of
    California, La Jolla, CA 92093, USA

Iyengar, E. R. R., CSIR, Central Salt & Marine Chemicals, Research
    Institute, Bhavnagar 364002, INDIA

Javor, B., c/o K. Nealson A-002, Scripps Institute of Oceanography,
    University of California, La Jolla, CA 92093, USA

Kelley, D. B., Dept. of Land, Air & Water Resources, University of
    California, Davis, CA 95616, USA

Kingsbury, B. W., 887 Industrial Road, San Carlos, CA, 94070 USA

Kirkham, M. B., Evapotranspiration Lab., Kansas State University,
    Manhattan, KS 66506, USA

Lanyi, J., Dept. of Physiology and Biophysics, University of
    California, Irvine, CA 92717, USA

Lauchli, A., Dept. of Land, Air & Water Resources, University of
    California, Davis, California 95616, USA

Lenilton, J., Conshelo Nacional de Desenvolvimiento, Cientifico e
    Tecnologico, Edificio CNPQ., AV W3 Norte Q. 507/B, Brasilia,
    BRASIL, S.A.

Lopez Gutierrez, F., Depto. de Biologia Celular, Centro de
    Investigaciones Biologicas, de Baja California, Guillermo
    Prieto 1042, La Paz, B.C.S., MEXICO

Makki, Y. M., College of Agriculture, King Faisal University,
    P. O. Box 380, Al-Hassa, SAUDI ARABIA

Madero, G., Conshelo Nacional de Desenvolvimiento, Cientifico e
    Tecnologico, Edificio CNPQ, AV W3 Norte Q 507/B, Brasilia, D.
    F., BRASIL, S.A.

Maeda Martinez, A., Depto. de Biologia Marina, Centro de
    Investigaciones Biologicas, de Baja California, 5 de Mayo No.
    9, La Paz, B.C.S., MEXICO

Maravilla, M. O., Depto. de Biologia Marina, Centro de
    Investigaciones Biologicas, de Baja California, 5 de Mayo No.
    9, La Paz, B.C.S., MEXICO

Mitsui, A., Div. of Biology & Living Resources, School of Marine &
    Atmospheric Sci., University of Miami, 4600 Rickenbacker
    Causeway, Miami FL 33149 USA

Mota-Urbina, J. C., Instituto Nacional de Investigaciones,
    Vaso de Ex Lago Texcoco (CIFREC), Forestales, MEXICO

Munoz Ley, E., Depto. de Biologia Marina, Centro de Investigaciones
    Biologicas, de Baja California, 5 de Mayo No. 9, La Paz,
    B.C.S., MEXICO

Munoz Ley, I., Depto. de Biotecnologia, Centro de Investigaciones
    Biologicas de Baja California, La Paz B.C.S., MEXICO

Nagel, C., Environmental Research Laboratory, University of
    Arizona, 765 W. Limberlost 32, Tucson, AZ 85705, USA

Neushul, M., Dept. of Biological Sciences, University of
    California, Santa Barbara, CA 93106, USA

Nissen, P., Scripps Institute of Oceanography, University of
    California, La Jolla, CA 92093, USA

Norlyn, J. D., Dept. of Land, Air & Water Resources, University of
    California, Davis, CA 95616, USA

Oswald, W. J., 615 Davis Hall, University of California, Berkeley,
    CA 94720, USA

Padilla, G., Depto. de Biologia Marina, Centro de Investigaciones
    Biologicas, de Baja California, 5 de Mayo No. 9 La Paz,
    B.C.S., MEXICO

Parra Hake, H., Centro de Investigaciones Forestales del Noroeste,
    Esquina con Republica, Ignacio Ramirez No. 580 N., La Paz,
    B.C.S., MEXICO

Pasternak, D., Research & Development Authority, Ben Gurion
    University of the Negev, P. O. Box 1025, Beer-Sheva, ISRAEL

Perez, F., Centro de Investigaciones Forestales del Noroeste,
    Esquerro No. 10, La Paz, B.C.S., MEXICO

Poljakoff-Mayber, A., Dept. of Botany, The Hebrew University of
    Jerusalem, Jerusalem, ISRAEL

Prisco, J. T., Depto. de Bioquimica y Biologia Molecular, C. P.
    1065, Fortaleza, Ceara, BRASIL

Pulich, Jr., W., Port Aransas Marine Lab, Marine Science Institute,
    University of Texas, Port Aransas, TX 78373, USA

Rains, S. W., Plant Growth Lab., University of California, Davis,
    CA 95616, USA

Ramirez Angulo, V., Depto. de Biotecnologia, Centro de
    Investigaciones Biologicas, de Baja California, Ocampo 312 La
    Paz, B.C.S., MEXICO

Reynoso Granados, T., Centro de Investigaciones Bilogicas, de Baja
    California, Guillermo Prieto No. 1042, APDO Postal No. 128, La
    Paz. B.C.S., MEXICO

Richards, R. A., Dept. of Agronomy & Range Science, University of
    California, Davis, CA 95616, USA

Risk, M, 711 Holiday Drive, No. 63, Galveston, TX 71550 USA

Rodriguez, G. A., CCML-UNAM, Apado. Postal No. 811, Mazatlan,
    Sinaloa, MEXICO

Romero, L. A., Depto. de Biotecnologia, Centro de Investigaciones
    Biologicas de Baja California, Ocampo, B.C.S., MEXICO

Rush, D. W., Dept of Land, Air & Water Resources, University of
    California, Hoagland Hall, Davis, CA 95616 USA

Sacher, R., Boyce Thompson Institute, Cornell University,
    Tower Road, Ithaca, NY 14853, USA

Salgado, R. M., Depto. de Biologia Marina, Centro de
    Investigaciones Biologicas, de Baja California, 5 de Mayo No.
    9 La Paz, B.C.S., MEXICO

Salman, A. J., Kuwait Institute for Scientific Research, P. O. Box
    24885, Safat, KUWAIT 2

Sanches Rueda, P., Depto. de Biologia Marina, Centro de
    Investigaciones Biologicas, de Baja California, 5 de Mayo No.
    9 La Paz, B.C.S., MEXICO

Sanchez Lopez, R., Depto. Biologia Celular, Centro de
    Investigaciones Biologicas, de Baja California, Guillermo
    Prieto 1042 La Paz, B.C.S., MEXICO

San Pietro, A., Dept. of Biology, Indiana University, Jordan Hall,
    Bloomington, IN 47405 USA

Schaller, C. W., Dept. of Agronomy & Range Science, University of
    California, Davis, CA 95616, USA

Shannon, M. C., US Salinity Lab., 4500 Glenwood Dr., Riverside, CA
    92501, USA

Shurkin, J., Stanford University, 2727 Midtown Ct. #23,
    Palo Alto, CA, USA

Sierra Beltran, A., Depto. de Biotecnologia, Centro de
    Investigaciones Biologicas, de Baja California, Ocampo 312 La
    Paz, B.C.S., MEXICO

Silva, E. A., Depto. de Biologia Marina, Centro de Investigaciones
    Biologicas, de Baja California, 5 de Mayo No. 9 La Paz,
    B.C.S., MEXICO

Somers, G. F., School of Life & Health Science, University of
    Delaware, Newark, DE 19711, USA

Sordo Cedeno, M., Depto. de Biotecnologia, Centro de
    Investigaciones, Biologicas de Baja California, La Paz B.C.S.,
    MEXICO

Staples, R. C., Boyce Thompson Institute for Plant Research, Tower
    Road, Ithaca, NY 14853, USA

Stoeckenius, W., Cardiovascular Research Institute, University of
    California, San Francisco, CA 94143, USA

Teas, H. J., Dept. of Biology, University of Miami, Coral Gables,
    FL 33124, USA

Terry, O. W., Marine Science Research, SUNY-Stony Brook, Stony
    Brook, NY 11794, USA

Valle Mercado, G., Campo Experimental Agricola, Carret.
    Transpeninsular KM, APDO. Postal 127, 208 CD Constitucion
    B.C.S., MEXICO

Walker, R. R., Div. of Horticultural Research, CSIRO, Merbein,
    Victoria 3505, AUSTRALIA

Weber, D. J., Dept. of Botany/Range Science, Brigham Young
    University, Provo, UT 84602, USA

White, D. H., Dept. of Chemical Engineering, University of Arizona,
    Tucson, AZ 85721, USA

Wrona, A. F., Dept. of Land, Air & Water Resources, University of
    California, Davis, CA 94616, USA

Wyn-Jones, R. G., Dept. of Biochemistry and Soil Sciences,
    University College of North Wales, Bangor, Gwynedd LL57,
    2UW Wales, ENGLAND

Zaborsky, O., ASRA, Room 1136, National Science Foundation, 1800
    'G' Street, N.W., Washington, D.C. 20550, USA